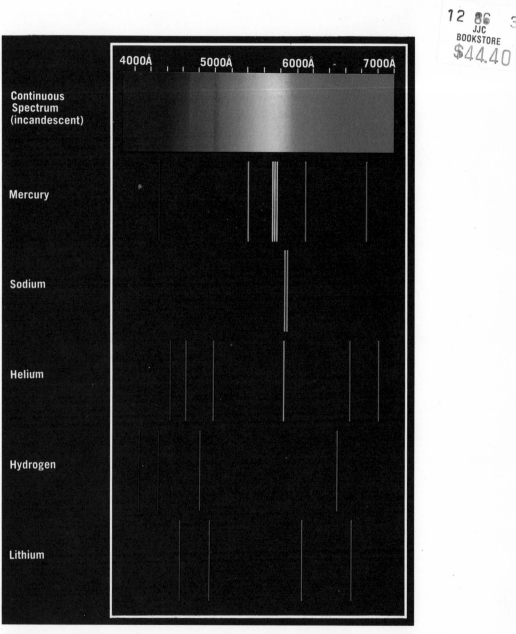

Color plate of continuous spectrum (incandescent) [*TOP*] and spectral lines for Mercury, Sodium, Helium, Hydrogen, and Lithium [*BOTTOM*].

Albert Einstein sailing at Saranac Lake, N.Y. in 1936. (Courtesy of the American Institute of Physics Niels Bohr Library.)

Introduction to
MODERN
PHYSICS

Elmer E. Anderson

Professor of Physics

The University of Alabama in Huntsville

SAUNDERS GOLDEN SUNBURST SERIES

SAUNDERS COLLEGE PUBLISHING

Philadelphia New York Chicago
San Francisco Montreal Toronto
London Sydney Tokyo Mexico City
Rio de Janeiro Madrid

Address orders to:
383 Madison Avenue
New York, NY 10017

Address editorial correspondence to:
West Washington Square
Philadelphia, PA 19105

This book was set in Times Roman by Progressive Typographers.
The editors were John Vondeling, Maryanne Miller, and Jay Freedman.
The art & design director was Richard L. Moore.
The text design was done by Nancy E. J. Grossman.
The cover design was done by Richard L. Moore.
The artwork was drawn by Danmark & Michaels.
The production manager was Tom O'Connor.
This book was printed by Fairfield Graphics.

> This book is dedicated to
>
> SVEN HARSTAD ANDERSON
> *and*
> DIANE CAYEY ANDERSON

Cover credit: Space-shuttle launch (NASA)

INTRODUCTION TO MODERN PHYSICS   ISBN 0-03-058512-0

©1982 by CBS College Publishing. All rights reserved. Printed in the United States of America. **Library of Congress catalog card number 81-53079.**

3 4 5   016   9 8 7 6 5 4 3

CBS COLLEGE PUBLISHING
Saunders College Publishing
Holt, Rinehart and Winston
The Dryden Press

Preface

The aim of this book is to provide an introduction to modern physics at the intermediate level. It is intended primarily for engineering and science sophomores or juniors who have completed courses in classical physics and introductory calculus. Certain sections of the book require some knowledge of differential equations. The more difficult sections, which may be omitted without destroying the continuity of the presentation, are marked with asterisks. Since there is more material here than what is normally covered in a one-semester course, the instructor has considerable flexibility in the choice of topics. For example, if it is desired to emphasize engineering applications, Chapters I and II may be omitted or treated lightly, and Chapters III and IV may be omitted entirely. The book contains more than one hundred worked examples and over 300 problems. As in the case of the text itself, the more challenging problems are marked with asterisks. A solutions manual is available to instructors.

Chapters I and II contain an introduction to the atomic theory of matter and the elementary particles as the building blocks of atoms. Chapters III and IV treat the special theory of relativity. In addition to the standard topics in kinematics and dynamics, brief treatments are given of four-vectors, relativistic collisions, particle creation and annihilation, and the relativistic Doppler effect. The quantum theory of light is developed in Chapters V and VI. After the student grasps the Planck quantum concept he is introduced to the Einstein transition probabilities and the operating principles of lasers. Next, the particle nature of light is examined in the photoelectric effect and Compton collisions, followed by the reconciliation of the wave and particle aspects of light through the concept of the wave packet. Chapter VII traces the development of the quantum theory of the atom from the Thomson atom to the Bohr model and the quantization of angular momentum. Particle waves are discussed in Chapter VIII and the theory is applied to the diffraction of particles from crystal lattices. The equation of motion for particle waves, the Schrödinger wave equation, is developed in Chapter IX. Special sections are devoted to the separation of variables, the eigenvalue problem and expectation values. A number of one-dimensional problems are solved by the Schrödinger method in Chapter X, including a brief treatment of quantum mechanical tunneling. Chapter XI is devoted to the quantum mechanical solution of the hydrogen atom, spatial quantization and the Zeeman effect. Quantum mechanics is applied to solids in Chapter XII, with emphasis upon the electron theory of metals, the band theory of metals, and semiconductors. Chapters XIII and XIV deal with the atomic nucleus. The basic theory is reviewed in the former, while the final chapter treats nuclear transformations and applications such as carbon dating, fission and fusion. The Appendices contain some important derivations and useful tables.

SI (mks) units have been used throughout, although other units have been used occasionally. Answers to odd-numbered problems are included at the back of the book.

Every effort has been made to minimize errors, but they will surely be found. The reader's assistance in pointing them out, as well as other comments or suggestions, will be greatly appreciated.

It is a pleasure to acknowledge the help of the following and to thank them for their indispensable roles in the preparation of this book: Sandy Armstrong, Melinda McCreless and Claire Doering for typing parts of the manuscript; Donald Cooke and Professors John Albright of Florida State University, David Cole of the University of Alabama, Allen Hermann of Tulane University, Carl Kocher of Oregon State University, Fred Lobkowicz of the University of Rochester, Ray Mikkelson of Macalester College, Curt Moyer of Clarkson College of Technology, Raymond Serway of James Madison University, William Walker of the University of Alabama, T. A. Wiggins of Pennsylvania State University, George A. Williams of the University of Utah, and E. L. Wolfe of Iowa State University for their critical reviews of the manuscript; Joan Garbutt, John Vondeling and Maryanne Miller of Saunders College Publishing for their encouragement and assistance; the Center for the History of Physics of the American Institute of Physics, the Lawrence Livermore National Laboratory, the Fermi National Accelerator Laboratory, and NASA's Marshall Space Flight Center for providing photographs; my students over the years for their enthusiasm and inspiration; and to my wife Diane for her tremendous support throughout this project in spite of the countless inconveniences she has endured.

ELMER E. ANDERSON
Huntsville, Alabama

Table of Contents

Introduction

The term "modern physics" usually refers to the rapid conceptual developments of twentieth century physics. In contrast, all of the physics that was so arduously conceived prior to 1900 is labeled "classical physics." Surely this does not mean that classical physics is old-fashioned and obsolete! No, much of what is still taught under the headings of mechanics, electricity and magnetism, heat, light, and sound has changed little since the late 19th century. Furthermore, many of the problems of our everyday world can be solved by classical physics without recourse to modern physics. The reason for this is that classical physics deals with the world of our physical perceptions, that is, the common-sense world of "ordinary" sizes, masses, and speeds, whereas the dramatic effects of modern physics generally require indirect verification. This fact should in no way detract from the great importance of the study of modern physics, for it is modern physics itself that provides us with clues about the limitations of classical physics and tells us when to use quantum physics or special relativity.

Moreover, modern physics gives us a theory of the atom and a new mechanics to use for solving problems in the atomic world. It also refines the physical intuition so that one can achieve a more sophisticated world-view which eliminates some of the errors or restrictions of the older physics. For example, the Newtonian view of absolute space and time is replaced by Einstein's postulate that such absolutes do not exist. The old axiom that matter can be neither created nor destroyed gives way in the face of the wholesale creation and annihilation of particle pairs. Even such an old standby as the conservation of energy does not hold in relativity unless it is modified. The arguments of ancient philosophers regarding continuity *vs.* discreteness are settled in modern physics, where we not only know the discrete sizes of atoms but we also have discovered the discrete nature of electric charge, angular momentum, energy, and magnetic flux. Particles and waves are no longer regarded as mutually exclusive in modern physics. Certainty is replaced by probabilities, and causality is open to serious question.

Admittedly, there have been some bold postulates made by the architects of modern physics, but these generally arose out of the compelling urge either to incorporate new experimental facts into known theory or to make existing theories consistent with each other. It is hoped that the reader will learn to appreciate the conceptual structure of modern physics as well as its application to specific physical problems.

1

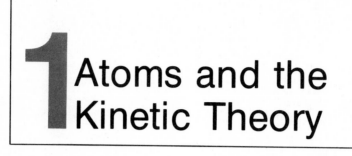

1 Atoms and the Kinetic Theory

1. THE ATOMIC THEORY OF MATTER

The concept of an ultimate particle or *atom* of matter was proposed as far back as the fifth century B.C. by Democritus and later by others of the Epicurean school of philosophers. However, little came of this until the development of quantitative experimental results in chemistry in the 17th and 18th centuries. The work of Proust, Gay-Lussac, Lavoisier, Dalton, and others led to the definition of the elements and to the concept of atomic masses.

It should be mentioned that hydrogen, the lightest element, was the first standard for relative atomic masses, and it was given the value of unity. Most of the light elements then had nearly integral values for their relative atomic masses. It was later found, however, that many of the heavier elements had non-integral relative masses and that fewer discrepancies would result if oxygen were used as the standard with its mass set at exactly sixteen. More recently, carbon-12 was adopted as the reference, and one-twelfth of its mass is defined as the *unified mass unit, u*. The approximate value of u is

$$u = 1.66 \times 10^{-27} \text{ kg.}$$

In 1811 Avogadro proposed the existence of molecules and suggested that each molecule of a gaseous element like hydrogen or oxygen contains two atoms. He then enunciated a principle which has henceforth been called Avogadro's law. It may be stated as follows: *equal volumes of gas at the same temperature and pressure contain the same number of molecules.* Avogadro made no estimate of the number of molecules contained in a given volume of gas, nor the size of a molecule, and it was about 50 years before the first calculation was made.

Avogadro's law holds for all elements and compounds, whether gas, liquid or solid. It is easily generalized by defining a kilogram-mole, or simply a "kilomole," as the amount of a substance contained in a number of kilograms of that substance numerically equal to its atomic mass. Thus, 32 kilograms of oxygen would be a kilomole of O_2, 28 kilograms of nitrogen would be a kilomole of N_2, and 18 kilograms of water contain 1 kilomole of H_2O. In each of these samples the number of molecules is the same, and that number is called *Avogadro's number*. Its approximate value is

$$N_A = 6.022 \times 10^{26} \text{ molecules/kmole.}$$

Once Avogadro's number is known, it immediately becomes possible to determine the mass of a single molecule or atom of a substance. For example, if 32 kilograms of oxygen contain N_A molecules of O_2, then each oxygen atom has a mass of

$$\frac{1}{2} \times \frac{32}{N_A} = 2.66 \times 10^{-26} \text{ kg.}$$

This is such an unbelievably small quantity that it is no wonder that evidence for the atomic theory of matter remained undetected so long.

Avogadro's number also permits estimates of the sizes of atoms, as the following example will show. A kilomole of copper has a mass of 63.54 kg. Since the density of copper is 8.92×10^3 kg/m³, then the kilomolar volume of copper is

$$V_M = \frac{63.54 \text{ kg}}{8.92 \times 10^3 \text{ kg/m}^3} = 7.12 \times 10^{-3} \text{ m}^3.$$

The space available for each atom is the small cubical volume

$$\frac{V_M}{N_A} = \frac{7.12 \times 10^{-3} \text{ m}^3}{6.02 \times 10^{26}} = 11.8 \times 10^{-30} \text{ m}^3.$$

The edge of this cube is obtained by taking the cube root of its volume, that is,

$$d \approx (12 \times 10^{-30})^{1/3} \text{ m} \approx 2 \times 10^{-10} \text{ m} = 2 \text{ Å,}$$

where Å is the Angstrom unit (1 Å $= 1 \times 10^{-10}$ m). Since the distance d represents the distance between two copper atoms, one can estimate the atomic radius of copper to be about 1 Å.

2. THE KINETIC THEORY OF GASES

Further evidence for the existence of molecules arose from the application of classical mechanics to a gas consisting of an enormously large collection of identical particles. This subject is known as the *kinetic theory;* it assumes that the particles of the gas are moving rapidly but in random directions and that they undergo successive collisions with each other and the walls of the container. Since no computer is large enough to keep track of the positions and velocities of all of the particles, it is necessary to resort to statistical averages in order to obtain useful results.

Consider a cubical box of edge L, as shown in Figure 1-1. It contains N molecules of gas at an equilibrium temperature T. If no forces act on the particles, other than the elastic forces during collisions, we can assume that complete chaos exists at the molecular level. That is, for a given molecule, all positions in the box and all directions of velocity are equally probable. We also assume that the walls are rigid and that the particles are hard spheres, so that all collisions between particles and with the walls are perfectly elastic. Further, we regard the molecules to be so small in comparison with their average distance of separation that, in spite of their great number, they fill a negligibly small fraction of the volume L^3.

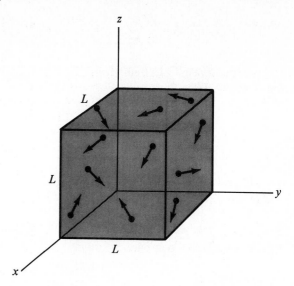

Figure 1–1

Molecules in a cubical box.

The velocity of a molecule may be represented at any instant by the components v_x, v_y, and v_z. The square of its velocity is

$$v^2 = v_x{}^2 + v_y{}^2 + v_z{}^2.$$

Then the mean-square speed of all of the N particles is

$$\overline{v^2} = \frac{\sum\limits_{i=1}^{N} v_i{}^2}{N} = \overline{v_x{}^2} + \overline{v_y{}^2} + \overline{v_z{}^2}, \tag{1.1}$$

where the bar indicates the average over N. When the gas is at equilibrium there is no preferred direction of motion. Thus all directions of the velocities are equally probable, and it follows that

$$\overline{v_x{}^2} = \overline{v_y{}^2} = \overline{v_z{}^2},$$

Then Equation 1.1 becomes

$$\overline{v^2} = 3\,\overline{v_x{}^2}. \tag{1.2}$$

If we denote the molecular mass by m, then a molecule whose x-component of velocity is v_x will have a momentum of mv_x in the x-direction. After it collides elastically with the wall at $x = L$, its momentum becomes $-mv_x$. Thus the change of momentum in the x-direction resulting from the collision is $2mv_x$. How frequently will this same molecule strike the wall at $x = L$? Its round-trip transit time will be $\dfrac{2L}{v_x}$, so it will strike the wall $\dfrac{v_x}{2L}$ times per second. Since the momentum change per second is by Newton's law the average force, this single molecule will exert a force on the wall given by

$$f = 2mv_x \frac{v_x}{2L} = \frac{mv_x{}^2}{L}. \tag{1.3}$$

Then the total force on the wall due to all N molecules will be

$$F_x = \sum_{i=1}^{N} f_i = \frac{m}{L} \sum_{i=1}^{N} v_{xi}^2 = \frac{mN}{L} \left(\frac{1}{N} \sum_{i=1}^{N} v_{xi}^2 \right) = \frac{mN}{L} \overline{v_x^2} = \frac{mN}{3L} \overline{v^2}, \quad (1.4)$$

where Equation 1.1 has been used. Since the force per unit area is the pressure on the wall, we have that

$$P = \frac{F}{L^2} = \frac{mN}{3L^3} \overline{v^2} = \frac{mN}{3V} \overline{v^2}, \quad (1.5)$$

where V is the volume of the box. It follows that

$$PV = \tfrac{1}{3} mN\overline{v^2} = \tfrac{2}{3} N \left(\tfrac{1}{2} m\overline{v^2} \right) = \tfrac{2}{3} N\overline{\epsilon_k} \quad (1.6)$$

where we have defined the average translational kinetic energy per molecule by

$$\overline{\epsilon_k} = \tfrac{1}{2} m\overline{v^2}. \quad (1.7)$$

It is instructive to compare Equation 1.6 with the ideal gas law,

$$PV = \nu RT. \quad (1.8)$$

Here $\nu = N/N_A$, the number of kilomoles of gas, T is the absolute temperature of the gas, and R is the kilomolar gas constant,

$$R = 8314 \text{ J/kmole-K} = 1.987 \text{ kcal/kmole-K}.$$

Then we have

$$\tfrac{2}{3} \nu N_A \overline{\epsilon_k} = \nu RT,$$

or

$$\overline{\epsilon_k} = \frac{3}{2} \left(\frac{R}{N_A} \right) T = \frac{3}{2} kT. \quad (1.9)$$

The constant $k = R/N_A$ is called the Boltzmann constant. Note that it is the "gas constant per particle." Its value is

$$k = \frac{R}{N_A} = 1.381 \times 10^{-23} \text{ J/K}. \quad (1.10)$$

Equation 1.9 expresses the fact that the *average kinetic energy per molecule of an ideal gas is proportional to the absolute temperature of the gas.*

Example 1-1
   The air in a warm room is at an equilibrium temperature of 27° C. What is the average kinetic energy of a molecule of air?

Solution
   The average kinetic energy per molecule is given by Equation 1.9,

$$\overline{\epsilon_k} = \tfrac{3}{2} kT = \tfrac{3}{2} (1.38 \times 10^{-23} \text{ J/K}) \times 300 \text{ K}$$

$$= 6.21 \times 10^{-21} \text{ J}$$

Example 1-2
   Under the conditions given in the preceding example, what is the root-mean-square speed of (a) a nitrogen molecule, (b) an oxygen molecule, and (c) a water molecule at equilibrium in the atmosphere of the room? The root-mean-square speed of a molecule is defined as $v_{\text{rms}} = (\overline{v^2})^{1/2}$.

Solution

The mass of a molecule is obtained by dividing its molecular weight A by Avogadro's number, N_A. That is, $m = A/N_A$. From Equation 1.7 we see that $\overline{v^2}$ is given by

$$\overline{v^2} = \frac{2\overline{\epsilon_k}}{m} = \frac{2\overline{\epsilon_k}N_A}{A} = \frac{2}{A} \times (6.21 \times 10^{-21} \text{ J}) \times (6.02 \times 10^{26})$$

$$= \frac{747.7}{A} \times 10^4 \text{ m}^2/\text{s}^2$$

(a) The diatomic molecule of nitrogen has a molecular weight of $A = 28$. Then $v_{\text{rms}} = \sqrt{\dfrac{747.7}{28}} \times 10^2 \text{m/s} = 516.8 \text{ m/s}.$

(b) Since an oxygen molecule has a molecular weight of 32, $v_{\text{rms}} = \sqrt{\dfrac{747.7}{32}} \times 10^2 \text{m/s} = 483.4 \text{ m/s}.$

(c) Since a water molecule has a molecular weight of 18, $v_{\text{rms}} = \sqrt{\dfrac{747.7}{18}} \times 10^2 \text{m/s} = 644.5 \text{ m/s}.$

3. SPECIFIC HEATS OF GASES

Equation 1.9 gives the average kinetic energy per molecule of an ideal gas as a function of temperature. This result of the kinetic theory was subjected to immediate experimental verification by using it to predict the heat capacity of an ideal gas. If no external forces act, the potential energy is zero and the total mechanical energy of a kilomole of gas is simply

$$U = N_A \overline{\epsilon_k} = \tfrac{3}{2}N_A kT = \tfrac{3}{2}RT. \qquad (1.11)$$

The amount of energy required to change the temperature of a gas one degree Kelvin is called the heat capacity. Then from Equation 1.11 it follows that the heat capacity per kilomole of an ideal gas at constant volume is

$$C_V = \left(\frac{\partial U}{\partial T}\right)_{V \text{ constant}} = \frac{3}{2} R = 2.98 \text{ kcal/kmole-K}. \qquad (1.12)$$

Kinetic theory thus predicts a constant value for all ideal gases. A glance at the experimental results shown in Table 1–1 confirms this value for the monatomic gases argon, helium, and mercury. At the same time, there is evidently something wrong, since all of the molecules that contain more than one atom have heat capacities that exceed the value of 2.98.

The solution to this problem lay in the proper identification of the degrees of freedom of the molecules. Each independent coordinate that is required to define the position or orientation of a mass in a mechanical system is known as a *degree of freedom*. Thus, a bead sliding on a wire has only one degree of freedom since just one coordinate is needed to specify its position. However, a ball rolling on a table would require two coordinates, say (x, y) or (r, θ). In order to specify the position of a gnat in a room one would need three coordinates, (x, y, z).

Referring to Equation 1.9, we see that the average translational kinetic energy of each molecule of a gas in thermal equilibrium is $\tfrac{3}{2}kT$. Assuming that each mole-

Table 1–1
Kilomolar Heat Capacities for Common Gases at Room Temperature
and 1 atm.

Gas	C_V(kcal/kmole-K)
A	2.98
He	2.98
Hg	2.98
CO	4.94
H_2	4.88
HCl	5.11
N_2	4.96
NO	5.02
O_2	5.03
CO_2	6.80
NH_3	6.65
SO_2	7.50

cule is a point mass with just three degrees of freedom along the x, y, and z axes, we note that *on the average there is $\frac{1}{2}kT$ of kinetic energy associated with each degree of freedom*. This principle, enunciated by Clausius, is known as the *equipartition of energy principle*.

Although a monatomic molecule can be treated very nearly as a point mass having only three translational degrees of freedom, a diatomic molecule should resemble a rigid dumbbell having two additional degrees of freedom. This can be understood by noting that three position coordinates and two angular coordinates would be required to specify both its position and its orientation. (See Figure 1–2.)

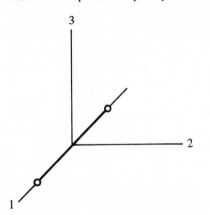

Figure 1–2

The three orthogonal rotation axes for a dumbbell molecule. Axis 1 is neglected when atoms are regarded as point masses.

Alternatively, one can argue that the dumbbell has three rotation axes but that the axis which passes through the centers of the two atoms should be discarded for point masses. At any rate, we find that diatomic molecules must possess at least five degrees of freedom. Assigning $\frac{1}{2}kT$ of energy to each in accordance with the equipartition theorem would result in a kilomolar heat capacity of

$$C_V = \tfrac{5}{2}R = 4.97 \text{ kcal/kmole-K.} \tag{1.13}$$

This result is in reasonably good agreement with values given in Table 1–1 for diatomic molecules.

One might ask whether additional degrees of freedom are possible. Indeed they are! No polyatomic molecule is truly rigid. The link holding the two atoms of a diatomic molecule should be regarded as a spring, not a rigid rod, so simple harmonic motion of the masses is possible. This provides two additional degrees of freedom; one is associated with the vibrational kinetic energy and the other with the elastic potential energy. Hence C_V becomes 6.95 kcal/kmole-K. In complex molecules many vibrational modes are possible, and the molar heat capacity can become quite large.

The explanation of the heat capacities of gases was a triumph of kinetic theory that gave considerable credibility to the atomic theory of matter. This success, however, was accompanied by another problem, namely, that of explaining the behavior of the heat capacity of a gas as a function of temperature. According to Equation 1.12 the heat capacity should have a constant value at all temperatures. That this is *not* the case is demonstrated by the sketch of C_V vs. T for hydrogen that is shown in Figure 1–3. Although kinetic theory could explain any one

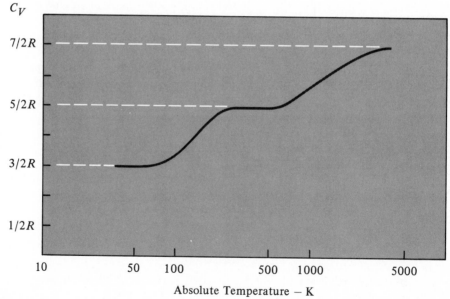

Figure 1–3

The kilomolar heat capacity at constant volume for H_2 versus the logarithm of absolute temperature. Below 100 K, the heat added goes into translational kinetic energy of the molecule. Between about 300 K and 600 K the molecule behaves like a rigid dumbbell. Above 600 K vibrations also occur. The hydrogen molecule dissociates at about 3200 K.

of the constant values $\frac{3}{2}R$, $\frac{5}{2}R$, or $\frac{7}{2}R$, it could not explain why the heat capacity would change from one value to another with increasing temperature. Nor could it explain those heat capacities that deviate greatly from the predicted values, such as $C_V = 6.15$ for chlorine. It remained for the advent of quantum mechanics and the concept of the quantization of energy to provide a satisfactory explanation. This subject will be discussed in subsequent chapters.

4. THE MAXWELL DISTRIBUTION OF VELOCITIES

In the previous sections we have referred to the root-mean-square speed of the molecules in a gas, but we have said nothing about the actual distribution of speeds. In a real system where the collisions are not all elastic, one would expect the speeds of the molecules to vary over a wide range at any instant; it is also reasonable to assume that the speed of a given molecule would vary considerably with time.

First consider the individual velocity components. At equilibrium we would expect the distribution of v_x to be symmetric about $v_x = 0$. Likewise, we require the distributions of v_y and v_z to be symmetric about $v_y = 0$ and $v_z = 0$, respectively. These assumptions are consistent with the discussion in Section 3. It was learned from experiments on the speeds of molecules that these distribution functions are indeed symmetric and that the shape of the distribution curve is the well-known Gaussian function shown in Figure 1–4. Hence, the probability that a

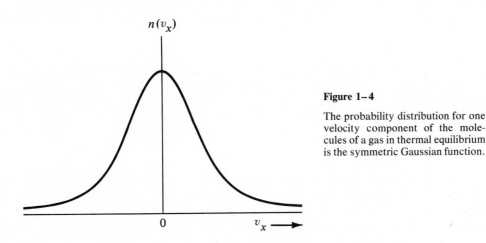

Figure 1–4

The probability distribution for one velocity component of the molecules of a gas in thermal equilibrium is the symmetric Gaussian function.

molecule will have an x-component of velocity given exactly by v_x is

$$P(v_x) = C \exp\left(-\frac{v_x^2}{2\sigma^2}\right), \tag{1.14}$$

where the values of C and σ will be determined later. C is related to the total number of molecules in the gas, while σ is a property of the distribution function called the *standard deviation*. About 68% of the values of v_x will be within $\pm \sigma$ of the mean value, $v_x = 0$. In like manner we may define the probability functions

$$P(v_y) = C \exp\left(-\frac{v_y^2}{2\sigma^2}\right) \tag{1.15}$$

and
$$P(v_z) = C \exp\left(-\frac{v_z^2}{2\sigma^2}\right). \tag{1.16}$$

Since the variables v_x, v_y, and v_z are completely independent, the three probabilities defined above are also independent. It follows that the probability of a velocity having a given set of components (v_x, v_y, v_z) is given by the product of the three probabilities. That is,

$$P(v) = P(v_x) \cdot P(v_y) \cdot P(v_z) = C^3 \exp\left(-\frac{v_x{}^2 + v_y{}^2 + v_z{}^2}{2\sigma^2}\right)$$

$$= C^3 \exp\left(-\frac{v^2}{2\sigma^2}\right). \tag{1.17}$$

Now the number of molecules whose velocities lie in the interval between v and $v + dv$ is given by the product of the probability and the volume of the spherical shell bounded by v and $v + dv$. Thus,

$$n(v)dv = P(v) \cdot 4\pi v^2 dv = 4\pi C^3 \exp\left(-\frac{v^2}{2\sigma^2}\right) v^2 dv. \tag{1.18}$$

If N is the total number of molecules, then[1]

$$N = \int_0^\infty n(v)dv = 4\pi C^3 \int_0^\infty \exp\left(-\frac{v^2}{2\sigma^2}\right) v^2 dv$$

$$= 4\pi C^3 \cdot \tfrac{1}{4}\sqrt{8\pi\sigma^6} = 2\pi C^3 \sigma^3 \sqrt{2\pi}.$$

We may now express the constant C as

$$C^3 = \frac{N}{(\sigma\sqrt{2\pi})^3}. \tag{1.19}$$

Substituting this value in Equation 1.18 yields

$$n(v)dv = \frac{2N}{\sigma^3\sqrt{2\pi}} \exp\left(-\frac{v^2}{2\sigma^2}\right) v^2 dv. \tag{1.20}$$

We may now obtain the total kinetic energy of the gas by multiplying the number of molecules having velocity v times the kinetic energy $\tfrac{1}{2}mv^2$ and then summing this quantity over all possible values of v. But the total kinetic energy is also obtained by multiplying Equation 1.9 by the total number of molecules N. Equating these two expressions for the total kinetic energy, we have:

$$\tfrac{3}{2}NkT = \int_0^\infty (\tfrac{1}{2}mv^2) \cdot n(v)dv = \frac{Nm}{\sigma^3\sqrt{2\pi}} \int_0^\infty \exp\left(-\frac{v^2}{2\sigma^2}\right) v^4 dv$$

$$= \tfrac{3}{2}Nm\sigma^2.$$

We may now express the quantity σ as

$$\sigma^2 = \frac{kT}{m}. \tag{1.21}$$

Inserting this into Equation 1.20, we finally obtain the Maxwell distribution of velocities,

$$n(v)dv = 4\pi N \left(\frac{m}{2\pi kT}\right)^{3/2} v^2 \exp\left(-\frac{mv^2}{2kT}\right) dv. \tag{1.22}$$

---

[1] See Appendix D for the appropriate integrals.

Sketches of this distribution are shown in Figures 1–5 and 1–6. The most probable speed,

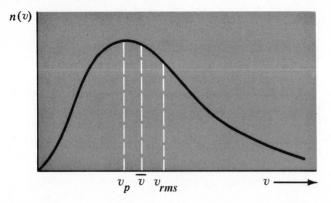

Figure 1–5

The Maxwell distribution of molecular speeds, showing the most probable speed, the mean speed, and the rms speed.

v_p, is indicated by the peak of the distribution curve. Also shown are the mean and root-mean-square speeds. The calculation of these expressions will be left for the problems, but the results are as follows:

$$v_p = \sqrt{\frac{2kT}{m}}; \qquad \bar{v} = \sqrt{\frac{8kT}{\pi m}}; \qquad v_{\text{rms}} = \sqrt{\frac{3kT}{m}} \tag{1.23}$$

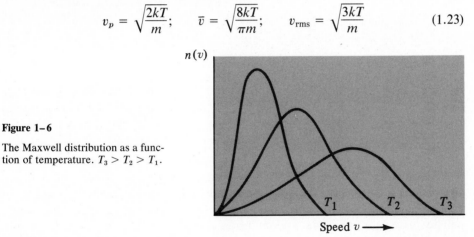

Figure 1–6

The Maxwell distribution as a function of temperature. $T_3 > T_2 > T_1$.

Example 1-3
Find the most probable speed, the average speed, and the rms speed for hydrogen molecules at a temperature of 1000 K.

Solution
Since all of these speeds require the value of kT/m, let us first calculate this.

$$\frac{kT}{m} = \frac{1.38 \times 10^{-23} \times 10^3}{2/6.02 \times 10^{26}} \ (\text{m/s})^2 = 4.15 \times 10^6 (\text{m/s})^2$$

$$\sqrt{\frac{kT}{m}} = 2.04 \times 10^3 \text{ m/s}$$

$$v_p = \sqrt{2} \times 2.04 \times 10^3 = 2{,}880 \text{ m/s}$$

$$\bar{v} = \sqrt{\frac{8}{\pi}} \times 2.04 \times 10^3 = 3{,}260 \text{ m/s}$$

$$v_{\text{rms}} = \sqrt{3} \times 2.04 \times 10^3 = 3{,}530 \text{ m/s}$$

5. BROWNIAN MOTION

The application of kinetic theory to a phenomenon that had puzzled scientists for a long time proved to be convincing evidence for the molecular theory of matter. In 1827 Robert Brown observed the random motions of microscopic pollen grains suspended in water. (See Figure 1–7.) At first this movement was

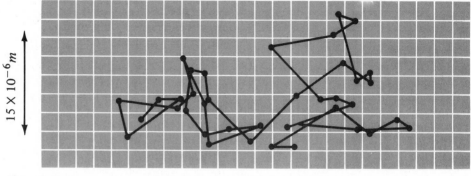

$15 \times 10^{-6} m$

Figure 1–7

The recorded positions of a solid particle in water at intervals of 30 seconds. (From J. Perrin, *Atoms*, D. Van Nostrand Co., Inc., 1923.)

regarded as a property of organic matter, but later it was also observed for small particles of inorganic matter suspended in both liquids and gases. The correct explanation of this *Brownian motion* was given by Einstein in 1905,[1] who concluded that the irregular movement of a suspended particle is due to fluctuations in the bombardment of the particle by the molecules of the fluid. Since even a microscopic particle is huge compared to a molecule, it is safe to assume that millions of molecules strike the particle at any instant and that the net force on the particle will fluctuate with time. The random movement of the suspended particles should resemble the random motions of the molecules themselves.

Indeed, a suspended particle is in thermal equilibrium with the molecules and its average kinetic energy is also $\frac{3}{2}kT$. (Recall Equation 1.9.) Einstein correctly perceived that the size of the fluctuations, that is, the amount of Brownian motion, must be related to the magnitude of Avogadro's number. He treated the fluctuations as a random-walk problem. In a random walk the direction of each successive step must be completely independent of any preceding steps. It turns out that the mean-square displacement from the starting point after N steps is given by

$$\overline{s^2} = N\ell^2, \tag{1.24}$$

---

[1] A. Einstein, *Investigations on the theory of the Brownian movement.* R. Furth (ed.), Dover Publications, Inc., New York, 1965.

where ℓ is the length of each step. That is, *the mean-square displacement from the starting point is proportional to the number of steps.* Applying this to the present problem, we let N be the number of collisions, so that ℓ is the average distance between collisions. It can then be shown that *the mean-square distance traveled is proportional to the elapsed time.* Einstein derived the following expression for the mean-square displacement of spherical particles of radius r suspended in a fluid having viscosity η:

$$\overline{s^2} = \frac{RTt}{3\pi\eta r N_A}, \tag{1.25}$$

where the elapsed time t is in seconds, R is the kilomolar gas constant, T is the absolute temperature, and N_A is Avogadro's number. In 1908, Jean Perrin verified Einstein's equation experimentally and thereby obtained a reasonably good value for Avogadro's number.[1]

SUMMARY

The kinetic theory of gases firmly established the atomic or molecular theory of matter that had long been advocated by chemists and by natural scientists of earlier eras. Avogadro's number, N_A, gives the number of molecules in a kilogram-mole, and its reciprocal is the unified mass unit, u. The approximate values of these two important constants are

$$N_A = 6.02 \times 10^{26} \text{ molecules/kmole},$$

$$u = 1.66 \times 10^{-27} \text{ kg}.$$

By the application of kinetic theory to a monatomic gas, the pressure of the gas was derived and an expression was obtained for the average translational kinetic energy per atom,

$$\overline{\epsilon_k} = \tfrac{3}{2}kT.$$

Utilizing the principle of the equipartition of energy, kinetic theory predicts that the heat capacity of an ideal gas should be $\tfrac{1}{2}R$ per degree of freedom. This prediction is confirmed reasonably well for a large number of gases. However, kinetic theory by itself is unable to explain the sometimes complex temperature dependence of the heat capacity.

The Maxwell distribution of molecular speeds was derived. In a collection of N molecules in equilibrium at temperature T, the number of molecules having speed v is given by

$$n(v) = 4\pi N \left(\frac{m}{2\pi kT}\right)^{3/2} v^2 \exp\left(-\frac{mv^2}{2kT}\right).$$

Brownian motion was mentioned in this chapter for two reasons. It provides striking evidence to the unaided eye of the random motions of molecules assumed in the kinetic theory, and it leads to a reasonably good value for Avogadro's number.

---

[1] For details of the experiments see Jean Perrin, *Atoms*, D. Van Nostrand Co., Inc., New York, 1923.

Additional
Reading

C. H. Blanchard, C. R. Burnett, R. G. Stoner, and R. L. Weber, *Introduction to Modern Physics* (2nd Ed.), Prentice-Hall, Inc., Englewood Cliffs, N.J., 1969, Chapter 3.

H. Boorse and L. Motz, editors, *The World of the Atom,* Basic Books, Inc., New York, 1966.

R. M. Morse, *Thermal Physics,* W. A. Benjamin, Inc., New York, 1964.

F. Reif, *Statistical Physics,* Berkeley Physics Course, Vol. 5, McGraw-Hill Book Co., New York, 1967.

P. A. Tipler, *Modern Physics,* Worth Publishers, Inc., New York, 1978, Chapter 2.

F. W. Van Name, Jr., *Modern Physics* (2nd Ed.), Prentice-Hall, Inc., Englewood Cliffs, N.J., 1962, Chapter 7.

PROBLEMS

1-1. What is the mass of an atom of gold?

1-2. Calculate the mass of an iron atom.

1-3. What is the mass of a molecule of carbon dioxide?

1-4. Determine the mass of a molecule of water.

1-5. An ideal gas is in thermal equilibrium at temperature T. If it is cooled to an equilibrium temperature of $T/2$, how is the average energy per molecule affected?

1-6. What is the volume of a kilomole of an ideal gas at a temperature of 0° C and at atmospheric pressure?

1-7. What is the particle density in molecules per liter for an ideal gas at standard conditions (0° C and 1 atm. of pressure)?

1-8. A gas cylinder having a volume of 0.5 m³ contains 1 kilomole of O_2 at a temperature of 28° C. (a) What is the pressure of the gas? (b) What is the particle density in the gas? (c) What is the mass per unit volume?

1-9. Using the equipartition of energy, obtain the specific heats of He and N_2 and explain why they differ.

1-10. Find the total kinetic energy per kilomole of an ideal gas at a temperature of 300 K.

1-11. Show that the most probable speed in a Maxwell distribution is $v_p = \sqrt{\dfrac{2kT}{m}}$.

1-12. Show that the average speed in a Maxwell distribution is $v = \sqrt{\dfrac{8kT}{\pi m}}$.

1-13. Show that the root-mean-square speed in a Maxwell distribution is $v_{\mathrm{rms}} = \sqrt{\dfrac{3kT}{m}}$.

1-14. Use Equation 1.23 to calculate the rms speeds of nitrogen, oxygen, and water vapor in equilibrium at 300 K. (Do these results agree with the values calculated in Example 1-2?)

1-15. Obtain the Maxwell distribution for translational kinetic energy, $E_K = \frac{1}{2}mv^2$, from Equation 1.22 and show that it may be written in the form $n(E_K)dE_K = \dfrac{2N}{\sqrt{\pi}(kT)^{3/2}} E_K{}^{1/2} \exp\left(-\dfrac{E_K}{kT}\right) dE_K$.

1-16. Show that the most probable kinetic energy of a gas molecule is $\frac{1}{2}kT$. (Hint: See Problem 1-15.)

1-17. Find the rms speed for carbon dioxide at an equilibrium temperature of 300 K.

1-18. At what temperature would v_{rms} for hydrogen equal 11.2 km/s, the escape speed from the earth's gravitational field?

1-19. A thermal neutron is one that is in equilibrium with gas atoms at room temperature. Calculate the kinetic energy and rms speed of neutrons in thermal equilibrium at 300 K. (The mass of a neutron is 1.675×10^{-27} kg.)

1-20. Find the rms speed of a helium atom at 4.2 K.

*1-21. (a) Using the hard-sphere model, show that a molecule of radius r and speed v would undergo $4\pi r^2 nv$ collisions per second in a gas having a particle density of n molecules per unit volume. (b) Find the mean free path, λ, that is, the average distance between collisions. (c) Find the mean free path in an ideal gas at standard conditions. (Assume a molecular radius of 1.8 Å.)

*1-22. In a container of gas at 300 K the pressure is reduced until the mean free path is 1.0 cm. If the molecules are regarded as spheres of radius 1.4 Å, what will be the final pressure? (See Problem 1-21.)

2 The First of the Elementary Particles

Although atoms are now known to be the fundamental particles of the chemical elements, one immediately asks, "Of what do atoms consist?" It was long suspected that electrical charge had something to do with atomic structure, but it was not at all clear just what role it played. Studies of the electrification of bodies show that many neutral bodies can be charged simply by friction and that charge can be transferred freely from one body to another. Yet there is no discernible change in the matter itself other than the presence or absence of charge. Thus charge was first thought of as some sort of continuous fluid that resided in matter.

1. DISCOVERY OF THE ELECTRON

During the 19th century considerable effort was devoted to the study of electrical discharges in rarified gases. All that was required for such research was a glass tube with an electrode at each end, a vacuum pump, and a high voltage source. At atmospheric pressure a large voltage is required to produce a discharge, but at a pressure of 2 or 3 mm of Hg a potential difference of 400 volts is generally sufficient to produce a glow discharge. Early workers suspected that the glow was caused by some kind of ray that traveled from one electrode to the other, and a number of ingenious tubes were devised in which obstructions could be manipulated in and out of the beam by the experimenter. (See Figure 2–1.) From the shadows produced by these obstructions it was correctly inferred that the rays emanated from the cathode. It was also learned that cathode rays are deflected by both electric and magnetic fields.

In 1897, J. J. Thomson[1] established the fact that cathode rays consist of negative particles, which were soon to be called electrons.[2] Thomson determined the ratio q/m, the ratio of the charge to the mass of the electron, by means of an apparatus of the type shown in Figure 2–2. Most of the electrons emitted from the cathode C strike the anode A_1. However, those electrons traveling along the axis

---

[1] J. J. Thomson, *Phil. Mag. 44,* 293 (1897).
[2] Apparently, Stoney first used this term for the average charge carried by an ion in solution. In 1874, he obtained a crude value for this charge by dividing the Faraday constant by an estimate of Avogadro's number.

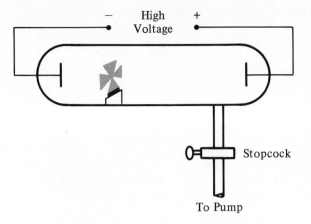

Figure 2–1

Low-pressure gas discharge tube. The opaque obstacle S is made of a magnetic metal and is hinge-mounted, so that it can be moved in and out of the beam by means of an external magnet.

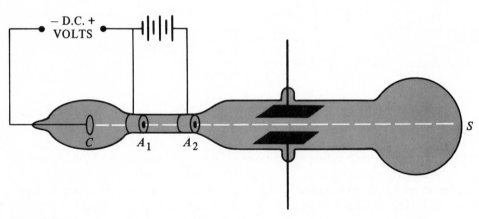

Figure 2–2

Apparatus for determining the ratio of charge to mass for cathode rays.

of the tube pass through a small hole in anode A_1 and are further accelerated and collimated by A_2. Since the tube is highly evacuated, the particles that leave A_2 will travel at nearly the constant speed v_x until they strike the fluorescent screen at S. Now, if a voltage is placed on the vertical deflecting plates in the center of the tube, the beam will be deflected as shown in Figure 2–3. The trajectory is parabolic while the beam is within the electric field of the plates, but it is again a straight line when it enters the field-free region beyond the plates. The vertical force on each electron in the beam is given by

$$F_y = q\mathscr{E}, \tag{2.1}$$

where q is the charge of the electron and $\mathscr{E}$ is the magnitude of the electric field between the plates in volts/meter. If we let m be the mass of the electron, its ver-

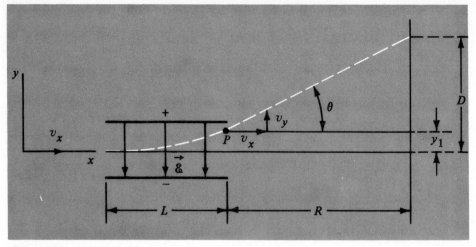

Figure 2–3

Deflection of cathode rays by means of a constant electric field.

tical acceleration will be

$$a_y = \frac{q\mathscr{E}}{m}. \tag{2.2}$$

After a time t, the electron will "fall" a distance

$$y = \frac{1}{2} a_y t^2 = \frac{q\mathscr{E}t^2}{2m}, \tag{2.3}$$

and it will have a vertical velocity given by

$$v_y = a_y t = \frac{q\mathscr{E}t}{m}. \tag{2.4}$$

It is now easy to determine the y-coordinate and the y-component of the velocity at point P. The time interval required to transit the plates is simply L/v_x. Then, at point P,

$$y_1 = \frac{q\mathscr{E}L^2}{2mv_x^2}, \tag{2.5}$$

$$v_{y_1} = \frac{q\mathscr{E}L}{mv_x}, \tag{2.6}$$

and $$\tan\theta = \frac{v_{y_1}}{v_x} = \frac{q\mathscr{E}L}{mv_x^2}. \tag{2.7}$$

Then, the total deflection on the screen may be expressed as

$$D = y_1 + R\tan\theta$$

$$= \frac{q\mathscr{E}L}{2mv_x^2}(L + 2R). \tag{2.8}$$

Thomson then determined the velocity v_x by adding a uniform magnetic field perpendicular to the electric field of the deflecting plates. If the magnetic field $\mathscr{B}$ is directed into the page, the beam of negative particles will be deflected downward by the force

$$\vec{F} = q\vec{v} \times \vec{\mathscr{B}}. \tag{2.9}$$

The magnetic field can be adjusted so that it cancels the electric force, that is, the net force on the electron beam is zero and the beam is accordingly undeflected. The condition for this is

$$q\mathscr{E} = qv_x\mathscr{B},$$

or
$$v_x = \mathscr{E}/\mathscr{B}. \tag{2.10}$$

Inserting this result in Equation 2.8, we obtain

$$D = (q/m)\frac{\mathscr{B}^2 L}{2\mathscr{E}}(L + 2R). \tag{2.11}$$

Since the ratio q/m is the only unknown quantity in Equation 2.11, it can be readily determined. The value obtained by Thomson was about 1×10^{11} coul/kg, whereas the accepted value today is about 1.76×10^{11} coulombs per kilogram.

Thomson's cathode ray tube was the precursor of those used in modern oscilloscopes and television receivers.

Example 2-1

A heated filament or cathode emits electrons by "boiling" them off. When the emitted electrons are collimated into a narrow beam and accelerated, the device is called an *electron gun*. The electrons in such a beam will normally have a wide distribution of speeds unless special steps are taken to produce a beam of particles having the same speed. One way to obtain such a mono-energetic beam is to use a *velocity*

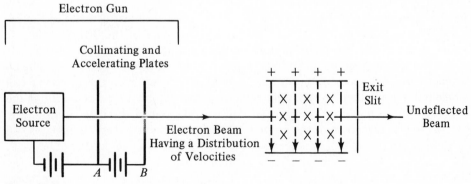

Figure 2-4

Velocity selector using crossed $\mathscr{B}$ and $\mathscr{E}$ fields. The $\mathscr{B}$ field goes into the page (indicated by the tail feathers X of an arrow). The $\mathscr{E}$ field vector is directed downward from the (+) plate to the (−) plate of the parallel deflecting plates.

selector consisting of crossed electric and magnetic fields, as shown in Figure 2–4. The electric field is directed downward in the figure. If it acted alone, it would deflect a positive particle downward and a negative electron upward. The magnetic field is directed into the plane of the paper. Since the direction of the force acting on a moving charge is given by Equation 2.9, it is evident from the definition of the vector cross product that a positive charge would be deflected upward and a negative charge downward. For a given set of values for $\mathscr{B}$ and $\mathscr{E}$ there is only one particle speed that will allow a particle to travel undeflected to the exit slit.

What voltage must be applied across the plates in order to select electrons having a speed of 6×10^5 m/s if the plates are 1 cm apart and the $\mathscr{B}$ field is 0.05 tesla?

Solution

From Equation 2.10,

$$\mathscr{E} = \mathscr{B}v_x = (0.05 \text{ tesla}) \times (6 \times 10^5 \text{ m/s}) = 3 \times 10^4 \text{ volt/m}$$

Then

$$V = \mathscr{E}d = (3 \times 10^4 \text{ volt/m}) \times (10^{-2}\text{m}) = 300 \text{ volts}.$$

2. THE QUANTIZATION OF ELECTRIC CHARGE

Although the q/m ratio was now known for the electron, it was still necessary to obtain an independent measurement of either the charge or the mass. After all, it was possible that cathode rays consisted of particles of different mass and charge such that the ratio always remained the same. Using a technique suggested by J. J. Thomson and H. A. Wilson, R. A. Millikan[1] measured the charge of the electron in 1909. In summary, the method was as follows. Minute droplets of oil will reach a terminal velocity of only a few millimeters per second when allowed to fall freely in air. When they are confined to a draft-free chamber and illuminated, such droplets can be observed with a telescope and their velocities may be easily measured. The mass of a given oil droplet can be calculated from its observed terminal velocity. Now, if this same droplet acquires a charge, it can be accelerated by means of an applied electric field until it achieves its terminal velocity in the upward direction. From the terminal velocities, the mass of the droplet, and the value of the electric field, the amount of charge on the drop can be determined.

A schematic diagram of Millikan's apparatus is shown in Figure 2–5. The oil droplets are produced by an atomizer. With a little practice the observer can single out a droplet and measure its velocity of free fall by means of a scale in the eyepiece of the telescope. Before the droplet falls out of the field of view, the electric field is turned on. If the droplet fails to respond to the electric field it is still uncharged. In that event, another drop that has acquired a charge should be used. An external radiation source greatly facilitates the acquisition of charge by the droplets. A careful experimenter can use the same droplet for an hour or more, making repeated measurements of the terminal velocities during free fall and with the electric field turned on.

---

[1] R. A. Millikan, *Phil. Mag. 19,* 209 (1910); *Phys. Rev. 32,* 349 (1911).

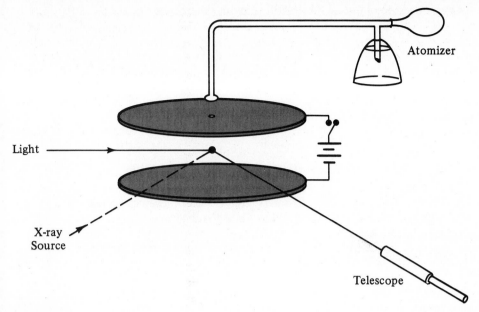

Figure 2–5

Schematic diagram of Millikan's oil drop apparatus.

DETERMINATION OF THE STOKES CONSTANT. The net downward force acting on the droplet during free fall is given by

$$F = mg - f_B - Kv, \qquad (2.12)$$

where mg is the weight of the droplet, f_B is the buoyant force due to the air, and Kv is the friction force, which is proportional to the speed of the droplet. According to Stokes' law,[1] the coefficient of the velocity in the friction force is given by

$$K = 6\pi\eta r, \qquad (2.13)$$

for a small sphere of radius r moving through a homogeneous fluid of viscosity η. Since the buoyant force is simply the weight of the air displaced by the volume of the droplet, we may write

$$mg - f_B = \tfrac{4}{3}\pi r^3(\rho_0 - \rho_A)g, \qquad (2.14)$$

where ρ_0 is the density of the oil and ρ_A is the density of air.

During free fall the terminal velocity v_g is reached when $F = 0$. Then,

$$mg - f_B = Kv_g, \qquad (2.15)$$

or

$$\tfrac{4}{3}\pi r^3(\rho_0 - \rho_A)g = 6\pi\eta r v_g. \qquad (2.16)$$

Solving for r, we obtain

$$r = 3\left[\frac{\eta v_g}{2g(\rho_0 - \rho_A)}\right]^{1/2}. \qquad (2.17)$$

[1] See, for example, F. W. Sears and M. W. Zemansky, *University Physics* (3rd Ed.), Addison-Wesley Publ. Co., Inc., Reading, Mass., 1964, p. 323.

Using this value of r, one can readily obtain the Stokes constant,

$$K = 18\pi \left[\frac{\eta^3 v_g}{2g(\rho_0 - \rho_A)} \right]^{1/2}. \qquad (2.18)$$

Before we proceed further, it should be pointed out that Millikan found that Equation 2.13 needed to be corrected for the extremely small spheres used in this experiment. He determined experimentally that it should be expressed as

$$K' = 6\pi\eta r \left(1 + \frac{b}{pr}\right)^{-1}, \qquad (2.19)$$

where p is the barometric pressure in cm of mercury, and b is an empirical constant whose value is 6.17×10^{-6}. Note that the radius of the droplet r (in meters) also appears in the correction factor. When this correction is applied, the approximate value of r is first calculated from Equation 2.17 and this value is used in the correction term. Then a better value for r may be obtained by multiplying Equation 2.17 by the factor $\left(1 + \frac{b}{pr}\right)^{-1/2}$.

Example 2-2

In a Millikan oil-drop experiment, the average time for a drop to fall 0.250 cm after reaching its terminal speed was 20.0 seconds. The oil used had a density of 896 kg/m³, and the viscosity of air at the ambient temperature was $\eta = 1.83 \times 10^{-5}$ N-s-m⁻². Find the radius of the drop.

Solution

The average free-fall speed was

$$v_g = \frac{2.50 \times 10^{-3}\text{m}}{20.0 \text{ s}} = 1.25 \times 10^{-4} \text{ m/s}.$$

Taking the density of air to be about 1 kg/m³, Equation 2.17 gives for r

$$r = 3 \left(\frac{1.83 \times 10^{-5} \times 1.25 \times 10^{-4} \text{ N/m}}{2 \times 9.8 \text{ m/s}^2 \times 895 \text{ kg/m}^3} \right)^{1/2}$$

$$= 1.08 \times 10^{-6} \text{ m}$$

$$= 1.08 \ \mu\text{m}.$$

Example 2-3

For the oil-drop of the previous example, find the corrected radius using Equation 2.19.

Solution

At a pressure of 76 cm of Hg,

$$\frac{b}{pr} = \frac{6.17 \times 10^{-6}}{76 \times 1.08 \times 10^{-6}} = 7.52 \times 10^{-2}.$$

Then the corrected radius is

$$r = 1.08(1.0752)^{-1/2} = 1.04 \ \mu\text{m}.$$

The correction is about 4% in this case.

DETERMINATION OF THE CHARGE. Now let us assume that the droplet has acquired a charge of q, and that it is being accelerated against gravity in an electric field $\mathscr{E}$ as shown in Figure 2–6. The net upward force acting on the

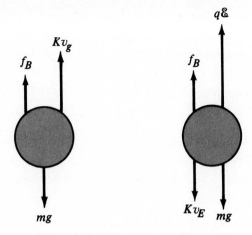

(a) Falling at speed v_g (b) Rising at speed v_E

Figure 2–6

Forces acting when terminal velocities are achieved.

droplet in this case is

$$F = q\mathscr{E} + f_B - mg - Kv. \qquad (2.20)$$

When the upward terminal velocity v_E is reached, then $F = 0$, and

$$q\mathscr{E} - Kv_E = mg - f_B. \qquad (2.21)$$

Combining Equations 2.15 and 2.21 yields

$$q\mathscr{E} - Kv_E = Kv_g,$$

or

$$q = \frac{K}{\mathscr{E}} (v_g + v_E). \qquad (2.22)$$

Since all of the quantities on the right-hand side of Equation 2.22 can be measured, q is readily calculated.

From the results of thousands of measurements, Millikan and his co-workers always found q to be an integral multiple of one value, which he took to be the charge of the electron, e. Thus electric charge is discrete and *not* continuous. It comes in units of e, whose value is approximately

$$e = 1.60 \times 10^{-19} \text{ coulomb.}$$

Using this value for e and the value for e/m, we then obtain for the mass of the electron

$$m = 9.11 \times 10^{-31} \text{ kg.}$$

The electronic mass and charge are such inconceivably small quantities that neither can be measured directly.

An important consequence of the new value for the electronic charge was that it led to the first good determination of Avogadro's number. Electrochemists had known for a long time that a definite quantity of electricity was required to electrodeposit a mole of any monovalent material. This quantity of electricity is called the faraday, and its value was known quite accurately. Avogadro's number may be determined simply by dividing the faraday unit by the electronic charge.

A convenient energy unit for calculations involving the electron and other elementary particles is based upon the value of the electronic charge. It is called the *electron volt,* and it simply represents the amount of kinetic energy that an electron receives when it is accelerated through a potential difference of one volt. That is,

$$1 \text{ eV} = 1.60 \times 10^{-19} \text{ coulomb} \times 1 \text{ volt} = 1.60 \times 10^{-19} \text{ J}.$$

Example 2-4

What electron velocities are produced by the following accelerating potentials: (a) 100 volts; (b) 500 volts; (c) 1000 volts?

Solution

(a) $100 \text{ eV} = 1.6 \times 10^{-17} \text{ J} = \frac{1}{2}mv^2$

$$v = \left(\frac{2 \times 1.6 \times 10^{-17} \text{ J}}{9.1 \times 10^{-31} \text{ kg}}\right)^{1/2} = 5.9 \times 10^6 \text{m/s}$$

(b) $$v = \left(\frac{2 \times 8.0 \times 10^{-17} \text{ J}}{9.1 \times 10^{-31} \text{ kg}}\right)^{1/2} = 1.3 \times 10^7 \text{ m/s}$$

(c) $$v = \left(\frac{2 \times 1.6 \times 10^{-16} \text{ J}}{9.1 \times 10^{-31} \text{ kg}}\right)^{1/2} = 1.9 \times 10^7 \text{ m/s}$$

At these energies, the classical expression for the kinetic energy is valid for electrons. In Chapter 4 we will see how to treat the relativistic effects that occur at higher energies.

Example 2-5

An electron enters the slit of the first accelerating plate A in Figure 2–4 with a speed of 4×10^6 m/s. If the potential difference between the two accelerating plates is 100 volts, what speed does the electron have when it leaves plate B?

Solution

If v is the final speed and v_0 is the initial speed, then

$$\frac{1}{2}m(v^2 - v_0^2) = 100 \text{ eV} = 1.6 \times 10^{-17} \text{ J}$$

$$v^2 = v_0^2 + \frac{2 \times 1.6 \times 10^{-17} \text{ J}}{9.1 \times 10^{-31} \text{ kg}} = (16 \times 10^{12} + 35.17 \times 10^{12})(\text{m/s})^2$$

$$v = 7.15 \times 10^6 \text{ m/s}$$

Example 2-6

An electron having a kinetic energy of 200 eV enters a uniform magnetic field of 0.01 tesla which is oriented perpendicular to the velocity of the electron. What is the radius of the electron's path?

Solution

The force on the electron due to the magnetic field is given by Equation 2.9. Since this force must provide the centripetal acceleration required for circular motion, we write

$$\frac{mv^2}{r} = ev\mathscr{B},\qquad(2.23)$$

or
$$p = mv = er\mathscr{B},\qquad(2.24)$$

where p is the momentum of the electron. Now the momentum and the classical kinetic energy are related by

$$E_K = \frac{1}{2}mv^2 = \frac{p^2}{2m}.\qquad(2.25)$$

Then Equation 2.24 becomes

$$\sqrt{2mE_K} = er\mathscr{B},$$

and
$$r = \frac{\sqrt{2mE_K}}{e\mathscr{B}}.\qquad(2.26)$$

$$r = \frac{(2 \times 9.1 \times 10^{-31}\ kg \times 200 \times 1.6 \times 10^{-19}\ J)^{1/2}}{1.6 \times 10^{-19}\ coul \times 0.01\ tesla}$$

$$r = \frac{7.63 \times 10^{-24}}{1.6 \times 10^{-21}}\ m = 4.77\ mm$$

Example 2-7

A cathode-ray tube has electrostatic deflecting plates that are 2 cm long and 0.5 cm apart. After leaving the plates, electrons must travel 25 cm to reach the fluorescent screen. What voltage must be applied to the plates in order to deflect 1000-eV electrons to a point on the screen that is 10 cm above the axis of the tube?

Solution

From Figure 2–3 we see that $L = 2$ cm, $R = 25$ cm and $D = 10$ cm. The kinetic energy is given by

$$E_K = \tfrac{1}{2}mv_x^2 = 1000\ eV = 1.6 \times 10^{-16}\ J.$$

We first need to solve for $\mathscr{E}$ in Equation 2.8.

$$\mathscr{E} = \frac{4DE_K}{eL(L + 2R)} = \frac{4 \times 10^{-1}\ m \times 1.6 \times 10^{-16}\ J}{1.6 \times 10^{-19}\ coul \times 2(2 + 50) \times 10^{-4}\ m} = 3.85 \times 10^4\ volts/m$$

$$V = d\mathscr{E} = 5 \times 10^{-3}\ m \times 3.85 \times 10^4\ volts/m = 193\ volts.$$

---

3. THE PROTON, THE NEUTRON, ANTIPARTICLES, AND SPIN

In the course of the studies of cathode rays described in Section 1, Goldstein first noted in 1886 that there were also rays traveling in the opposite direction. These rays were found to carry positive charge, so they became known as "positive rays." Since they could be deflected by both electric and magnetic fields, it was also possible to measure the speeds and the q/m ratios of these particles by the methods used for electrons. The most innovative approach to the study of positive rays, however, was Thomson's parabola method.[1] It turned out that the par-

---

[1] See, for example, J. D. Stranathan, *The Particles of Modern Physics,* The Blakiston Co., Philadelphia, 1942, Chapter 5.

ticle speeds and the q/m ratios were always much smaller than for electrons. The largest q/m ratio was found for hydrogen, and it agreed with the value obtained for the hydrogen ion during electrolysis. This is, of course, the q/m value for the *proton,* although that name was not given to the hydrogen nucleus until 1920 by Lord Rutherford. The proton has the same magnitude of charge as the electron and its mass is about

$$m_p = 1.673 \times 10^{-27} \text{ kg.}$$

It remains to this day a mystery why there is such a large disparity in the masses of the electron and the proton, while their charges are identical in magnitude so far as we can determine.

As early as 1920 Rutherford predicted the existence of a heavy particle comparable to the proton in mass but possessing no charge. This particle, called the *neutron,* was finally identified in 1932 by Chadwick.[1] Since the neutron cannot be accelerated by electric or magnetic fields like charged particles, it must be studied indirectly by its effects on other particles. The approximate value for the mass of the neutron is

$$m_n = 1.675 \times 10^{-27} \text{ kg.}$$

Note that this value is greater than the sum of the electron and proton masses.

In the same year that the neutron was identified, Carl Anderson[2] discovered the positive electron or *positron.* It is the antiparticle to the electron, that is, it is identical to the electron except for the sign of its charge. Its discovery led to the prediction that there should also be a negative proton or antiproton. As we shall see later, this particle and others have now been found, and the list of subatomic particles has grown in number and in complexity.

There is another property of the elementary particles introduced above that should be mentioned at this time. In applying the law of the conservation of angular momentum to interactions involving these particles, it soon becomes apparent that they possess intrinsic angular momenta that cannot be accounted for by classical mechanics. An orbital angular momentum can be described simply by the vector product $\vec{r} \times \vec{p}$, where $\vec{r}$ is the position vector and $\vec{p}$ is the linear momentum of the particle. But an elementary particle has an additional angular momentum for which it is impossible to even define $\vec{r}$ or $\vec{p}$! Since its origin is not known, we assume that it derives from the particle spinning on its own axis. Hence, we call it spin angular momentum or simply *spin.* The electron, proton, neutron, positron, and antiproton are all classed as spin-$\frac{1}{2}$ particles. The unit of spin will be discussed in a later chapter.

4. DISCOVERY OF X-RAYS AND NATURAL RADIOACTIVITY

In 1895 Roentgen[3] who was studying cathode rays in electrical discharge tubes, discovered that a new penetrating radiation was produced at the point

[1] J. Chadwick, *Proc. Roy. Soc.* (London) *A136,* 692 (1932).
[2] C. D. Anderson, *Science 76,* 238 (1932).
[3] W. K. Roentgen, *Electrician* (London) *36,* 415, 850 (1896).

where the cathode rays struck the plate or the glass wall of the tube. He detected the new rays, which he called x-rays, by means of the fluorescence they produced in some salts that were accidentally in the vicinity of his apparatus. His persistence as an investigator led to the early discovery of many of the properties of x-rays. For example, he noted that most materials are transparent to x-rays but that lead glass is quite opaque. He determined the relative opacities of various thicknesses of different substances. (He even observed the shadows of the bones of his hand.) He found that x-rays discharge electrified bodies, and that they affect photographic plates, but that they are not deflected by a magnetic field. He also learned that targets (anodes) of heavy metals are more efficient than those of light metals.

Roentgen's discovery stimulated such a burst of scientific research that virtually hundreds of papers on x-rays were published in the year 1896 alone! In a relatively short time it was established that x-rays are electromagnetic waves having shorter wavelengths than ultraviolet rays. As a result of unsuccessful diffraction experiments it was believed that x-ray wavelengths must be of the order of one Ångstrom (10^{-8} cm). After Thomson's discovery of the electron in 1897, it was immediately evident that the origin of x-rays is connected with the impact of an electron with the target. That is, the electron's kinetic energy is converted partially to heat and partially to x-radiation. The shorter the stopping time, the greater the energy of the x-ray. Since heavy atoms are effective in stopping electrons, it is easy to see why heavy metals are the most efficient anodes for x-ray tubes.

Just a few months after Roentgen's discovery, but before the origin of x-rays was understood, another accidental discovery was made by Henri Becquerel.[1] Noting the fact that the fluorescence of the cathode ray tube was accompanied by the emission of x-rays, he speculated that perhaps all fluorescing objects produce x-rays. His experiments involved a number of salts whose fluorescence can be activated by exposing them to sunlight. As luck would have it, one of these was a uranium salt, and Becquerel discovered an emanation from it that, like x-rays, blackened protected photographic plates and discharged an electrometer. This radiation was emitted while the uranium salt was kept in the dark, so its origin has no connection whatsoever with the property of fluorescence. Becquerel called the unknown rays *radioactivity*. It was soon learned that uranium, thorium, and the two newly-discovered elements radium and polonium are all radioactive.

When they are subjected to a transverse magnetic field, the radioactive rays, unlike x-rays, separate into three distinct groups (see Figure 2–7) designated as α-rays, β-rays, and γ-rays. From the nature of the deflections, it was concluded that α's carry positive charge, β's are negative, and γ's are uncharged. Researchers using Thomson's method for measuring q/m readily established that β-particles are high-speed electrons and that α-particles have the same charge and mass as an ionized helium atom. Rutherford[2] later proved conclusively that α-particles are helium nuclei, by collecting α-rays in a helium-free chamber. Knowing the number of α-particles entering the chamber per second, he later detected helium gas in the chamber and showed that its growth rate equalled the rate of the incoming particles. Gamma rays displayed all of the properties of x-rays,

[1] H. Becquerel, *Compt. rend. 122* (1896).
[2] E. Rutherford and H. Geiger, *Proc. Roy. Soc.* (London) *A81,* 141, 162 (1908).

Carl D. Anderson shortly after his discovery of the positron in 1932. (Courtesy of the American Institute of Physics Niels Bohr Library, W. F. Meggers Collection.)

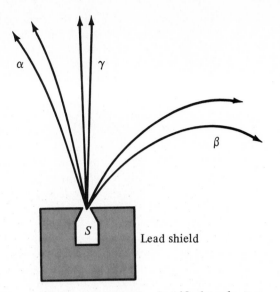

Figure 2–7

Magnetic deflection of the radiation from the radioactive source S. The magnetic field is directed into the plane of the page.

though with greater penetrating power, so they were properly classified as electromagnetic radiation.

Further discussion of x-rays and radioactivity will be deferred until later chapters.

SUMMARY

The discovery of the electron established two very important facts, namely, (1) electric charge appears in discrete units (that is, it is "quantized"), and (2) subatomic particles do indeed exist. The charge-to-mass ratios of a large number of positive ions were measured and the proton, the deuteron, and the alpha particle were soon identified. The neutron was predicted and it was thought by some that the electron, the proton, and the neutron (when found) would complete the family of subatomic particles. However, the first antiparticle, the positron, was detected in the same year that the neutron was discovered, and the list of new particles has been steadily growing ever since.

Additional Reading

C. H. Blanchard, C. R. Burnett, R. G. Stoner, and R. L. Weber, *Introduction to Modern Physics* (2nd Ed.), Prentice-Hall, Inc. Englewood Cliffs, N.J., 1969, Chapter 4.

J. B. Hoag and S. A. Korff, *Electron and Nuclear Physics* (3rd Ed.), D. Van Nostrand Co., Princeton, 1948.

R. A. Millikan, *Electrons + and −, Protons, Photons, Neutrons, and Cosmic Rays,* University of Chicago Press, Chicago, 1935.

H. Semat and J. R. Albright, *Introduction to Atomic and Nuclear Physics* (5th Ed.), Holt, Rinehart and Winston, Inc., New York, 1972, Chapter 1.

J. D. Stranathan, *The Particles of Modern Physics,* The Blakiston Co., Philadelphia, 1942.

P. A. Tipler, *Modern Physics,* Worth Publishers, Inc., New York, 1978, Chapter 3.

F. W. Van Name, Jr., *Modern Physics* (2nd Ed.), Prentice-Hall, Inc., Englewood Cliffs, N.J., 1962, Chapter 1.

PROBLEMS

2-1. The faraday is the quantity of electricity required to electrodeposit a kilogram-mole of a monovalent metal. Calculate its value.

2-2. What is the speed of an electron that has been accelerated through a potential difference of 50 volts?

2-3. The electric field between two parallel plates that are one centimeter apart has the constant value of 2000 volts/m. If an electron is released with zero velocity midway between the plates, with what kinetic energy and speed will it strike one of the plates?

2-4. What value of constant electric field would be required to stop electrons traveling at a speed of 1×10^7 m/s within a distance of 2 cm? 1 cm? 5 mm?

2-5. (a) How many coulombs of electricity would be required to electroplate 10 g of silver? (b) If the plating current is 0.3 ampere, how long would this take?

2-6. What is the kinetic energy of an electron that is bent into a circular path of radius 1.5 cm by a perpendicular magnetic field of 10^{-3} tesla?

2-7. Show that Equation 2.26 may be written in the form

$$r = \frac{2E_K}{(e/m)\mathscr{B}p}$$

2-8. A beam contains protons, deuterons, and alpha particles, all having the same energy of 10^4 eV. The beam enters a mass spectrometer that has a perpendicular magnetic field of 0.1 tesla. What are the radii of the orbits of the three kinds of particles?

2-9. Suppose that protons, deuterons, and alpha particles are accelerated from rest through a potential difference of 10^4 volts and are then sent into the mass spectrometer of the previous problem. Find the radii of the orbits. Compare with the answers of Problem 2-8.

2-10. A 1000-eV electron enters a uniform magnetic field perpendicular to the field lines. If the field strength is 0.01 tesla, what is the radius of the electron's path while it is in the field?

2-11. A 50-eV electron enters a chamber where only the earth's magnetic field acts upon it. Assuming a value of $\mathscr{B} = 5 \times 10^{-5}$ tesla acting perpendicular to the electron's velocity, what would be the radius of the orbit?

2-12. A velocity selector (see Figure 2–4) has a potential difference of 1000 volts across its electrostatic deflecting plates, which are 2 cm apart. If the $\mathscr{B}$ field is 0.05 tesla, what are the speeds of electrons and protons that will pass through the device undeflected?

2-13. In Millikan's apparatus, an oil droplet of density $\rho = 896$ kg/m³ and radius 2.35×10^{-4} cm has an excess charge of 10 electrons.
(a) Find its free-fall terminal velocity with the electric field off. (b) What value of electric field is required to produce zero net force on the droplet? (Take the density of air as 1.0 kg/m³ and the viscosity of the oil as 1.83×10^{-5} N-s-m⁻².)

*2-14. A cathode-ray tube utilizing magnetic deflection has a magnetic field of 0.005 tesla, which is uniform over a distance of 1 cm and zero elsewhere. If the fluorescent screen is 20 cm beyond the edge of the deflecting field, what is the deflection on the screen of a beam of electrons having a kinetic energy of 1000 eV?

*2-15. A discharge tube contains 10^{-7} kilomole of helium gas at a pressure of 10^{-3} atmosphere and a temperature of 300 K. What electric field strength in volts/meter would be required to give a singly ionized helium atom an additional energy of 10 eV between collisions? Use 1.25 Å for the radius of the helium atom.

3 Special Relativity: Space and Time

The special theory of relativity is much more than a conceptual revolution of interest only to philosophers of science. The consequences of the postulates of the special theory of relativity to dynamical systems and to the interactions between matter and energy are so far-reaching that scarcely any area of contemporary physics is free of them, and the relativistic equations now play a significant role in the research activities of many scientists. Therefore, an understanding of the special theory of relativity is extremely important for the student of modern physics.

1. CLASSICAL PRINCIPLE OF RELATIVITY

The theory of relativity is concerned with the way in which observers who are in a state of relative motion describe physical phenomena. The idea of an absolute state of rest or motion has long been abandoned, since an observer "at rest" in an earth-bound laboratory is sharing the motion of the earth about its axis, the earth about the sun, the solar system through the Milky Way, and so on. It is also common knowledge that one can perform *simple* experiments with bouncing balls, oscillating springs, or swinging pendula in a laboratory fixed on the earth or in a smoothly running truck moving at constant velocity and obtain identical results in both sets of experiments.[1] We describe this fact by saying that the laws of motion are *covariant,* that is, they retain the same form when expressed in the coordinates of either frame of reference.

Consider two frames having relative translational velocity V in their common x-directions. It is evident that the coordinates are related by

$$x' = x - Vt$$
$$y' = y$$
$$z' = z$$
$$t' = t,$$

---

[1] This is true only for experiments that are insensitive to the rotation of the earth.

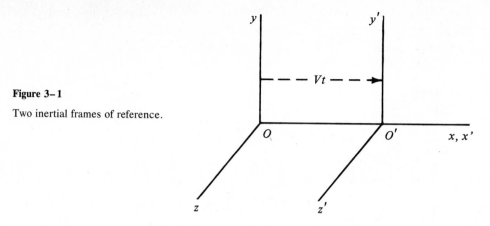

Figure 3–1

Two inertial frames of reference.

provided that $t = t' = 0$ when the origins coincide. Differentiating with respect to time, we find

$$u'_x = u_x - V$$

$$u'_y = u_y$$

$$u'_z = u_z,$$

where $u_x = dx/dt$, and so forth. This again checks with our everyday experience with velocities. Differentiating once more, we have

$$a'_x = a_x$$

$$a'_y = a_y$$

$$a'_z = a_z,$$

where $a_x = du_x/dt$, and so forth. Then,

$$\vec{F'} = m\vec{a'} = m\vec{a} = \vec{F}. \tag{3.1}$$

Equation 3.1 says that Newton's laws have the same form in both coordinate frames, that is, they are covariant. Note that the coordinates and velocities are different in the two frames but each observer always knows how to obtain the other's values. Frames in which Newton's laws are covariant are called *inertial frames*. Inertial frames are *equivalent* in the sense that there is no mechanical experiment which can distinguish whether either frame is at rest or in uniform motion; hence, *there is no preferred frame.* This is known as the *Galilean* or *classical principle of relativity,* and the coordinate transformation given above is called a *Galilean transformation.* Strictly speaking, the earth is not an inertial frame because of its rotation and its orbital motion, but it can often be treated as an inertial frame without serious error.

As a consequence of the principle of relativity, observers in different inertial frames would all discover the same set of mechanical laws, namely, Newton's laws of motion. By way of illustration, consider the following thought experiment. An observer, riding on a flat car which is moving at constant velocity past an observer on the ground, fires a ball vertically upward from a small cannon and notes its maximum height as well as the total time of flight for it to return to the muzzle of the cannon. The observer on the ground also measures the maximum height and the time of flight. Although the two observers will not agree on the shape of the trajectory of the ball, they *will* agree on the height, the time, the calculated muzzle velocity, and the value of g, the gravitational acceleration.

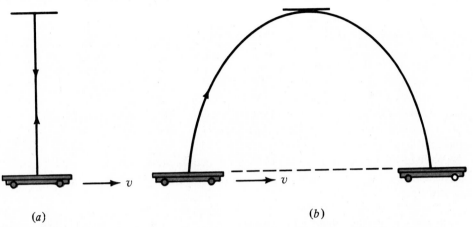

(a)   (b)

Figure 3–2

A ball fired vertically upward from a flatcar moving with speed V relative to the ground will have the trajectory shown in (a) for an observer on the car. An observer at rest on the ground will see the trajectory shown in (b).

The classical principle of relativity is widely used in the everyday world; as long as the speeds of the moving objects are small compared to the speed of light, the error is negligible. For example, consider the following problem.

Example 3-1

An aircraft has a take-off speed of 200 km/hr relative to the deck of an aircraft carrier. If the carrier is moving in the same direction at 50 km/hr with respect to the water, what is the aircraft's speed relative to the water?

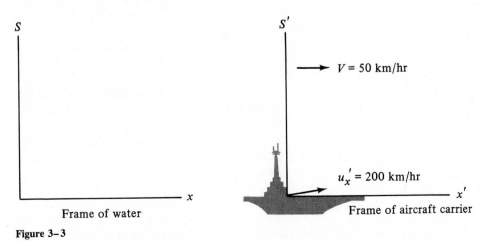

Frame of water   Frame of aircraft carrier

Figure 3–3

Reference frames for Example 3-1.

Solution

The speed of the aircraft relative to the water is

$$u'_x + V = 200 \text{ km/hr} + 50 \text{ km/hr} = 250 \text{ km/hr}.$$

2. ELECTROMAGNETIC THEORY AND THE GALILEAN TRANSFORMATION

Long after the Galilean principle of relativity was well established, Maxwell formulated his electromagnetic field equations, out of which arose a finite and constant velocity for light in vacuum. In spite of the great success of Maxwell's equations in describing the behavior of electromagnetic phenomena in space and time, it was quite disturbing to find that they are *not* covariant under a Galilean transformation. Thus, the classical transformation, which permits a covariant description of mechanical forces, does *not* hold for electromagnetic forces.

Example 3-2

Using the classical (Galilean) principle of relativity, what would an observer on earth measure for the speed of a laser pulse fired toward the observer from a rocket having a speed V relative to the earth? The speed of the laser pulse in the frame of the rocket (the primed frame) is c.

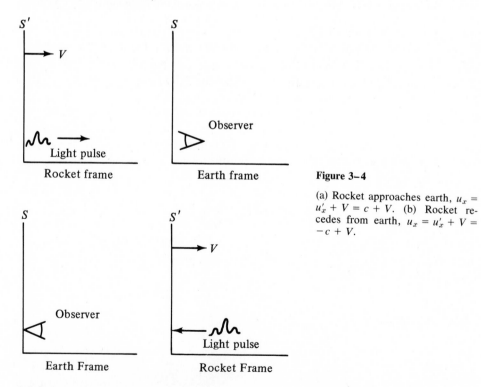

Rocket frame   Earth frame

Earth Frame   Rocket Frame

Figure 3–4

(a) Rocket approaches earth, $u_x = u'_x + V = c + V$. (b) Rocket recedes from earth, $u_x = u'_x + V = -c + V$.

Solution

The situation, apparently analogous to that of Example 3-1, is shown in Figure 3–4. Hence, the Galilean transformation yields the result that the speed of light is $V +$

c when the rocket approaches the observer and $V - c$ when the rocket recedes from the observer.

If the above results *were* true, then Maxwell's prediction that the speed of light in free space should be c would hold true only for an observer in the rest frame of the light source. On the other hand, if the speed of light is really c for all observers, then the Galilean transformation *cannot* be correct for electromagnetic waves.

Another serious difficulty arose from electromagnetic theory, in connection with a medium which was postulated to sustain the wave motion. This ethereal substance was called "the ether," and was endowed with infinite elasticity and inertia but zero mass. These contradictory properties were required in order to sustain the transverse vibrations of light waves and yet prevent any longitudinal vibrations. Furthermore, the ether had to pervade all of space without inhibiting the movements of celestial bodies, since it was known that these bodies move in a non-viscous medium.

It was hoped that the ether might provide the necessary reference frame for measuring absolute motion. However, these hopes were destined to be unfulfilled, as the following summary will show:

(a) As early as 1728 it was known that the observed position of a star varies slightly throughout the year.[1,2] This phenomenon, which is called the *aberration* of starlight, is due to the fact that the observed direction of a light ray depends upon the relative velocity of the light source and the observer. Since the earth moves in its orbit at about 30 km/sec, a telescope should be tilted toward a star at an angle $\theta \sim 30/300{,}000 = 1 \times 10^{-4}$ radian or 20 seconds of arc. Six months later,

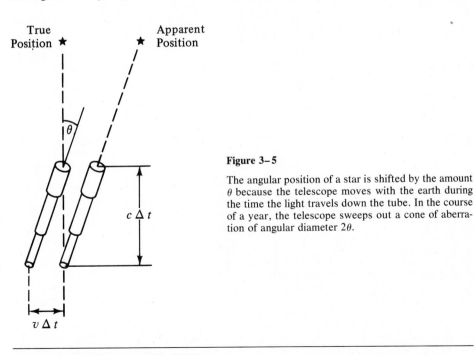

True
Position ★

★ Apparent
Position

Figure 3–5

The angular position of a star is shifted by the amount θ because the telescope moves with the earth during the time the light travels down the tube. In the course of a year, the telescope sweeps out a cone of aberration of angular diameter 2θ.

[1] J. Bradley, *Phil. Trans. 35,* 637 (1728).
[2] A. B. Stewart, "The Discovery of Stellar Aberration," *Sci. Amer.,* p. 100 (March 1964).

the telescope should be tilted the same amount in the opposite direction. See Figure 3–5. An important conclusion can be drawn from the fact that aberration of the correct amount is experimentally observed, namely, that the *earth moves through the ether*. The ether, then, was assumed fixed with respect to the stars.

(b) A number of experiments were devised to measure the earth's motion through the ether, that is, to detect an ether wind with respect to the earth. The most famous of these was the Michelson-Morley experiment,[1] which will be described below. This experiment was repeated many times but always with a null result, that is, no ether wind was detected. Another famous but unsuccessful attempt to detect an ether wind was the experiment of Trouton and Noble.[2] They sought to detect a torque on a charged capacitor due to its motion through the ether.

(c) All attempts to explain the absence of an ether wind while retaining the ether concept were also refuted. It could not be argued that the earth carries the ether along with it since this would contradict the observed aberration of starlight. Some proposed that the velocity of light adds vectorially to the velocity of the source. This theory (historically called the emission theory) would easily account for the null result of the Michelson-Morley experiment, but it contradicts a known fact about wave motion, namely, that the velocity of a wave depends only upon the properties of the medium. Perhaps the most convincing evidence for rejecting the emission theory comes from the study of the periods of binary stars as proposed by De Sitter.[3] The relative velocity (to an earth observer) of a distant star performing circular motion will vary sinusoidally with time. Therefore, the measured Doppler shift[4] of the light emitted by such a star would show the same sinusoidal variation if the speed of light were constant. On the other hand, if the emission theory were true, the speed of the emitted light would be greater as the star approached and less as the star receded. In this event the measured Doppler shifts would no longer show a sinusoidal variation, but the curve would be distorted as in Figure 3–6(a). An actual plot of the Doppler shifts measured for the components of the binary star Castor C is shown in Figure 3–6(b).[5] The absence of distortion in the curves is strong evidence for the validity of the constancy of the speed of light. Many such binaries have been studied, and there is no indication of distortion other than that due to known eccentricities in the orbits. More recent experiments which refute the emission theory of light are measurements of the velocity of light in the form of gamma rays emitted from positron annihilation[6] and from the decay of π° mesons.[7]

A third attempt to explain the null result of the Michelson-Morley experiment

---

[1] A. A. Michelson, *Amer. J Sci.* (3), 22, 20 (1881); A. A. Michelson and E. W. Morley, ibid. *34*, 333 (1887). These results were confirmed by R. J. Kennedy, *Proc. Nat. Acad. 12*, 621 (1926), and K. K. Illingworth, *Phys. Rev. 30*, 692 (1927).

[2] F. T. Trouton and H. R. Noble, *Phil. Trans. Roy. Soc. A202*, 165 (1903); *Proc. Roy. Soc.* (London) *72*, 132 (1903).

[3] W. De Sitter, *Proc. Amsterdam, Acad. 15*, 395 (1913).

[4] The mathematical details of the Doppler shift, which will be discussed in the next chapter, are not required for the present argument.

[5] A. H. Joy and R. F. Sanford, *Astrophys. J. 64*, 250 (1926).

[6] D. Sadeh, *Phys. Rev. Letters 10*, 271 (1963).

[7] T. Alväger, F. J. M. Farley, J. Kjellman, and I. Wallin, *Phys. Letters 12*, 260 (1964).

was the Lorentz-FitzGerald contraction theory. It was postulated that a length is contracted by a factor $(1 - v^2/c^2)^{1/2}$, where v is the component of the velocity parallel to the length and c is the velocity of light. This contraction is just the right amount to explain the null Michelson-Morley result, although it was not derived from first principles. The Lorentz-FitzGerald contraction was extremely difficult

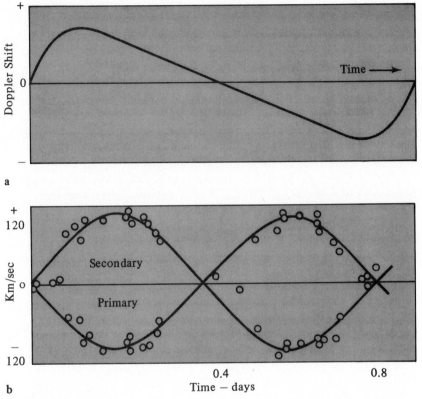

Figure 3-6

The measured Doppler shifts due to the orbital velocity of a star performing circular motion would appear as in (a) if the speed of light were to increase as the star approached and to decrease as the star receded from the observer. Shown in (b) are the relative velocities of the components of the binary star Castor C as determined from the Doppler shifts of their spectra. The absence of distortion is strong evidence for the constancy of the speed of light. (The data are from A. H. Joy and R. F. Sanford, *Astrophysical Journal 64*, 250 [1926].)

to refute,[1] and its physical significance did not become clear until after the work of Einstein.

3. THE MICHELSON-MORLEY EXPERIMENT

As mentioned in the previous section, if the ether is assumed to be at rest in the frame of the "fixed" stars, then the earth's motion through the ether should result in an ether wind for an earth observer. Michelson[2] proposed to detect this

---

[1] R. J. Kennedy and E. M. Thorndike, *Phys. Rev. 42*, 400 (1932).
[2] See previous references and R. S. Shankland, *Amer. J. Phys. 32*, 16 (1964).

wind by looking at the interference pattern produced by two coherent beams of light, where the optical paths of the two beams are respectively parallel to and perpendicular to the earth's motion through the ether. If the two beams are initially in phase and if their optical paths differ by zero or an integral number of wavelengths, they will still be in phase when they are brought together again. In general, the optical paths of the two beams will differ by an arbitrary amount, and either destructive or constructive interference can occur. The arrangement used by Michelson is shown in Figure 3–7. One of the mirrors of the interferometer is usually tilted slightly so that a pattern of alternately bright and dark lines, called fringes, is viewed in the eyepiece. Each dark fringe corresponds to a path difference of an odd number of half wavelengths in the two arms of the instrument. Slight variations in either path produced while observing through the eyepiece will result in an apparent motion of the fringe pattern. Shifts corresponding to a fraction of a wavelength can be easily detected.

Suppose the interferometer is set up so that one light path, say ℓ_2, is collinear with the earth's motion through the ether with velocity v. Then, assuming a Galilean transformation, the time for one round trip along that path is

$$t_2 = \frac{\ell_2}{c - v} + \frac{\ell_2}{c + v} = \frac{2c\ell_2}{c^2 - v^2} = \frac{2\ell_2}{c} (1 - \beta^2)^{-1} \sim \frac{2\ell_2}{c} (1 + \beta^2),$$

where $\beta = v/c$. Since the velocity of the earth in its orbit is about 30 kilometers per second, $\beta \sim 10^{-4}$ and terms higher than β^2 are neglected in the expansion. For the path that is perpendicular to the relative motion the situation is analogous to

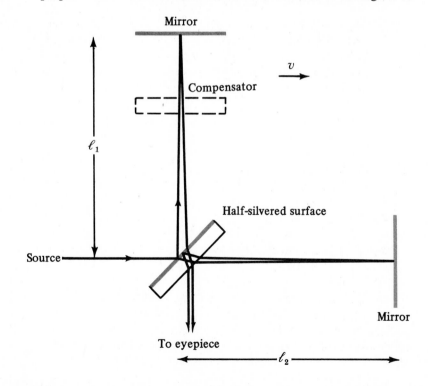

Figure 3–7

Michelson interferometer with arm ℓ_2 parallel to the earth's motion.

the problem of rowing a boat across a river having a current, v. If the rower's speed is c, his path is actually of length

$$ct = \sqrt{\ell_1{}^2 + v^2 t^2},$$

where t is the time for one transit. Then, his velocity for a direct crossing is $\ell_1/t = \sqrt{c^2 - v^2}$, and the time for a round trip transverse to the current is

$$t_1 = \frac{2\ell_1}{\sqrt{c^2 - v^2}} = \frac{2\ell_1}{c}\,(1 - \beta^2)^{-1/2} \sim \frac{2\ell_1}{c}\,(1 + \tfrac{1}{2}\beta^2).$$

The phase difference between the two beams as they enter the eyepiece is

$$\Delta\phi_1 = (2\pi/\lambda) \cdot (\text{path difference}) = (2\pi/\lambda) \cdot c(t_2 - t_1)$$

$$= (2\pi/\lambda) \cdot [2(\ell_2 - \ell_1) + \beta^2(2\ell_2 - \ell_1)].$$

By rotating the interferometer through 90°, the roles of the two beams are interchanged, and the phase difference for this case is

$$\Delta\phi_2 = \frac{2\pi}{\lambda} \cdot [2(\ell_1 - \ell_2) + \beta^2(2\ell_1 - \ell_2)].$$

Thus, the effect of rotating the instrument through 90° is to enable the observer to detect a total phase difference of $\Delta\phi_1 + \Delta\phi_2$. This corresponds to a fringe shift of

$$\frac{1}{2\pi}\,(\Delta\phi_1 + \Delta\phi_2) = \frac{\beta^2}{\lambda}\,(\ell_1 + \ell_2). \tag{3.2}$$

For $\ell_1 \sim \ell_2 = 10$ meters and $\lambda \sim 5000$ Å, Equation 3.2 predicts a shift of about 0.4 fringe. Such a shift could be easily detected, as Michelson's instrument could resolve 0.01 of a fringe at that wavelength.

The Michelson-Morley experiment was repeated many times, at all hours of the day and in all portions of the earth's orbit, yet no fringe shift has ever been observed. The null result requires us either to accept the non-physical Lorentz-FitzGerald contraction or to reject the ether theory, since the ether cannot be fixed with respect to the earth as well as the distant stars.

4. THE POSTULATES OF SPECIAL RELATIVITY

The failure of such efforts as the Michelson-Morley and the Trouton-Noble experiments to discover a preferred frame for the electromagnetic equations of Maxwell suggested that the latter must conform to a principle of relativity. However, the fact that the Galilean principle of relativity—which was known to be valid for classical mechanics—failed for Maxwell's equations was a considerable source of frustration to physicists at the turn of the century. After a critical examination of the concepts of space, time, and simultaneity, Einstein[1] discarded the Galilean principle of relativity and postulated in its stead a principle of relativity for all physical laws. His postulates may be stated as follows:

---

[1] A. Einstein, *Ann. Physik 17*, 891, (1905).

(1) There is no preferred or absolute inertial system. That is, all inertial frames are equivalent for the description of all physical laws.

(2) The speed of light in vacuum is the same for all observers who are in uniform, rectilinear relative motion and is independent of the motion of the source. Its free space value is the universal constant c given by Maxwell's equations.

Since the first postulate says that Maxwell's equations as well as Newton's laws are covariant, a new set of relativistic transformation equations must be obtained to replace those of Section 1. The second postulate follows from the first, since, if an observer were to measure, for example, the value $c - v$ for the speed of light, he would then have a means of determining the relative speed v, which would violate the principle of relativity.

Maxwell's equations enable one to express the universal constant c in terms of the fundamental constants of electromagnetic theory. Thus, in m.k.s. units, $c = (\epsilon_0\mu_0)^{-1/2}$, and its value is very nearly 3×10^8 meters per second. Frequent attempts have been made to determine c to as many significant figures as possible. Some of the methods employed have involved direct measurements of distance and transmission time; others have been concerned with precise measurements of λ and ν in order to calculate c from their product, and still others have used interferometry. There is a large amount of literature dealing with this experimental problem, and a number of good summaries exist which are replete with references.[1,2,3] The currently accepted value for the speed of light is given in the table inside the front and back covers.

Before deriving the new relativistic transformation, let us look briefly at the concept of time. According to classical physics, $t = t'$, and it was tacitly assumed that clocks in different inertial frames could be easily synchronized. The concept of simultaneity was not questioned. Einstein, however, realized that the process of synchronization requires the transmission of signals at a finite velocity which can be at most the velocity of light in free space. Furthermore, simultaneity is a meaningful concept only in a frame in which light sources, clocks, and observers are all at rest. By way of illustration, suppose observer O has measured two equal distances x_1 and x_2 from his reference position and installs identical flash lamps S_1 and S_2 as shown in Figure 3–8. He can fire these flash lamps simulta-

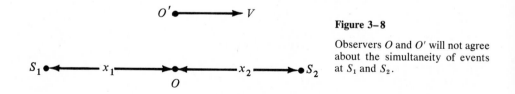

Figure 3–8

Observers O and O' will not agree about the simultaneity of events at S_1 and S_2.

neously by means of matched cables or radio operated relays, and he will assert that the flashes reach him simultaneously. However, to a second observer O', who is in a state of relative motion parallel to the line joining S_1 and S_2, the flashes will not appear to be simultaneous because of the finite distance through which he

[1] J. F. Mulligan, *Amer. J. Phys.* **20,** 165 (1952).
[2] J. F. Mulligan and D. F. McDonald, *Amer. J. Phys.* **25,** 180 (1957).
[3] J. H. Sanders, *The Velocity of Light,* Pergamon Press, New York, 1965.

moves during the transit time of the light waves. Thus, if the flashes occur just as O' passes O, then O' will claim that S_2 flashed earlier than S_1.

Since simultaneity cannot be defined independently of spatial coordinates for observers who are in relative motion, their clocks are not synchronized and their measured times are different. As a consequence of this, taken together with the constancy of the observed speed of light in either frame, we conclude that the statement $t' = t$ is no longer valid.

5. THE EINSTEIN TRANSFORMATION[1]

Although the Galilean transformation is not the correct expression of the principle of relativity, the new transformation should reduce to the Galilean transformation for small relative velocities, since the latter gives correct results for Newton's laws in our macroscopic world. Also, the transformation must be linear in order to make the intervals between events independent of the choice of origin in the space-time coordinate system. The simplest transformation that will satisfy both of these requirements will have the form

$$\left.\begin{array}{l} x' = C(x - vt) \\ y' = y \\ z' = z, \end{array}\right\} \qquad (3.3)$$

where $C \to 1$ as $v \to 0$, and where v is parallel to the common direction of the x and x' axes. Since the transformed time must, in general, depend upon both time and space coordinates, the simplest form to try is

$$t' = At + Bx, \qquad (3.4)$$

where $A \to 1$ and $B \to 0$ as $v \to 0$.

Consider two observers in different inertial frames having relative velocity v along their common x-directions. At the instant that their origins coincide, a light is flashed by the unprimed observer and a spherical wave emanates from his origin, the equation of which is

$$c^2t^2 = x^2 + y^2 + z^2.$$

However, by Einstein's postulates, the primed observer must also see a spherical wave emanating at speed c from *his* origin as its center. If he were not to see a spherical wave but, say, an ellipsoidal wavefront, he would have a means of determining which frame was moving with velocity v. The equation of the wave front in the primed system is

$$c^2t'^2 = x'^2 + y'^2 + z'^2.$$

Then we have

$$x^2 + y^2 + z^2 - c^2t^2 = x'^2 + y'^2 + z'^2 - c^2t'^2. \qquad (3.5)$$

If the new transformation satisfies Equation 3.5, it will be consistent with Einstein's postulates. The derivation will be left as Problem 3-3, so here we will sim-

---

[1] Frequently called the "Lorentz transformation."

ply give the Einstein transformation as follows:

$$\left.\begin{array}{l} x' = \gamma(x - \beta ct) \\ y' = y \\ z' = z \\ ct' = \gamma(ct - \beta x) \end{array}\right\} \tag{3.6}$$

where

$$\beta = v/c$$

and

$$\gamma = (1 - \beta^2)^{-1/2}.$$

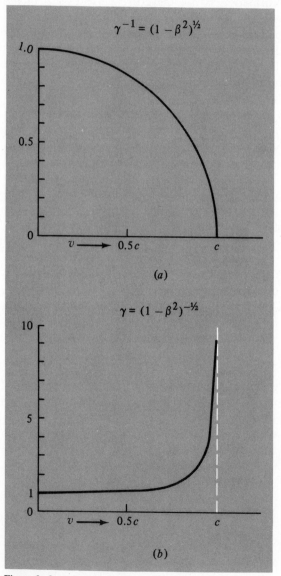

(a)

(b)

Figure 3–9

(a) Plot of γ^{-1} vs. v and (b) plot of γ vs. v.

6. RELATIVISTIC KINEMATICS

We define an event as a physical occurrence that is localized in space and time. That is, to an observer who is at rest in the frame of the event, the time of the event can be measured by means of a single clock and the position of the event can be determined by means of a meter stick. We call the time interval in the rest frame the *proper time interval, τ_0.*

If the event consists of the measurement of the length of an object in its rest frame, the coordinates of the two ends of the object can be measured simultaneously by an observer in that frame. We call the length of the object measured in this fashion its *proper length, L_0.* To an observer in any inertial frame other than that of the event, a measurement of either a length or a time will require the use of *both* a meter stick and a clock. We will now look in detail at the way lengths and time intervals transform under an Einstein transformation.

a. Preservation of Lengths Perpendicular to the Direction of Relative Motion

Consider the following thought experiment. The lengths of two meter sticks are known to be exactly the same in a frame where they are both at rest. If they are now placed in relative motion in a direction perpendicular to their lengths, as in Figure 3–10, what do observers O and O' conclude about their lengths? O can install synchronized flash cameras at each end of his meter stick which he can trigger just as O' passes him. O' will agree that the flash pictures were simultaneous since he is at all times equidistant from the two flash lamps. Since both observers agree that the flashes were simultaneous, they must agree with the result shown on the photographs. That is, the result is an *absolute* one, not a relative one. Now the experiment can be repeated with the flash cameras in the primed frame, and once again both observers will agree that the flashes were simultaneous. The same result must be obtained in both sets of experiments. If, for example, the stick in the unprimed frame were found to be shorter in the first set of pictures, then the stick in the unprimed frame must be found to be shorter in the second set of pictures. However, such a result would contradict the principle of relativity, since in that case motion to the right would stretch a stick and motion to the left would shrink it. Therefore, we conclude that the lengths of the two sticks must remain unchanged. This result is consistent with the symmetrical conditions of the experiments.

b. Contraction of Lengths Parallel to the Direction of Relative Motion

Let the two meter sticks of Figure 3–10 now be placed in relative motion parallel to their common x-axes and the photographs repeated as before. The first set of photographs, taken by O, would look like Figure 3–11 (a), and the second, taken by O', would look like Figure 3–11 (b). In the first, O would say that his picture shows the "moving" stick was contracted. O' would explain the photographs by claiming that the flashes fired by O were *not* simultaneous. In fact, he would say

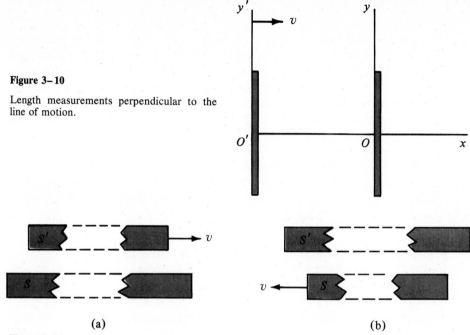

Figure 3–10

Length measurements perpendicular to the line of motion.

(a)   (b)

Figure 3–11

Length measurements parallel to the line of motion. In (a) the photographs are simultaneous to observer O in S; in (b) they are simultaneous to observer O' in S'.

that the leading end of his meter stick was photographed *before* it reached the coincidence position and the lagging end of his meter stick was photographed *after* it had passed, the position of coincidence. Thus, both observers accept the photographs; but since they do not agree that the flashes were simultaneous, they have different explanations for them. In the second experiment the roles are reversed and now observer O' says that his picture shows the meter stick in S is contracted by virtue of its "motion." We cannot argue, as we did in the previous thought experiment, that the sticks should appear equal in length since *there is no agreement on simultaneity* in the present case. We must conclude that the stick that is moving with respect to the observer appears to be contracted.

This contraction may be demonstrated analytically as follows. Suppose observer O' has a meter stick situated so that the end points are at x_1' and x_2' for any time t' in his frame. By the Einstein transformation, an observer in S will obtain the result

$$x_1' = \gamma(x_1 - \beta c t_1)$$
$$x_2' = \gamma(x_2 - \beta c t_2),$$

from which we obtain

$$x_2' - x_1' = \gamma(x_2 - x_1) - \beta\gamma c(t_2 - t_1).$$

But $x_2' - x_1' = L_0$, the proper length, since the stick is at rest in S'. Furthermore, if x_2 and x_1 are measured simultaneously in S, then $x_2 - x_1 = L$, the length of the stick as observed in S. Thus, for $t_1 = t_2$,

$$L_0 = \gamma L \quad \text{or} \quad L = \sqrt{1 - \beta^2}\, L_0. \tag{3.7}$$

The same result can be obtained by using the inverse transformation, but the algebra requires a few more steps. To illustrate this we write

$$x_1 = \gamma(x_1' + \beta c t_1')$$

$$x_2 = \gamma(x_2' + \beta c t_2')$$

$$x_2 - x_1 = \gamma(x_2' - x_1') + \beta \gamma c(t_2' - t_1').$$

But the time transformations must be used here since the simultaneous measurement of x_1 and x_2 in frame S will not appear simultaneous in frame S'. Then,

$$ct_1' = \gamma(ct_1 - \beta x_1)$$

$$ct_2' = \gamma(ct_2 - \beta x_2),$$

from which

$$c(t_2' - t_1') = \gamma c(t_2 - t_1) - \beta \gamma(x_2 - x_1).$$

Putting $t_1 = t_2$ and eliminating $(t_2' - t_1')$ from the above equations,

$$x_2 - x_1 = \gamma(x_2' - x_1') - \beta^2 \gamma^2 (x_2 - x_1),$$

which reduces immediately to Equation 3.7.

This result is identical with the Lorentz-FitzGerald contraction, but it now has a theoretical basis in the postulates of Einstein. Note that the proper length is the *maximum* length that can be measured by any observer.

c. Time Dilation

Suppose an observer in S measures a proper time interval τ_0 by a single clock and that the coordinates of the events marking the start and the finish of the interval are (x_0, t_1) and (x_0, t_2). An observer in the inertial frame S' will observe the coordinates of the events as follows:

$$ct_1' = \gamma(ct_1 - \beta x_0)$$

$$ct_2' = \gamma(ct_2 - \beta x_0)$$

$$c(t_2' - t_1') = \gamma c(t_2 - t_1) - \beta \gamma(x_0 - x_0).$$

Letting $t_2' - t_1' = \tau$ and $t_2 - t_1 = \tau_0$,

$$\tau = \gamma \tau_0 \quad \text{or} \quad \tau_0 = \sqrt{1 - \beta^2}\, \tau. \tag{3.8}$$

As in the case of Equation 3.7, the inverse transformations can be used to derive Equation 3.8, but additional algebraic steps would be required to eliminate x_1' and x_2'. Note that the proper time interval is the *minimum* time interval that any observer can measure between two events. This phenomenon is called *time dilation* and provides the basis for the statement that "moving clocks run slow." If the two events are two "ticks" of a clock, Equation 3.8 says that the minimum interval between ticks will occur for the clock in the rest frame of the observer; that is, the proper clock will run fastest.

It is possible to perform thought experiments in order to compare measurements of time intervals by clocks that are in relative motion. In the course of such experiments it is assumed that any number of clocks in the same rest frame as an observer can by synchronized to any desired degree of precision. It is also assumed that clocks in different inertial frames can be photographed or compared visually from either frame at the instant that the clocks pass the same point in space. Consider the following example. Let observer O' send a light flash a distance ℓ to a mirror at rest in his frame, S', at the instant that he arrives at the origin of the unprimed frame where observer O is stationed. O' sees the light reflected directly back to his position, and he records the proper time interval for the round trip by means of a single clock. The proper time interval is $\tau_0 = 2\ell/c$. In the unprimed frame the light beam is reflected from a point to the right of observer O, as shown in Figure 3–12, since the mirror is moving to the right with velocity v. The

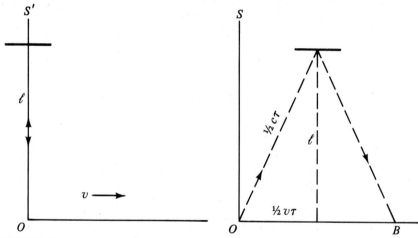

Figure 3–12

Comparison of time intervals measured in two frames.

time at which the beam returns is measured by observer B on a clock which had previously been synchronized with the one used by observer O. If the elapsed time in the unprinted frame is τ, then from the figure we have

$$(\tfrac{1}{2}c\tau)^2 = \ell^2 + (\tfrac{1}{2}v\tau)^2$$

$$\tau^2 = \left(\frac{2\ell}{c}\right)^2 + (\beta\tau)^2$$

$$(1 - \beta^2)\tau^2 = \tau_0^2$$

$$\tau = \gamma\tau_0,$$

as in Equation 3.8. Observer O thus concludes that the clock used by O' runs slow. Had O' photographed the clocks at O and at B as he passed those points, his photos would confirm the time interval reported by O. However, O' would claim that the clocks used by O and B were *not* synchronized but that the clock used by B was fast by the amount $\beta^2\tau$. Therefore, according to O', the time interval in the unprimed frame should be

$$\tau - \beta^2\tau = (1 - \beta^2)\tau = (1 - \beta^2)\gamma\tau_0 = \sqrt{1 - \beta^2}\,\tau_0,$$

and O' would also conclude that "moving clocks run slow."

Before concluding this section, mention should be made of the visual appearance of objects traveling at relativistic speeds, since misconceptions are prevalent in the literature. A visual image formed at the retina or a camera film is produced by photons that have *arrived* simultaneously, but they did not necessarily *leave* the object simultaneously. Thus, photographs of high speed objects must be corrected for the transit time of the photons to the various points of the image in order to determine the effects of the Lorentz contraction.[1]

Example 3-3

A space crew has a life-support system that will last 1000 hours. What minimum speed would be required for safe travel between two space stations that are separated by a fixed distance of 8.0×10^{11} km?

First solution

In the rest frame of the space stations, the distance separating them is the proper length, $L_0 = 8.0 \times 10^{11}$ km. If we denote the speed of the rocket relative to the space stations by v, then the travel time is given simply by

$$t = \frac{L_0}{v}.$$

Also, $t = \gamma t_0$, where $t_0 = 1000$ hours. Equating these two expressions for t,

$$\gamma t_0 = \frac{L_0}{v}$$

$$\gamma v = \frac{L_0}{t_0} = \frac{8.0 \times 10^{14} \text{ m}}{1000 \times 3600 \text{ s}}.$$

Dividing by c,

$$\beta\gamma = \frac{\beta}{\sqrt{1 - \beta^2}} = 0.74.$$

Squaring,

$$\frac{\beta^2}{1 - \beta^2} = 0.55$$

$$\therefore \beta = 0.6.$$

Second solution

In the rest frame of the rocket, the astronauts determine the distance between the space stations to be the contracted distance, $L = L_0/\gamma$. Since this distance must be

---

[1] See V. T. Weisskopf, *Physics Today* 13, 24 (1960); also, G. D. Scott and M. R. Viner, *Amer. J. Phys.* 33, 534 (1965).

covered in the proper time t_0, the least average speed must be

$$v = \frac{L}{t_0} = \frac{L_0}{\gamma t_0}.$$

$$\therefore \beta\gamma = \frac{L_0}{ct_0} = 0.74,$$

and $\beta = 0.6$ as before.

7. RELATIVISTIC VELOCITY ADDITION

By means of the Einstein transformation given in Equation 3.6, we can read-ily obtain expressions for the velocity components that each observer will mea-sure in his own frame. Thus, first taking the differentials

$$dx' = \gamma(dx - \beta c \, dt)$$
$$dy' = dy$$
$$dz' = dz$$
$$dt' = \gamma \left(dt - \frac{\beta}{c} \, dx \right),$$

we define

$$\left. \begin{aligned}
u'_x &= \frac{dx'}{dt'} = \frac{u_x - \beta c}{1 - \dfrac{u_x \beta}{c}} \\[2em]
u'_y &= \frac{dy'}{dt'} = \frac{u_y}{\gamma \left(1 - \dfrac{u_x \beta}{c} \right)} \\[2em]
u'_z &= \frac{dz'}{dt'} = \frac{u_z}{\gamma \left(1 - \dfrac{u_x \beta}{c} \right)}.
\end{aligned} \right\} \tag{3.9}$$

Note that β, the relative velocity of the frames, is in the common x-direction in Equations 3.9. The numerators of these expressions are just the velocity transfor-mations as given by the Galilean principle of relativity; these are the transforma-tions that are normally used in the everyday world. The denominators represent the correction due to the special theory of relativity; they all approach unity as β gets small. These equations can be used to obtain the relative velocity of one body with respect to the other when two bodies are moving with respect to the labora-tory frame. For example, suppose two electrons are fired in opposite directions, each with speed V with respect to the laboratory. Let the S frame be the rest frame of the electron moving in the negative x-direction, and the S' frame be the rest frame of the laboratory. Then, in the S frame the laboratory is moving to the right with velocity $\beta c = V$; in the S' frame, $u'_x = V$ is the velocity of the other electron.

Since we seek u_x, we must use the inverse transformation,

$$u_x = \frac{u'_x + \beta c}{1 + \frac{u'_x \beta}{c}} = \frac{V + V}{1 + \frac{V^2}{c^2}}. \tag{3.10}$$

If V is large, such as $0.9c$, then $u_x = (1.80/1.81)c$, which is very close to, but still less than, the velocity of light in free space. An observer in the laboratory will assert that the electrons are separating at a relative velocity of $1.80c$. This is not in conflict with the theory of relativity, since neither electron is moving at a speed greater than c in any frame of reference. If the particles in this example were photons fired by two lasers, Equation 3.10 tells us that relative to each photon, the other would have a velocity of

$$u_x = \frac{2c}{2} = c,$$

which is the result required by Einstein's second postulate.

Example 3-4

A stream of electrons has velocity components $u_x = 0.9c$, $u_y = 0.3c$, $u_z = 0$, with respect to the rest frame of the electron gun from which the electrons are ejected. Find the velocity components and the total velocity of the electrons with respect to an observer approaching the electron gun at relative velocity $V = 0.8c$ in the negative x-direction (see Figure 3–13).

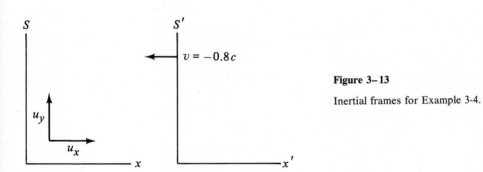

Figure 3–13

Inertial frames for Example 3-4.

Solution

Since $\beta = 0.8$, then $\gamma = 1.67$. Using Equation 3.9,

$$u'_x = \frac{u_x - V}{1 - \frac{u_x V}{c^2}} = \frac{0.9c + 0.8c}{1 + (0.9) \cdot (0.8)} = \frac{1.70c}{1.72} = 0.988c$$

$$u'_y = \frac{u_y}{\gamma \left(1 - \frac{u_x V}{c^2}\right)} = \frac{0.3c}{(1.67) \cdot (1.72)} = 0.105c$$

$$u'_z = 0$$

$$u' = \sqrt{(u'_x)^2 + (u'_y)^2 + (u'_z)^2} = 0.994c$$

*8. THE SPACE-TIME FOUR-VECTOR

Recall that the Einstein transformation was derived from the condition that observers in different inertial frames will agree about the magnitude of the quantity s^2, where

$$s^2 = x^2 + y^2 + z^2 - c^2t^2.$$

In three-dimensional space, the sum

$$x^2 + y^2 + z^2$$

is the square of the magnitude (or length) of the vector whose components are (x, y, z). The fact that the length of this vector is preserved or *invariant* as the vector is moved about in space is expressed mathematically by the statement that

$$x^2 + y^2 + z^2 = r^2,$$

where r is a constant.

In like manner, we may now conceive of a four-dimensional space in which the components of a four-vector are (x, y, z, ict), where the i is required in order to obtain the minus sign when the components are squared. The square of the magnitude of this vector is,

$$s^2 = x^2 + y^2 + z^2 - c^2t^2 = r^2 - c^2t^2. \tag{3.11}$$

The magnitude or "length" of the four-vector is called the *interval* between the two events that are determined by the two end points of the vector. In special relativity the interval s between two events is invariant. Note that the distance r will not be the same to different observers, but *the interval s will be the same for all observers*.

It is possible to obtain a convenient expression for the constant value of s by using the invariance principle itself. When $s^2 = 0$, there is only one frame in which the two events may be observed by a single clock at rest. In that frame $r = 0$, and the time interval is the proper time interval, τ_0. Therefore, $s^2 = -c^2(\tau_0)^2$ in all inertial frames, or

$$s^2 = r^2 - c^2t^2 \equiv -c^2(\tau_0)^2. \tag{3.12}$$

This may be rearranged in the form

$$t^2 = (\tau_0)^2 + \frac{r^2}{c^2}, \tag{3.13}$$

which shows that the proper time interval is indeed the minimum time interval recorded by any observer.

It is also instructive to rewrite Equation 3.13 in the form

$$(\tau_0)^2 = t^2 - \left(\frac{r}{c}\right)^2. \tag{3.14}$$

When $(\tau_0)^2 > 0$, τ_0 is *real* and the interval is called a *time-like interval*. In this case one can always find an inertial frame traveling at a velocity less than c such that the two events occur at the *same point in space* but at different times. Two events that are separated by a time-like interval can never be regarded as simultaneous by any observer.

On the other hand, when $(\tau_0)^2 < 0$, τ_0 is imaginary and the interval is called a *space-like interval*. Here one can always find an inertial frame traveling at a velocity less than c such that the two events occur simultaneously but at different points in space. No observer can eliminate the space interval between the events regardless of his state of relative motion. The imaginary proper time interval corresponds to the transit time for a light pulse between the two spatial points.

There is one additional special case, namely, when $s^2 = 0$. This means that the two events can be connected by a light ray and so there is no interval.

Example 3-5

What is the proper time interval between two events if it is known that in some inertial frame the events are separated by 3×10^6 km and occur 12 seconds apart?

Solution

Using Equation 3.14,

$$(\tau_0)^2 = t^2 - \left(\frac{r}{c}\right)^2$$

$$= (12)^2 - \left(\frac{3 \times 10^9 \text{ m}}{3 \times 10^8 \text{ m/s}}\right)^2 = 144 - 100$$

$$= 44$$

$$\tau_0 = \sqrt{44} = 6.63 \text{ s}$$

---

9. RELATIVITY AND "COMMON SENSE"

After such a brief introduction to special relativity one is tempted to say, "This violates common sense! Does the world *really* behave this way?" We must remember that common sense is man's perception of the world about him based upon a limited range of experiences. When one's experience is enlarged to include high relative velocities, then he is compelled to answer, "Yes, the world really is like this."

Studies made of short-lived muons have confirmed[1] the Einstein transformations of length and time, Equations 3.6. The muons, whose mean lifetime in their own rest frames is well-known, are observed to travel a much greater distance than that which is consistent with their known speed and mean lifetime. However, if we consider the slowing of the muon "clocks" as observed from our frame, we find that the distance traveled is consistent with their measured speed and the dilated lifetime (see Problem 3-8). Another way of studying the effect of relativistic speeds on the lifetime of unstable particles is to trap the particles in circular orbits in a magnetic field (see Example 4-2). In all such experiments the increase in the mean lifetime of the particles is in direct agreement with the prediction of special relativity.

The reader who can easily accept the evidence of these experiments might still be troubled by a statement such as, "A moving clock runs slow." When carried to its logical extreme this concept leads to the so-called "twin paradox." One

---

[1] See D. H. Frisch and J. H. Smith, *Amer. J. Phys.* **31**, 342 (1963), or J. H. Smith, *Introduction to Special Relativity*, W. A. Benjamin, New York, 1965. section 3-5.

twin takes a rapid rocket trip into space. His clock, as well as his own biological clock, runs slow from the viewpoint of the earth twin. At the end of the trip, then, when the two twins stand side by side, the traveling twin would have aged less than the earth-bound twin. The paradox arises when we consider the events from the frame of the rocket ship. The twin in the rocket claims that the earth-bound clocks run slow and, accordingly, the earth-bound twin will be younger at the end of the journey. Who is correct?

At first glance it may appear as if the two twins undergo equivalent experiences, but this is not the case. The twin who remains on earth knows that he is at rest with respect to the earth. On the other hand, the traveling twin has undergone three major accelerations—lift-off, turnaround, and re-entry—and he has occupied at least three different inertial frames. Both twins readily agree about which one did the traveling, so this example does not have the symmetry characteristic of our previous examples, which had constant relative motion between two frames.

Then which twin is older? It is generally agreed that the earth-bound twin is older after the trip. The reason for this really comes from general relativity, where we learn that large accelerations produce effects equivalent to strong gravitational fields. Since clocks slow down in strong gravitational fields, they must also run slow when accelerated.

SUMMARY

Einstein's two postulates in the special theory of relativity are that (1) all inertial frames are equivalent for the description of all physical laws, and (2) the speed of light in vacuum is the same for all observers. The Einstein transformation follows directly from these postulates. Hence, any physical law that retains the same mathematical form under this transformation automatically satisfies the principle of relativity and is said to be *covariant*. Two important kinematical effects that derive from the theory of relativity are the contraction of lengths parallel to the direction of relative motion and the dilation of time. These are expressed mathematically by the equations

$$L_0 = \gamma L$$

and

$$\tau = \gamma \tau_0,$$

where $\gamma = (1 - \beta^2)^{-1/2}$ and $\beta = v/c$. Here L_0 is called the *proper length* and τ_0 is called the *proper time* interval. A proper length is one that is measured in the rest frame of the object or events. The observer has plenty of time to lay out a tape measure or set up an interferometer or other method to make the measurement as accurate as he wishes. A proper time interval is measured by a single clock (or a set of synchronized clocks) in the rest frame of the events. The proper length is the maximum length obtainable in any inertial frame, and the proper time interval is the minimum interval for the events.

The relativistic formula for the addition of velocities is quite different from ordinary vector addition for speeds greater than one-tenth the speed of light, but it reduces to the classical value for ordinary speeds. The speed of light was shown to be the limiting velocity for transmitting information.

The concept of the four-vector was introduced by treating time as a fourth dimension on an equal footing with the three dimensions of Euclidean geometry. The invariance of the "length" of a four-vector was shown to be a useful property when solving physical problems. In space-time, the length of the four-vector is called the *interval* between two events. The square of the interval is $s^2 = r^2 - c^2t^2$. When $r^2 > c^2t^2$, the interval is said to be space-like; when $r^2 < c^2t^2$, the interval is *time-like*. When the interval is space-like, there is an inertial frame in which the events are simultaneous, but there is no frame in which the events occur at the same place. On the other hand, when the interval is time-like, there is no frame for which the events are simultaneous, but there is one frame in which the events occur at the same place.

Additional Reading

A. Einstein, *The Meaning of Relativity.* Princeton University Press, Princeton, N.J., 1946.

G. Gamow, "Mr. Tompkins in Wonderland," in *Mr. Tompkins in Paper Back,* Cambridge University Press, New York, 1965.

R. Resnick, *Introduction to Special Relativity,* John Wiley & Sons, New York, 1968.

R. Skinner, *Relativity,* Blaisdell Publishing Co., Waltham, Mass., 1969.

J. H. Smith, *Introduction to Special Relativity,* W. A. Benjamin, New York, 1965.

E. F. Taylor and V. A. Wheeler, *Spacetime Physics,* W. H. Freeman & Co., San Francisco, 1966.

V. T. Weisskopf, "The Visual Appearance of Rapidly Moving Objects," in *Physics in the Twentieth Century: Selected Essays,* The MIT Press, Cambridge, Mass., 1972. (See also G. D. Scott and M. R. Viner, *Amer. J. Phys. 33,* 534, 1965.)

PROBLEMS

3-1. An airplane is landing on an aircraft carrier, which is heading into the wind at a speed of 10 m/s with respect to the water. The wind speed is 20 m/s with respect to the water. The plane requires a landing speed of 50 m/s with respect to the air.

   Assuming that the plane lands into the wind, what is the landing speed of the plane with respect to the carrier? What is the landing speed of the plane with respect to the water?

3-2. Show that the Lorentz-FitzGerald contraction can account for the null result of the Michelson-Morley experiment.

3-3. By means of Equations 3.3, 3.4, and 3.5, determine the Einstein transformation,

$$x' = \gamma(x - \beta ct)$$
$$y' = y$$
$$z' = z$$
$$ct' = \gamma(ct - \beta x),$$

where

$$\beta = v/c \quad \text{and} \quad \gamma = (1 - \beta^2)^{-1/2}.$$

3-4. An observer in S gives the coordinates of an event as ($x = 10$ km, $y = 5$ km, $z = 10$ km, $t = 1 \times 10^{-5}$ s). What are the coordinates (x', y', z', t') of this event to an observer in S' if the relative speed of S and S' is 0.98c along their common x-directions? Assume that $t = t' = 0$ when $x = x' = 0$.

3-5. A meter stick lies at rest in the xy-plane at an angle of 45° with the x-axis. What are the length and orientation of the stick to an observer moving at relative velocity βc in the x-direction?

3-6. If the galaxy has a diameter of 10^5 light-years, what will the diameter appear to be to a cosmic ray particle traveling at the relative speed $\beta = 0.99$? How long will the trip take as measured by a clock riding with the particle?

3-7. A truck-and-trailer having a total length at rest of 20 m approaches a covered bridge that is 18 m long. (a) At what speed must the truck travel in order for its length to appear equal to the bridge length, to an observer standing on the ground? (b) How long is the bridge as viewed from the truck?

3-8. A cosmic ray muon moves vertically through the atmosphere at a relative speed of $\beta = 0.99$. Its mean lifetime in its own rest frame is 2.22 microseconds. (a) What is the mean lifetime measured by an observer on earth? (b) What mean path length will be "seen" by the muon and by the earth observer?

3-9. A space voyager returned to earth after a trip that required a time span of one year by his clock. However, he discovered that five years had elapsed on earth while he was gone. What was his average speed during his voyage?

3-10. The distance to the nearest star is about four light-years. At what relative speed must an astronaut travel in order to reach the neighborhood of that star in a proper time of six months? (A light-year is the distance light travels in one year, that is, 9.46×10^{15} m.)

3-11. In the rest frame of an electron, another electron approaches at a speed v. What is the relative velocity of an observer who measures equal and opposite velocities for the electrons?

3-12. (a) Derive the result expressed by Equation 3.7 by allowing one observer to measure how long it takes for a fixed length in a moving frame to pass him. (b) Show that Equation 3.8 can be obtained if an observer in S measures the coordinates of two events which occur at the same point x_0' in S'.

3-13. A radioactive atom is moving with respect to the laboratory at a speed of $0.3c$ in the x-direction. If it emits an electron having a speed of $0.8c$ in the rest frame of the atom, find the velocity of the electron with respect to a laboratory observer when: (a) the electron is ejected in the x-direction, (b) it is ejected in the negative x-direction, and (c) it is ejected in the y-direction.

3-14. An observer in S places a laser so that the beam makes an angle θ with his x-axis. What angle does an observer in S' see if the relative velocity between S and S' is βc in their common x-directions? (This is the relativistic equation for the aberration of starlight.) What does the S' observer see when $\cos \theta \leq \beta$?

3-15. Consider three inertial frames such that S and S' have relative speed $\beta_1 c$ along their common x-directions and S' and S'' have relative speed $\beta_2 c$ along their common x-directions. Find the expression for transforming velocities from S to S''. Express βc, the relative speed of S'' with respect to S, in terms of $\beta_1 c$ and $\beta_2 c$.

*3-16 What synchronization error will an observer moving at relative speed βc detect in two stationary clocks separated by a distance L along the path of the motion?

*3-17. A chart recorder shows that there is a 10-second interval between two events that occur at the same point in a given frame. (a) What time interval between the events is recorded by an observer whose velocity relative to the first frame is such that $\gamma = 5$? (b) What spatial separation between the events is determined by the observer in (a)?

*3-18. An observer in frame S notes that two events are separated by 6 km and 8 microseconds. (a) How fast must frame S' travel relative to S in order for the two events to appear simultaneous in S'? (b) What is the distance between the events in S'?

*3-19. An observer in frame S notes that two events are separated by 6 km and 33.3 microseconds. (a) How fast must frame S' travel relative to S in order for the two events to occur at the same place in S'? (b) What is the time interval between the events in S'?

*3-20. A beacon sends out a pulse of light once each millisecond, as measured by a clock in its rest frame. To an observer in a spaceship, the beacon gets 100 km closer between each pair of flashes. (a) What is the speed of the spaceship relative to the beacon? (b) What is the time interval between pulses observed in the spaceship? (c) What distance is traveled by the ship between pulses, according to an observer in the rest frame of the beacon?

*3-21. Two clocks that are 10 m apart in S' are synchronized in that frame. An observer in S claims that the clocks are out of synchronism by 2.5×10^{-8} seconds. What is the relative speed between S and S'?

*3-22. Show that the wave equation,

$$\nabla^2 \phi(x, y, z, t) = \frac{1}{c^2} \frac{\partial^2}{\partial t^2} \phi(x, y, z, t),$$

is covariant under an Einstein transformation.

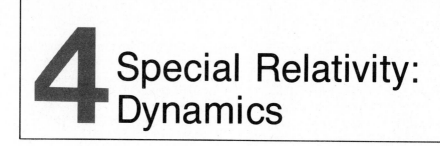

4 Special Relativity: Dynamics

In the previous chapter, we learned how the Einstein transformation enables observers in different inertial frames to interpret one another's measurements of position, time, and velocity. Now let us consider the dynamics of motion, that is, let us explicitly include the mass of an object and see what we can learn about its momentum and kinetic energy.

1. RELATIVISTIC CHANGE OF MASS

First, consider the following thought experiment. Two identical volley balls, having rest mass m_0 when measured in the same rest frame, are given to two observers in different inertial frames that have relative velocity βc in their common x-directions. Each observer is instructed to throw his volley ball so that it will make a head-on collision with the other volley ball at the instant that the line joining the two observers is perpendicular to their common x-axes (see Figure 4–1). Each observer claims that his own volley ball moves parallel to his own y-axis both before and after the collision and that it undergoes a momentum change of $2m_0 u_y$. Moreover, each claims that the other volley ball has an x-component of velocity before and after the collision. Since Newton's laws are covariant (that is, form-invariant) and momentum is conserved, either observer can write

$$2m_0 u_y = (2m_0 u_y)', \qquad \text{provided that } u_y \ll c.$$

Then, $$m_0 u_y = m_0' u_y' = m_0' u_y \sqrt{1 - \beta^2}$$

and $$m_0 = m_0' \sqrt{1 - \beta^2}.$$

If we simply denote m_0', the mass in the *other* frame, by the symbol m, we may write

$$m = \gamma m_0. \tag{4.1}$$

What this experiment tells us is as follows: Each observer claims that *his*

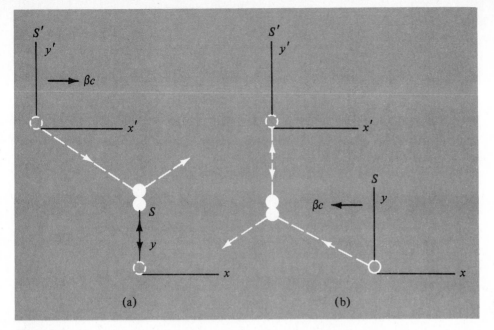

Figure 4–1

The collision of two identical volley balls: (a) as observed in S, (b) as observed in S'. The y-axes of S and S' are collinear at the instant of impact.

volley ball has mass m_0 and velocity u_y, whereas the *other* volley ball has mass γm_0 and a velocity in the y-direction given by u_y/γ. The inevitable conclusion is that a body of rest mass m_0 has its mass *increased* by the factor γ relative to an observer in a different inertial frame.

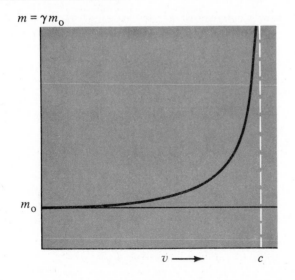

Figure 4–2

The relativistic mass m as a function of speed v. Since $m = \gamma m_0$, this curve is identical to the plot of γ vs. v in Figure 3–9(b) except for the constant factor m_0.

2. EQUIVALENCE OF MASS AND ENERGY

After squaring both sides of Equation 4.1 and removing fractions, the result may be written as

$$m^2c^2 - m^2v^2 = m_0^2c^2.$$

Since we must now regard both m and $\vec{v}$ as variables, differentiation leads to the expression

$$2mc^2dm - 2mv^2dm - 2m^2\vec{v} \cdot d\vec{v} = 0,$$

or

$$c^2dm = v^2dm + m\vec{v} \cdot d\vec{v}. \tag{4.2}$$

Also, from Newton's second law,

$$\vec{F} = \frac{d\vec{p}}{dt} = \frac{d}{dt}(m\vec{v}) = m\frac{d\vec{v}}{dt} + \vec{v}\frac{dm}{dt}.$$

If this force produces a displacement $d\vec{s}$ in the time interval dt, the incremental change in kinetic energy dT is,

$$dT = \vec{F} \cdot d\vec{s} = m\frac{d\vec{s}}{dt} \cdot d\vec{v} + \vec{v} \cdot \frac{d\vec{s}}{dt}dm = m\vec{v} \cdot d\vec{v} + v^2dm. \tag{4.3}$$

From Equations 4.2 and 4.3 we have

$$dT = c^2\,dm,$$

which can be integrated as follows:

$$\int_{v=0}^{v} dT = \int_{m_0}^{m} c^2\,dm.$$

Then,

$$T = (m - m_0)c^2 = (\gamma - 1)m_0c^2. \tag{4.4}$$

We note that the kinetic energy of a body is equal to the product of its mass change times the square of the velocity of light. This result can be compared with the classical expression for the kinetic energy by writing

$$T = \frac{m_0c^2}{\sqrt{1 - \beta^2}} - m_0c^2 = m_0c^2[(1 - \beta^2)^{-1/2} - 1]$$

$$= m_0c^2(1 + \tfrac{1}{2}\beta^2 + \cdots - 1) \qquad \text{(for small } \beta)$$

$$= m_0c^2(\tfrac{1}{2}\beta^2 + \cdots)$$

$$\sim \tfrac{1}{2}m_0v^2.$$

Thus, if β is small enough to permit discarding all terms higher than β^2 in the binomial expansion, Equation 4.4 reduces to the classical result.

Still more information can be gleaned from Equation 4.4 by writing it in the form

$$E = mc^2 = T + m_0c^2, \tag{4.5}$$

where we now call E the total energy and m_0c^2 the *rest mass energy*. Equations 4.4 and 4.5 express the equivalence of mass and energy, and provide the basis for restating the conservation laws of classical mechanics. Neither mass nor the quantity $(T + V)$ need be conserved separately in relativity[1]; instead, the quantity mc^2 is conserved. Example 4-2 below illustrates what is meant by the equivalence of mass and energy.

Example 4-1

At what speed will the relativistic increase of mass amount to 1%?

Solution

$$0.01 = \frac{m - m_0}{m_0} = \frac{(\gamma - 1)m_0}{m_0} = \gamma - 1$$

Then

$$\gamma = 1.01$$

$$\gamma^2 = 1.0201 = \frac{1}{1 - \beta^2}$$

$$\beta^2 = \frac{\gamma^2 - 1}{\gamma^2} = \frac{0.0201}{1.0201} = 0.01970$$

$$\beta = \frac{v}{c} = 0.140$$

Therefore, when $v = c/7$ the relativistic increase of mass is 1%; when $v = c/10$ the increase is 0.5%. For speeds less than $c/10$ the mass may normally be assumed to have the constant value m_0.

Example 4-2

Two identical balls of putty collide head-on and stick together. Assume that each ball has a rest mass of m_0 and a kinetic energy T as seen by a particular observer. What is the total energy E before the collision and after the collision?

Solution

Before the collision the total rest mass is $2m_0$, the total kinetic energy is $2T$, and the total energy E is

$$E = 2m_0c^2 + 2T.$$

After the collision the new body of rest mass M_0 is at rest with respect to the observer's frame (because the total momentum is zero in that frame). Since the kinetic energy is now zero, the total energy E is simply

$$E = M_0c^2 + \text{Heat}.$$

The kinetic energy that existed before the collision has been transformed into heat and rest mass of the composite body. The rest mass is now equal to

$$M_0 = 2m_0 + \left(\frac{2T - \text{Heat}}{c^2}\right)$$

---

[1] A potential energy V may be added to each side of Equation 4.4, and then Equation 4.5 becomes $E = mc^2 = T + V + m_0c^2$.

3. RELATIVISTIC MOMENTUM

The relativistic momentum of a mass m with velocity $\vec{v}$ is simply

$$\vec{p} = m\vec{v} = \gamma m_0 \vec{v},$$

and its magnitude may be written as

$$p = mv = \frac{mc^2 v}{c^2} = \frac{Ev}{c^2} = \frac{\beta E}{c}. \tag{4.6}$$

Using Equations 4.5 and 4.6, we find that

$$E^2(1 - \beta^2) = (m_0 c^2)^2$$

$$E^2 \left[1 - \left(\frac{pc}{E}\right)^2 \right] = (m_0 c^2)^2$$

or

$$E^2 = (pc)^2 + (m_0 c^2)^2. \tag{4.7}$$

Further,

$$(T + m_0 c^2)^2 = (pc)^2 + (m_0 c^2)^2$$

and

$$(pc)^2 = T^2 + 2T m_0 c^2. \tag{4.8}$$

Equations 4.7 and 4.8 provide useful relationships between the momentum of a body and its energy.

It is worth pointing out that the kinetic energy of a body, according to Equation 4.4, will increase rapidly as the velocity increases. In the limit as $v \rightarrow c$, γ would become infinite and hence the kinetic energy would also be infinite. We conclude that c must represent a limiting velocity that cannot be achieved by any particle that has a rest mass. Another way of stating this is to say that only those entities for which a rest frame exists can have a rest mass. Since an entity traveling at the speed of light has *no* rest frame (its speed will be c to all observers), it can have no rest mass. A zero rest mass particle *can*, however, have a momentum and kinetic energy. From Equation 4.7 we see that the kinetic energy and momentum of a particle of zero rest mass are $T = pc = E$ and $p = E/c$, respectively. The right triangle in Figure 4–3 is a convenient way to represent Equation 4.7. It is evident from the figure that $E = pc$ when $m_0 = 0$ and that $E \sim pc$ whenever $pc \gg m_0 c^2$.

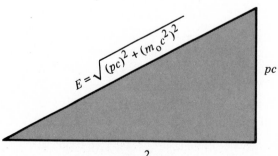

Figure 4–3

Right triangle showing the relation of the total energy, relativistic momentum, and rest mass energy as expressed by Equation 4.7.

Example 4-3

A particle of rest mass m_0 has a speed $v = 0.8c$. Find its relativistic momentum, its kinetic energy, and its total energy.

Solution

First note that

$$\beta = v/c = 0.8 \quad \text{and} \quad \gamma = (1 - \beta^2)^{-1/2} = (1 - 0.64)^{-1/2} = (0.36)^{-1/2} = \tfrac{5}{3}.$$

Then the relativistic mass is

$$m = \gamma m_0 = \tfrac{5}{3} m_0.$$

The relativistic momentum is given by

$$p = mv = \gamma m_0 v = \tfrac{5}{3} \times \tfrac{4}{5} \times m_0 c = \tfrac{4}{3} m_0 c.$$

The total relativistic energy may be obtained from Equation 4.6 as follows:

$$E = \frac{pc}{\beta} = \frac{5}{4} \times \frac{4}{3} m_0 c^2 = \frac{5}{3} m_0 c^2.$$

Alternatively, Equation 4.7 could have been used to find E, since

$$E^2 = (pc)^2 + (m_0 c^2)^2 = (\tfrac{4}{3} m_0 c^2)^2 + (m_0 c^2)^2 = \tfrac{25}{9}(m_0 c^2)^2.$$

Then $\quad E = \tfrac{5}{3} m_0 c^2.$

The kinetic energy is now easily obtained using Equation 4.5,

$$T = E - m_0 c^2 = \tfrac{2}{3} m_0 c^2.$$

Example 4-4

Find the velocity and the momentum of a particle having rest mass m_0 whose kinetic energy equals its rest mass energy.

Solution

We are given that $T = m_0 c^2$, where m_0 is the rest mass of the particle. Then, using Equations 4.5 and 4.1,

$$E = T + m_0 c^2$$

$$\gamma m_0 c^2 = 2 m_0 c^2$$

$$\gamma = 2$$

$$\beta = \frac{\sqrt{\gamma^2 - 1}}{\gamma} = \sqrt{\frac{3}{2}}$$

$$\therefore v = \sqrt{3}\, c/2$$

and

$$p = mv = \gamma m_0 v = \sqrt{3}\, m_0 c.$$

Example 4-5

Derive an expression for the radius of the orbit of a particle of mass m_0 and charge q moving with kinetic energy T at a relativistic velocity in a transverse magnetic field of intensity B. Express your answer in terms of q, B, c, T, and m_0.

Solution

The force on a particle of charge q and velocity $\vec{v}$ in a magnetic field $\vec{B}$ is

$$\vec{F} = q\vec{v} \times \vec{B}.$$

When $\vec{v} \perp \vec{B}$ and $\vec{B}$ is a uniform, constant field over a sufficiently large region of space, the particle will follow a circular path in a fixed plane perpendicular to the magnetic field. The radius of this path is determined by the condition that the centripetal acceleration must be provided by the magnetic force. That is,

$$\frac{mv^2}{r} = qvB,$$

or

$$p = qrB.$$

Using the relativistic momentum, Equation 4.8, we find

$$Brqc = \sqrt{T^2 + 2Tm_0c^2}, \tag{4.9}$$

where T is in joules.

In practice, the energies are usually expressed in MeV rather than in joules. In order to express T and m_0c^2 in MeV in Equation 4.9, we must multiply the left-hand side by 6.24×10^{12} MeV/joule. For electrons and protons, then, Equation 4.9 becomes

$$Br = \frac{1}{300} \sqrt{T^2 + 2Tm_0c^2}, \tag{4.10}$$

where B is in teslas, r is in meters, and the energies are in MeV.

Example 4-6

What magnetic field strength would be required to confine 100-MeV protons to an orbital radius of 10 meters?

Solution

The kinetic energy is $T = 100$ MeV, the proton rest mass energy is (see the list inside the front cover) $m_0c^2 = 938$ MeV, and $r = 10$ meters. Using Equation 4.10,

$$B = \frac{10^{-3}}{3} \sqrt{(100)^2 + 2(100)(938)} = \frac{10^{-2}}{3} \sqrt{100 + 1876}$$

$$= \frac{10^{-2}}{3} (44.45) = 0.15 \text{ tesla}$$

*4. THE MOMENTUM-ENERGY FOUR-VECTOR

Equation 4.7 expresses the fundamental relationship between momentum and energy in special relativity. In order to gain a better understanding of this relationship, let us rewrite Equation 4.7 as

$$p^2 - \frac{E^2}{c^2} = -m_0^2c^2, \tag{4.11}$$

where each term has been divided by c^2. By comparing this result with Equation 3.12, it is immediately evident that we have here all of the necessary ingredients for defining a new four-vector. The first three components of this four-vector are simply (p_x, p_y, p_z), since $p^2 = p_x^2 + p_y^2 + p_z^2$ as in classical mechanics. The fourth component contains an i as in the case of the space-time four-vector. The magnitude of this four-vector is a scalar invariant, as it should be, and we see from Equation 4.11 that the magnitude squared is equal to $-m_0^2c^2$, where m_0 is the invariant mass.

What we have gleaned from Equation 4.11 may be summarized as follows. There is only one inertial frame in which an object can be at rest. In that frame $p = 0$, and the relativistic energy E is simply the rest mass energy m_0c^2. In every other inertial frame $p \neq 0$, and E is the sum of the rest mass energy and the kinetic energy. However, the difference

$$p^2 - \frac{E^2}{c^2}$$

is *always* equal to the quantity $-m_0^2c^2$.

The Einstein transformation enables us to obtain the components of the momentum-energy four-vector in any inertial frame in the same manner that we have done for the space-time four-vector. The corresponding equations are obtained simply by replacing the quantities x, y, z, and ct with p_x, p_y, p_z, and E/c, respectively. The result is:

$$\left. \begin{aligned} p_x' &= \gamma\left(p_x - \frac{\beta E}{c}\right) \\ \frac{E'}{c} &= \gamma\left(\frac{E}{c} - \beta p_x\right) \end{aligned} \right|$$

(4.12)

For those readers who are familiar with matrix algebra, Equation 4.12 may be obtained by solving the following matrix equation:

$$\begin{pmatrix} p_x' \\ p_y' \\ p_z' \\ \frac{iE'}{c} \end{pmatrix} = \begin{pmatrix} \gamma & 0 & 0 & i\beta\gamma \\ 0 & 1 & 0 & 0 \\ 0 & 0 & 1 & 0 \\ -i\beta\gamma & 0 & 0 & \gamma \end{pmatrix} \begin{pmatrix} p_x \\ p_y \\ p_z \\ \frac{iE}{c} \end{pmatrix}$$

(4.13)

The following examples will illustrate the use of Equations 4.11 and 4.12 in solving problems.

Example 4-7

An object having a mass m_0 in its own rest frame is moving in the negative x-direction at a velocity of $0.8c$ relative to the laboratory.

(a) What are the components of the momentum-energy four-vector in the laboratory frame?

(b) What are the components of the momentum-energy four-vector in the frame of an observer who is traveling in the positive x-direction at a speed of $0.6c$ relative to the laboratory?

Solution

(a) The x-component of $\vec{p}$ is

$$p_x = mu_x = \gamma'm_0u_x,$$

where
$$\gamma' = \left(1 - \frac{u_x^2}{c^2}\right)^{-1/2} = \frac{5}{3}$$

so
$$p_x = -\frac{5}{3} \cdot \frac{4}{5} m_0 c = -\frac{4}{3} m_0 c;$$

$$p_y = p_z = 0.$$

Using Equation 4.6,

$$\frac{E}{c} = \frac{p_x c}{u_x} = \frac{-\frac{4}{3} m_0 c^2}{-0.8c} = \frac{5}{3} m_0 c.$$

(b) Since $\beta = 0.6$, we find that $\gamma = \frac{5}{4}$. Then, from Equation 4.12,

$$p'_x = \gamma\left(p_x - \frac{\beta E}{c}\right) = \frac{5}{4}\left(-\frac{4}{3} m_0 c - 0.6 \times \frac{5}{3} m_0 c\right) = -\frac{35}{12} m_0 c$$

$$p'_y = p'_z = 0$$

$$\frac{E'}{c} = \gamma\left(\frac{E}{c} - \beta p_x\right) = \frac{5}{4}\left(\frac{5}{3} m_0 c + 0.6 \times \frac{4}{3} m_0 c\right) = \frac{37}{12} m_0 c$$

Let us check these results in order to insure that the invariance property given by Equation 4.11 is satisfied. In (a),

$$p^2 - \frac{E^2}{c^2} = \frac{16}{9} m_0^2 c^2 - \frac{25}{9} m_0^2 c^2 = -m_0^2 c^2$$

In (b),

$$p'^2 - \frac{E'^2}{c^2} = \left[\left(\frac{35}{12}\right)^2 - \left(\frac{37}{12}\right)^2\right] m_0^2 c^2 = \frac{1}{144}[1225 - 1369] m_0^2 c^2$$

$$= -m_0^2 c^2$$

Alternate solution for part (b)
From Equation 3-9,

$$u'_x = \frac{u_x - v}{1 - \frac{u_x v}{c^2}} = \frac{-0.8c - 0.6c}{1 + (0.8)(0.6)} = -\frac{1.40}{1.48} c = -\frac{35}{37} c$$

$$\gamma' = \left[1 - \left(\frac{35}{37}\right)^2\right]^{-1/2} = \frac{37}{(37^2 - 35^2)^{1/2}} = \frac{37}{12}$$

$$\therefore p'_x = \gamma' m_0 u'_x = \left(\frac{37}{12}\right) m_0 \left(-\frac{35}{37} c\right) = -\frac{35}{12} m_0 c$$

and

$$\frac{E'}{c} = \frac{p'_x c}{u'_x} = \gamma' m_0 c = \frac{37}{12} m_0 c$$

Example 4-8

An object having a mass m_0 in its own rest frame is moving in the positive y-direction at a velocity of 0.8c relative to the laboratory.

(a) What are the components of the momentum-energy four-vector in the laboratory frame?

(b) What are the components of the momentum-energy four-vector in the frame of an observer who is traveling in the positive x-direction at a speed of 0.6c relative to the laboratory?

Solution

(a) The y-component of $\vec{p}$ is

$$p_y = mu_y = \gamma' m_0 u_y,$$

where

$$\gamma' = \left(1 - \frac{u_y^2}{c^2}\right)^{-1/2} = \frac{5}{3}.$$

Thus,

$$p_y = \frac{5}{3} \cdot \frac{4}{5} m_0 c = \frac{4}{3} m_0 c;$$

$$p_x = p_z = 0.$$

Using Equation 4.6,

$$\frac{E}{c} = \frac{p_y c}{u_y} = \frac{5}{3} m_0 c.$$

(b) Since $\beta = 0.6$, we find that $\gamma = \frac{5}{4}$. Then

$$u'_x = \frac{u_x - v}{1 - \frac{u_x v}{c^2}} = \frac{0 - 0.6c}{1 - 0} = -0.6c$$

and

$$u'_y = \frac{u_y}{\gamma} = \frac{0.8c}{\frac{5}{4}} = 0.64c.$$

Then,

$$(u')^2 = (u'_x)^2 + (u'_y)^2 + (u'_z)^2 = (0.36 + 0.4096 + 0)c^2$$

$$= 0.7696c^2$$

$$u' = 0.8773c.$$

It follows that

$$\gamma' = \left(1 - \frac{u'^2}{c^2}\right)^{-1/2} = (1 - 0.7696)^{-1/2} = 2.083.$$

The components of the momentum-energy four-vector are then

$$p'_x = \gamma' m_0 u'_x = -1.250 m_0 c$$

$$p'_y = \gamma' m_0 u'_y = 1.333 m_0 c$$

$$p'_z = 0$$

$$\frac{E'}{c} = \gamma' m_0 c = 2.083 m_0 c$$

Let us use Equation 4.11 to check these results:

$$(p_x')^2 + (p_y')^2 - \left(\frac{E'}{c}\right)^2 = [(1.250)^2 + (1.333)^2 - (2.0833)^2]m_0^2c^2$$

$$= -m_0^2c^2.$$

Note that $p' = \gamma'm_0u' = (2.0833)m_0 (0.8773c) = 1.8276m_0c$, and indeed

$$(p_x')^2 + (p_y')^2 + (p_z')^2 = (1.5625 + 1.7778 + 0)m_0^2c^2$$

$$= 3.3403m_0^2c^2$$

$$= (1.8276m_0c)^2 = (p')^2.$$

*5. RELATIVISTIC COLLISIONS AND THE C-FRAME

In the majority of experiments involving relativistic velocities, the observer is at rest in the laboratory and the events under study are taking place between particles that, for the most part, are moving at high velocities with respect to the laboratory. The physical nature of these events, however, depends upon the amount of energy available to do work in the zero-momentum frame of the particles. (The zero-momentum frame is the center-of-mass frame of classical mechanics.) If two particles approach each other and exchange momentum and energy, we call the event a *scattering* event. If the particles are not the same in *number* or in *kind* *after* the collision, we say that a *reaction* has occurred. In either case, it is often convenient to do the theoretical analysis in the zero-momentum frame and then transform the results to the laboratory frame in order to check against experiments.

It has been our practice to regard the laboratory frame as the unprimed frame and, accordingly, the energy and momentum of a particle would be E and p to a laboratory observer. An observer in any other frame S' would measure E' and p' for the same particle. Now let us designate a special frame, the C-frame, which corresponds to the center of mass in non-relativistic mechanics, by means of asterisks instead of primes. That is, in S^*, our dynamical variables are E^*, T^*, and p^*. *The C-frame is that particular S' frame in which the total momentum of the particles p^* is zero.* In some special cases the C-frame will be simply the laboratory frame, but generally, the C-frame will have relative velocity β^*c with respect to the laboratory. A simple example will show how β^* is determined.

Example 4-9

Consider two identical particles of rest mass m_0, one particle having momentum p_a and kinetic energy T, and the other being at rest in the laboratory. (See Figure 4-4.) Find an expression for β^*, the relative velocity of the C-frame with respect to the laboratory.

Solution

In the laboratory or L-frame, the total momentum is

$$p = p_a + p_b = p_a,$$

and the total energy is

$$E = E_a + E_b = T_a + m_0c^2 + m_0c^2 = T_a + 2m_0c^2.$$

Now, using Equation 4.12 we can write

$$p^* = \gamma^* \left(p - \frac{\beta^* E}{c} \right),$$

where p^* is the total momentum in the C-frame. Since $p^* = 0$, by definition, we have the result that

$$pc = \beta^* E,$$

or

$$\beta^* = \frac{pc}{E} = \frac{p_a c}{T_a + 2m_0c^2}. \qquad (4.14)$$

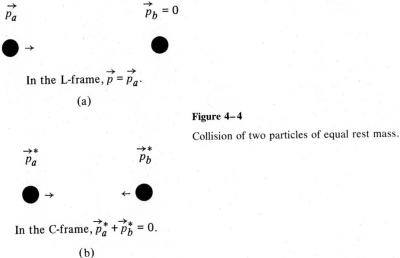

$\vec{p}_a$

$\vec{p}_b = 0$

In the L-frame, $\vec{p} = \vec{p}_a$.

(a)

Figure 4–4

Collision of two particles of equal rest mass.

$\vec{p}_a^*$

$\vec{p}_b^*$

In the C-frame, $\vec{p}_a^* + \vec{p}_b^* = 0$.

(b)

Example 4-10

Show that the total energies in the C-frame and the L-frame are related by

$$E = \gamma^* E^*. \qquad (4.15)$$

Solution

From the second expression in Equation 4.12,

$$E^* = \gamma^*(E - \beta^* p_x c).$$

Substituting $p_x c = \beta^* E$ from Equation 4.14,

$$E^* = \gamma^*(E - \beta^{*2}E) = \gamma^*(1 - \beta^{*2})E.$$

Example 4-11

In Example 4-9, if the kinetic energy of particle a were equal to its rest mass energy, what would be the values of β^* and γ^*?

Solution

We are given that $T_a = m_0c^2$. It follows that

$$E_a = T_a + m_0c^2 = 2m_0c^2.$$

Using Equation 4.7, $(p_ac)^2 = E_a{}^2 - (m_0c^2)^2$, and

$$p_ac = \sqrt{(2m_0c^2)^2 - (m_0c^2)^2} = \sqrt{3}\, m_0c^2.$$

From Equation 4.14,

$$\beta^* = \frac{\sqrt{3}\, m_0c^2}{3m_0c^2} = \frac{\sqrt{3}}{3} = 0.58,$$

Then,

$$\gamma^* = \sqrt{\tfrac{3}{2}} = 1.22.$$

Example 4-12

Find the total energy E^* in the C-frame and the kinetic energies $T_a{}^*$ and $T_b{}^*$ for the conditions given in Example 4-11.

Solution

The given quantities are as follows:

$$T_a = m_0c^2 \qquad\qquad p_a = \sqrt{3}\, m_0c$$

$$E_a = 2m_0c^2 \qquad\qquad p_b = 0$$

$$T_b = 0 \qquad\qquad p = p_a + p_b = \sqrt{3}\, m_0c$$

$$E_b = m_0c^2$$

$$E = E_a + E_b = 3m_0c^2$$

From Equation 4.15,

$$E^* = \frac{E}{\gamma^*} = \sqrt{\frac{2}{3}}\,(3m_0c^2) = \sqrt{6}\, m_0c^2.$$

Then,

$$T^* = E^* - 2m_0c^2 = (\sqrt{6} - 2)m_0c^2.$$

In the C-frame, since $p^* = 0$, $p_a{}^*$ and $p_b{}^*$ must be equal in magnitude but opposite in direction. This means that the kinetic energy must be divided equally between the particles since their masses are equal. Therefore,

$$T_a{}^* = T_b{}^* = \tfrac{1}{2}T^* = \tfrac{1}{2}(\sqrt{6} - 2)m_0c^2 = 0.22m_0c^2.$$

If, during a collision, kinetic energy is not conserved, the collision is said to be *inelastic*. The kinetic energy that "disappears" is converted to mass or potential energy and heat. This process may be illustrated by the following example.

Example 4-13

A particle of rest mass m_0 moving with velocity β_1c collides inelastically with and sticks to a stationary particle of rest mass M_0. Find the velocity of the composite particle.

Solution

In the L-frame the momentum is

$$p = \frac{m_0\beta_1c}{\sqrt{1 - \beta_1{}^2}} = \beta_1\gamma_1m_0c.$$

The total energy in the L-frame is

$$E = mc^2 + M_0c^2 = \gamma_1 m_0 c^2 + M_0 c^2.$$

Then, from Equation 4.14, the relative velocity of the C-frame is

$$\beta^* = \frac{pc}{E} = \frac{\beta_1 \gamma_1 m_0 c^2}{\gamma_1 m_0 c^2 + M_0 c^2} = \frac{\beta_1 \gamma_1 m_0}{\gamma_1 m_0 + M_0}.$$

Since the composite particle remains at rest in the C-frame, the required velocity is β^*c.

*6. THE CREATION AND ANNIHILATION OF PARTICLES

An important consequence of the equivalence of mass and energy is that under certain circumstances particles can be created or destroyed, the mass difference being accounted for by the disappearance or appearance of energy. Whether or not such a transformation occurs in a given physical situation is governed by the requirements of conservation laws and the availability of competing reactions or processes. The conservation laws with which we shall be concerned here include not only the classical requirements on the total linear momentum and the total angular momentum, but also the conservation of total *relativistic energy*, which includes the rest masses of all particles in the system. The classical conservation laws for mass and energy *separately* are no longer considered valid.

In addition, we will accept and make use of the conservation laws for total charge and total angular momentum for a system of particles. These laws are very simple to apply. For example, if a charged particle is created or destroyed, it must be accompanied by another particle of equal but opposite charge so that there is no net change in the amount of charge in the universe. Since a positron is a "positively charged electron," the creation or destruction of a positron-electron pair results in no net change in total charge. The conservation law for spin angular momentum is also easy to apply, though the concept of spin is a subtle one. Spin is an additional angular momentum which is characteristic of a given class of particles. It behaves like an ordinary angular momentum, but since its origin is not understood,[1] it is simply called an *intrinsic* angular momentum. Just as the classical angular momentum may be represented by a vector along the axis of rotation, we represent the spin by a hypothetical vector along the "spin axis." Thus, we may speak of a particle with its spin "up" or "down." When there is no orbital angular momentum in a collision or a reaction, the conservation law then means simply that the net sum of the up and down spins after the event must equal the net sum of the up and down spins before the event. Electrons, positrons, protons, and neutrons each have $\frac{1}{2}$ unit of spin angular momentum. Hence, they are all known as spin-$\frac{1}{2}$ particles.

Let us first consider the conversion of electromagnetic energy to rest mass energy. A common example of this is known as "pair-production," the creation of positron-electron pairs by the annihilation of gamma ray photons in the presence of heavy nuclei. A quantum[2] of energy $h\nu$ is associated with a photon of frequency

---

[1] That is, we do not know how to visualize a linear momentum and an axis such that the spin would equal $\vec{r} \times \vec{p}$. (See Chapter 2, section 3.)

[2] The quantum theory of radiation will be discussed in the next chapter.

ν, and this photon carries a momentum given by $E/c = h/\lambda$. A photon has no charge, but it has 1 unit of spin. Since the electron and positron each require 0.51 MeV of rest mass energy, as well as some kinetic energy, it is evident that the threshold photon energy for this process is slightly over 1 MeV. Photons having energy less than 1.02 MeV cannot produce particle pairs, whereas a photon whose energy exceeds 1.02 MeV might, under the right conditions, produce a positron-electron pair. The energy excess over the 1.02 MeV required for the rest masses is converted to kinetic energy of the particles. Note that charge conservation is satisfied, since no net charge is created. Spin conservation requires that the spins of the electron and positron be parallel so that the total spin will be 1 unit, thus conserving the spin of the photon.

The inverse process, the annihilation of a particle-antiparticle pair, can also occur. At least two photons are always required in order to conserve linear momentum, but it is interesting to see that the actual number of photons emitted is determined by the conservation of spin. In the case of electron-positron annihilation there are two possibilities for the total spin of the initial state. If the spins are parallel, the total spin of the system is unity; if they are antiparallel, the system has zero spin. Photons, on the other hand, have a spin of 1, so the production of two photons would result in a total spin of either 0 (antiparallel) or 2 (parallel). In the case in which the particle pair has a total spin of 1, an odd number of photons is required. Since a single photon will not permit conservation of linear momentum, at least three photons must be emitted. Therefore, two photons will be emitted in electron-positron annihilation when the particle spins are antiparallel, and three photons will generally be emitted when the particle spins are parallel.

The previous cases have dealt with the conversion of photon energy to rest mass energy and vice versa. It is also possible to use the kinetic energy of high-speed particles to create new particles. In order to achieve this, there must be sufficient kinetic energy in the C-frame to provide for the masses of the new particles. Consequently, the experimentalist, who is in the L-frame, must have a means of obtaining projectiles of very high energy. This is sometimes achieved in the case of heavy particle production by using two colliding beams of high-energy projectiles in order to optimize the energy available in the C-frame. For instance, if two particles of equal mass and velocity could be directed toward each other in a head-on collision, the C-frame as a whole would have no kinetic energy, and the energy available for conversion to mass in the C-frame would be maximized.

By way of illustration, consider the production of a proton-antiproton pair by bombarding protons with protons. Since the proton has a charge and a spin angular momentum, the conservation laws require that its antiparticle (having opposite charge and spin) must be created simultaneously. That is, the reaction may be represented by the expression

$$p + p \rightarrow p + p + p + \bar{p},$$

where the bar over the particle symbol indicates an antiparticle. Note that before the collision there were two particles, but after the collision there are four particles of equal rest mass.

Example 4-14

Find the threshold kinetic energy in the laboratory for the production of a proton-antiproton pair by bombarding stationary protons with protons.

Solution

After the collision the *minimum* total energy in the C-frame is

$$E^* = 4M_0c^2,$$

assuming all four particles are at rest in this frame. Before the collision, the total energy in the C-frame was

$$E^* = 2M_0c^2 + T^*.$$

Therefore, the threshold kinetic energy in the C-frame is

$$T^* = 2M_0c^2.$$

From Equation 4.4,

$$T^* = (\gamma^* - 1)2M_0c^2.$$

Therefore,
$$\gamma^* = 2,$$

and from Equation 4.15,

$$E = \gamma^*E^* = 2E^* = 8M_0c^2.$$

Since $E = T + 2M_0c^2$, we find that the threshold kinetic energy in the L-frame for this reaction is

$$T = E - 2M_0c^2 = 6M_0c^2.$$

*7. THE DOPPLER EFFECT

The apparent change of frequency resulting from the relative motion between source and observer is known as the Doppler effect. It should be recalled that in the case of a sound wave the magnitude of the frequency shift is *not* the same for a moving source as it is for a moving observer, even when the relative velocity is the same in both instances. If the equations derived for sound were also true for light, the Doppler shift would provide us with a means of determining which frame is *"really"* at rest and which is moving. In other words, we could discover an absolute frame, the ether frame, in violation of the principle of relativity.

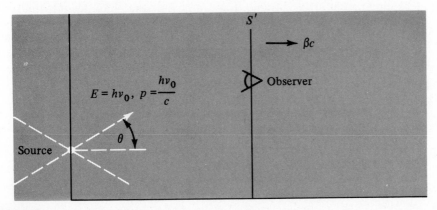

Figure 4–5

Relative motion of observer and light source.

Since we have adopted the principle of relativity, we must discard the classical Doppler equations in favor of a new one that is consistent with our postulates. The relativistic Doppler equation may be obtained from the Einstein transformation of the energy-momentum four-vector, Equation 4.12. We need only incorporate the Planck expression for the energy of a photon, $E = h\nu$, and the relativistic expression for the momentum, $p = h\nu/c$, where ν is the frequency. Consider a monochromatic source at rest in the unprimed frame as shown in Figure 4–5. An observer receding to the right at velocity βc will detect a frequency that is lower than the proper frequency, ν_0.

Using the second relation in Equation 4.12,

$$E' = \gamma(E - \beta p_x c)$$

$$h\nu' = \gamma(h\nu_0 - \beta h\nu_0 \cos \theta)$$

$$\nu' = \frac{\nu_0(1 - \beta \cos \theta)}{\sqrt{1 - \beta^2}}. \tag{4.16}$$

Now consider an earlier time when the observer in S' is approaching the source in S from the left. The observer sees light emitted at the angle $\theta = \pi$, and he measures the frequency

(Approaching observer
$$\nu' = \frac{\nu_0(1 + \beta)}{\sqrt{1 - \beta^2}} = \frac{\nu_0\sqrt{1 + \beta}}{\sqrt{1 - \beta}}. \tag{4.17}$$

As he passes the source he sees light emitted at the angle $\theta = \pi/2$, and he measures

(Transverse Doppler effect)
$$\nu' = \frac{\nu_0}{\sqrt{1 - \beta^2}}. \tag{4.18}$$

As he recedes to the right, the observer in S' measures light emitted at the angle $\theta = 0$, and he measures

(Receding observer)
$$\nu' = \frac{\nu_0(1 - \beta)}{\sqrt{1 - \beta^2}} = \frac{\nu_0\sqrt{1 - \beta}}{\sqrt{1 + \beta}}. \tag{4.19}$$

Now suppose we regard the observer as stationary in the S-frame and the source as moving with the S'-frame. Instead of Equation 4.16, we now must use the inverse transformation,

$$\nu = \frac{\nu_0(1 + \beta \cos \theta')}{\sqrt{1 - \beta^2}}. \tag{4.20}$$

Again we can treat three cases:

(Approaching source)
$$\theta' = 0: \quad \nu = \frac{\nu_0\sqrt{1 + \beta}}{\sqrt{1 - \beta}}. \tag{4.17a}$$

(Transverse effect)
$$\theta' = \pi/2: \quad \nu = \frac{\nu_0}{\sqrt{1 - \beta^2}}. \tag{4.18a}$$

(Receding source)
$$\theta' = \pi: \quad \nu = \frac{\nu_0\sqrt{1 - \beta}}{\sqrt{1 + \beta}}. \tag{4.19a}$$

Note that we obtain the same result for a relative velocity of approach, regardless of which frame is assumed to be moving, and likewise for the other cases.

These results are consistent with the principle of relativity, in that it is not possible to distinguish between motion of the source and motion of the observer. A relative velocity of separation decreases the observed frequency and a relative velocity of approach increases the observed frequency. Note that the transverse effect is simply an example of the phenomenon of time dilation, that is, that "moving" clocks run slowly.

An important dynamical effect in the theory of relativity is the increase of mass with speed. It follows that c appears as the upper limit for the speed of any object that has a rest mass. The Einstein expression for the equivalence of mass and energy, $E = mc^2$, accounts for the conversion of kinetic energy to rest mass, and vice versa, during collisions. Mass and energy are not conserved separately. In relativity, the conservation of total relativistic energy, which includes rest mass energy, replaces the separate conservation theorems for mass and mechanical energy in classical physics. The momentum-energy four-vector consists of the three momentum components (p_x, p_y, p_z) plus the total relativistic energy as the fourth component. The magnitude of this four-vector is the invariant rest mass term, $-m_0^2 c^2$. Relativistic collisions are introduced using the concept of the zero momentum frame of reference.

SUMMARY

N. Ashby and S. C. Miller, *Introduction to Special Theory of Relativity,* Physical Biological Sciences Misc., Blacksburg, Virginia, 1966.

R. Resnick, *Introduction to Special Relativity,* John Wiley & Sons, New York, 1968.

"Resource Letter SRT-1 on Special Relativity Theory," *Amer. J. Physics 30,* 462 (1962).

R. Skinner, *Relativity,* Blaisdell Publishing Co., Waltham, Mass., 1969.

J. H. Smith, *Introduction to Special Relativity,* W. A. Benjamin, New York, 1965.

Special Relativity Theory: Selected Reprints, American Association of Physics Teachers, New York, 1963.

E. F. Taylor and V. A. Wheeler, *Spacetime Physics,* W. H. Freeman & Co., San Francisco, 1966.

Additional Reading

PROBLEMS

4-1. At what relative speed will the kinetic energy of a body equal two times its rest mass energy?

4-2. At what relative speed will an object appear to have doubled its rest mass?

4-3. Show that if all of the mass of one atomic mass unit of matter were converted to energy, the amount of energy released would be 931.5 MeV. (1.0 u is equal to 1.66×10^{-27} kg.)

4-4. (a) What is the maximum speed for which the classical expression $\frac{1}{2}m_0 v^2$ will yield an error in the kinetic energy no greater than one percent? (b) What is the kinetic energy of an electron moving at this speed?

4-5. Find the expression for the angular frequency ω for the particle of Example 4-5. How does this compare with the classical cyclotron angular frequency?

4-6. A cosmic ray proton having a kinetic energy of 5×10^9 electron-volts approaches the earth on a trajectory that is perpendicular to the earth's magnetic field. What is the radius of curvature of the path in a region where the strength of the magnetic field is 4×10^{-5} tesla?

4-7. If the total relativistic energy of a particle is five times its rest mass energy, find (a) its velocity relative to the laboratory and (b) its momentum in terms of its rest mass energy E_0 and the constant c.

4-8.   What would be the error (in percent) in a calculation of the momentum and kinetic energy if the classical expressions were used for a particle having a relative speed of $0.6c$?

4-9.   What electron velocities are produced by the following accelerator energies: (a) 50 keV; (b) 1 MeV; (c) 10 MeV?

4-10.   Through what potential difference must an electron "fall" in order to achieve a speed of $0.95c$ with respect to the laboratory? What is the mass of the electron in the laboratory frame? What is its momentum? How could you verify experimentally that it actually has that mass and speed?

4-11.   Find the momentum and the kinetic energy of the electrons in Example 3-4 of Chapter 3 (see Figure 3–13) as observed from the S' frame.

4-12.   In the rest frame of an electron, a positron is observed coming directly toward the electron with momentum given by $pc = 4$ MeV. Find the relative velocity of the C-frame, the total energy in the C-frame, and the kinetic energy of each particle in the C-frame. (Use 0.51 MeV for the rest mass energy of the electron and the same value for the positron.)

4-13.   In Example 4-13, all of the kinetic energy that existed in the C-frame *before* the collision is converted to mass and heat since there is no kinetic energy in the C-frame *after* the collision. Find the kinetic energy in the C-frame before the collision if $\beta_1 = 0.8$ and $M_0 = 1.8m_0$.

4-14.   A 1 GeV proton is moving toward an alpha particle that is at rest in the L-frame. (a) What is the velocity of the proton in the L-frame? (b) What is the velocity of the C-frame (zero momentum frame) relative to the L-frame? (c) What is the total kinetic energy in the C-frame *before* the collision?

4-15.   Show that the conservation laws cannot be satisfied for the conversion of an isolated photon into a positron-electron pair. Let the energy of the photon be $h\nu$ and take its momentum to be $h\nu/c$.

4-16.   A positive pion can be produced through the reaction $p + p \rightarrow p + n + \pi^+$ by bombarding protons at rest with high-energy protons. Find the minimum kinetic energy for the incident protons (in the L-frame) required to initiate this reaction. Take the π^+ rest mass energy to be about 140 MeV.

4-17.   Assuming that electrons and positrons having equal energies collide head-on, what laboratory kinetic energy per particle is required to produce the reaction

$$e^+ + e^- \rightarrow \pi^+ + \pi^-?$$

Use 140 MeV for the rest mass energies of the pions.

*4-18.   The light emitted by hydrogen atoms that are traveling at a velocity of $0.1c$ is observed in the laboratory from the forward and backward directions by a means of a mirror.[1] Calculate the wavelength separation between the forward and backward beams for the spectral line of wavelength $\lambda_0 = 4861$ Å.

*4-19.   Assume that the stars of a binary pair revolve about their center of mass with a linear velocity of 6000 km/sec. If one of the stars emits a spectral line of wavelength 6000 Å, what wavelength will an observer measure for this line when

---

[1] This is the experiment of H. E. Ives and G. R. Stilwell, *J. Opt. Soc. Amer.* 28, 215 (1938) and 31, 369 (1941), but with an exaggerated atomic velocity.

the emitting star is receding, approaching, and moving transverse to the line of sight if (a) the observer is in the rest frame of the center of mass, and (b) the observer is moving with speed $0.6c$ with respect to the center of mass frame of the binary?

*4-20. In its own rest frame, a π^0 meson decays into two photons of equal energy and equal but opposite momenta. If a π^0 has a velocity v_0 in the laboratory, what are the highest and lowest photon frequencies that can be produced when it decays?

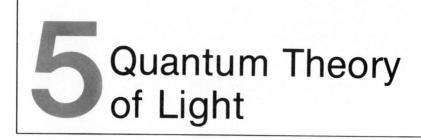

5 Quantum Theory of Light

1. BLACKBODY RADIATION

At the turn of the century Max Planck introduced the idea that light and matter could interact only through the exchange of discrete amounts of energy, called quantum units or quanta. The additional contributions of Einstein, Bohr, and others led to the development of the quantum theory of radiation as we understand it today. It all began with the search for an explanation of the spectral distribution of the radiant energy emitted by a heated body. By the term *spectral distribution,* we refer to the relative amount of energy associated with each wavelength interval of the emitted radiation. It has been known for a long time that the color of a heated object changes to a dull red at about 1100 K and that the color of the visible light emitted shifts toward the blue end of the spectrum as the temperature rises further. Experimental curves of the distribution of energy with wavelength for a given equilibrium temperature show the same general characteristics, regardless of the material of the body. Hence, it is natural to define an ideal *blackbody,* which is a perfect absorber (and emitter) of radiation. Since it reflects no light at all, it must appear perfectly black unless it is *emitting* light in the visible region of the spectrum. If a pulse of visible light, made up of a narrow band of frequencies, were incident upon such a blackbody, all of the light would be absorbed and would then be reradiated in all directions at greatly reduced intensity and with a different spectral distribution. It turns out that a blackbody that is in thermal equilibrium with its surroundings has a constant spectral distribution of radiated energy, which is characteristic of all blackbodies maintained at that same temperature.

A study of the radiation from an ideal blackbody can be approximated experimentally by observing the light emerging from a small hole in an isothermal enclosure, such as a hollow block of carbon. If the hole is sufficiently small, the energy radiated through it will have a negligible effect upon the equilibrium state in the cavity. For such a cavity, a typical plot of the spectral distribution of energy for several absolute temperatures is shown in Figure 5–1. The curves shown were obtained by Lummer and Pringsheim in 1900, although their general characteristics were known earlier. It should be noted that (1) the short wavelength "cutoff" advances toward the origin as the temperature increases, (2) raising the temperature

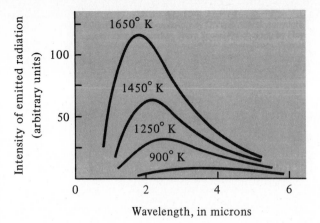

Figure 5–1

Spectral distribution of blackbody radiation for several different temperatures.

increases the energy of all spectral components, and (3) the peak of the curve shifts to shorter wavelengths as the temperature increases. The shift of the peak of the curve was found to obey the following empirical relationship, commonly called *Wien's displacement law:*

$$\lambda_p T = \text{constant}, \tag{5.1}$$

where the symbol λ_p refers to the value of the wavelength corresponding to the peak of the curve. A thermodynamic expression also exists which relates the total power radiated per unit area of a blackbody to its absolute temperature. This is known as the Stefan-Boltzmann law and is expressed mathematically as

$$E(T) = \sigma T^4, \tag{5.2}$$

where $\sigma = 5.6699 \times 10^{-8}$ watts m^{-2} K^{-4}. Thus the total energy radiated in a given time by a heated object is proportional to the fourth power of its absolute temperature. The monochromatic emissive power, $E(\lambda, T)$, is the power radiated per unit area at a given wavelength. This is related to the total power radiated per unit area by the integral,

$$E(T) = \int_0^\infty E(\lambda, T) \, d\lambda. \tag{5.3}$$

Wien proposed an empirical form for the monochromatic emissive power, $E(\lambda, T)$, by constructing a mathematical function to fit the experimental blackbody curves. That is, there was no attempt to relate the emitted radiation to physical processes taking place within the radiating body. Not long thereafter, a theoretical model was developed by Rayleigh and Jeans; this model was later refined by Planck to obtain the correct radiation law.

Example 5-1

(a) What will be the wavelength shift of the peak of the blackbody radiation curve when a star whose surface temperature is T_1 Kelvin cools to a temperature of T_2? (b) What is the fractional wavelength change when $T_1 = 5800$ K and $T_2 = 4350$ K?

Solution

(a) From Equation 5.1 we see that

$$\lambda_{p1}T_1 = \lambda_{p2}T_2.$$

Then
$$\lambda_{p2} = (T_1/T_2)\,\lambda_{p1}$$

$$\Delta\lambda_p = \lambda_{p2} - \lambda_{p1} = (T_1/T_2 - 1)\,\lambda_{p1}.$$

(b) Setting $T_1 = 5800$ K and $T_2 = 4350$ K, we obtain

$$\frac{\Delta\lambda_p}{\lambda_p} = \left(\frac{5800}{4350} - 1\right) = \frac{5800 - 4350}{4350} = \frac{1450}{4350} = \frac{1}{3}.$$

Example 5-2

If an object is heated from 100° C to 200° C, by what factor will the energy lost by thermal radiation increase?

Solution

Equation 5.2 tells us that the ratio of the power radiated at the two temperatures is

$$\frac{E_2}{E_1} = \left(\frac{T_2}{T_1}\right)^4,$$

where T_1 and T_2 must be expressed in Kelvin. Then, the radiation loss will increase by the factor

$$\frac{E_2}{E_1} = \left(\frac{473\ \text{K}}{373\ \text{K}}\right)^4 = 2.6.$$

2. THE RAYLEIGH-JEANS THEORY

The simplest model of a radiating body is to regard it as a collection of a large number (on the order of 10^{23}) of linear oscillators performing simple harmonic motion. Since the particles undergoing the oscillations are, in general, charged particles, they will radiate electromagnetic waves. In the case of a cavity as discussed above, at thermal equilibrium the electromagnetic energy density inside the cavity will equal the energy density of the atomic oscillators situated in the cavity walls. When the walls are raised to a higher temperature, the following events take place: more energy is put into existing oscillator modes by increasing their amplitudes, new modes corresponding to stiffer spring constants (higher frequencies) are excited, and the radiation density in the cavity is increased until a new equilibrium point is reached.

The number of oscillator modes per unit volume corresponding to wavelength λ can be readily calculated. This number is known as Jeans' number, and it is given by

$$g(\lambda) = \frac{8\pi}{\lambda^4}. \qquad (5.4)$$

If one knew the average energy per oscillator mode for a given wavelength and equilibrium temperature, then the energy per unit volume, $I(\lambda, T)$, would be known. The radiated power $E(\lambda, T)$ is related to the energy density $I(\lambda, T)$ by the constant factor $c/4$. (See Problem 5-19). Then,

$$E(\lambda, T) = \frac{c}{4} I(\lambda, T) = \frac{c}{4} \cdot g(\lambda) \cdot \langle \epsilon \rangle = \frac{2\pi c}{\lambda^4} \langle \epsilon \rangle, \qquad (5.5)$$

where $\langle \epsilon \rangle$ is the average energy per oscillator. It was at this point that the Rayleigh-Jeans theory went awry, for it assumed the classical value for the average energy per oscillator, namely, kT. Although this theory was in reasonable agreement with the curves of Figure 5–1 for very long wavelengths, it made disastrous predictions for short wavelengths. In fact, it is readily apparent that the radiated power in Equation 5.5 becomes infinite as the wavelength approaches zero, which is in sharp contrast with the experimental facts. This failure was such a crushing blow to classical physics that it has often been referred to as "the ultraviolet catastrophe."

*3.   CALCULATION OF JEANS' NUMBER

Our aim is to determine how many oscillator modes exist in a given wavelength interval in a cubical cavity of edge L. First let us consider the vibrational modes in a one-dimensional system.

In a one-dimensional vibrating system such as a stretched string, standing waves can be set up as illustrated in Figure 5–2. Each standing wave mode is char-

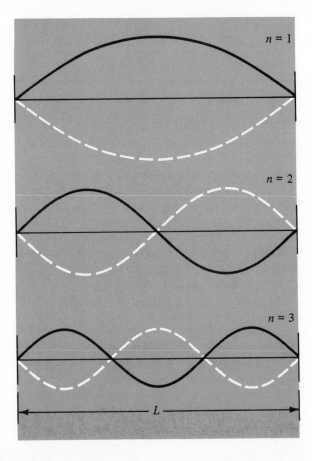

$n = 1$

$n = 2$

$n = 3$

L

Figure 5–2

The standing waves for $n = 1, 2,$ and 3 for a string of length L fixed at both ends. The nodes (points of zero amplitude) remain fixed. The antinodes (loops) vary from one extreme position to the other sinusoidally in time.

acterized by the number of loops (antinodes) contained in the vibration pattern. Each loop corresponds to a half wavelength. For a string stretched in the x-direction, the displacement is given by $u(x, t) = C(\sin k_x x) (\cos \omega t)$. At some time t a "snapshot" of the standing wave would show that $u(x) = C' \sin k_x x$. Since the string is clamped at both ends we have the boundary conditions that $u(x) = 0$ at $x = 0$ and at $x = L$. It follows that

$$k_x L = n_x \pi, \tag{5.6}$$

where n_x is an integer. Since $k_x = 2\pi/\lambda$ this becomes

$$n_x = \frac{2L}{\lambda}. \tag{5.7}$$

Extending this idea to a cubical volume, we may now visualize elastic waves in a solid, or sound waves in a room, or electromagnetic waves in a cavity. The displacement of the wave is given by

$$u(x, y, z, t) = c \sin k_x x \cdot \sin k_y y \cdot \sin k_z z \cdot \cos \omega t, \tag{5.8}$$

where $u = 0$ for $x = 0, L; y = 0, L; z = 0, L$. With these boundary conditions, we see that

$$k_x = \frac{n_x \pi}{L}, \qquad k_y = \frac{n_y \pi}{L}, \qquad k_z = \frac{n_z \pi}{L}. \tag{5.9}$$

Then
$$k^2 = k_x{}^2 + k_y{}^2 + k_z{}^2$$

$$= \left(\frac{\pi}{L}\right)^2 (n_x{}^2 + n_y{}^2 + n_z{}^2), \tag{5.10}$$

where n_x, n_y and n_z are all integers. Now define n in terms of these three integers by

$$n^2 = n_x{}^2 + n_y{}^2 + n_z{}^2. \tag{5.11}$$

Since the wave number $k = 2\pi/\lambda$, we may combine Equations 5.10 and 5.11 to write,

$$n^2 = \left(\frac{2\pi}{\lambda}\right)^2 \cdot \left(\frac{L}{\pi}\right)^2 = \frac{4L^2}{\lambda^2}. \tag{5.12}$$

Each set of integers (n_x, n_y, n_z) determines a point in *lattice space,* and each such point occupies unit volume in that space. (Each unit cube has eight points at its corners, but each lattice point is contained in eight such cubes.) Since there is a one-to-one correspondence between the volume of the space, the number of lattice points contained in that volume, and the number of allowed modes of a given wavelength, *we can count modes by merely calculating the volume in the lattice space.* Thus, the number of modes of wavelength λ in the first octant of a sphere of radius n is

$$N = \frac{1}{8} \cdot \frac{4}{3} \pi n^3 = \frac{\pi}{6} \left(\frac{2L}{\lambda}\right)^3. \tag{5.13}$$

Then the number of modes in the wavelength interval between λ and $\lambda + d\lambda$ is

$$|dN| = \frac{4\pi L^3}{\lambda^4} d\lambda. \tag{5.14}$$

Since L^3 is the volume of the cavity, the number of modes per unit volume in the same wavelength interval is

$$\frac{|dN|}{L^3} = \frac{4\pi}{\lambda^4}\, d\lambda. \tag{5.15}$$

In the case of transverse waves there are actually twice this number of modes, since there are two degrees of freedom associated with the two senses of polarization of each wave. Then, we finally obtain

$$g(\lambda)d\lambda = \frac{2|dN|}{L^3} = \frac{8\pi}{\lambda^4}\, d\lambda, \tag{5.16}$$

where $g(\lambda)$ is known as *Jeans' number,* the number of oscillator modes per unit volume per wavelength interval.

Example 5-3
   Express Jeans' number in terms of frequency instead of wavelength.

Solution
   Conversion between the parameters λ and ν is easily done if one remembers that

$$g(\lambda)\, d\lambda = g(\nu)\, d\nu. \tag{5.17}$$

Then, $$g(\nu) = \left|\frac{d\lambda}{d\nu}\right| \cdot g(\lambda) = \frac{c}{\nu^2}g(\lambda) = \frac{8\pi\nu^2}{c^3}. \tag{5.18}$$

---

4. PLANCK'S QUANTUM THEORY OF RADIATION

   Being well aware of the shortcomings of both the Rayleigh-Jeans and the Wien radiation laws, Planck examined the mathematical statistics of these theories to ascertain what changes, if any, might result in a reasonable description of the experimental radiation curve. As a result of this work, he was led to certain assumptions about the nature of the electromagnetic oscillators that are in equilibrium with the energy density within the blackbody cavity. These postulates, which have become the foundation of the quantum theory of radiation, are as follows:

   (1) The amount of energy emitted or absorbed by an oscillator is proportional to its frequency. Calling the constant of proportionality h, we then write the change in oscillator energy as

$$\Delta\epsilon = h\nu. \tag{5.19}$$

   (2) An oscillator cannot have an arbitrary energy, but must occupy one of a discrete set of energy states given by

$$\epsilon_n = nh\nu, \tag{5.20}$$

where n is an integer or zero. It was assumed that the ground state corresponded to the zero energy state. The value Planck gave for the constant of proportionality, h, was 6.55×10^{-34} Joule-second. He obtained this value by fitting curves such as those shown in Figure 5–1 with his radiation law, which will be derived below. However, the value for h is now known to be approximately

$$h = 6.626 \times 10^{-34} \text{ J-s}.$$

Planck's constant is a universal constant that plays an important role in all quantum phenomena.

The previous picture of a continuum of oscillator energy states is now replaced by a discrete set of "quantized" levels. Furthermore, the amount of energy emitted or absorbed is also quantized, since each quantum must correspond to the energy difference between two states of a given oscillator. Each quantum of electromagnetic energy $h\nu$ is called a *photon*. The photon also carries momentum given by Equation 4.7 as

$$p = \frac{h}{\lambda}. \tag{5.21}$$

The absorption of a photon of frequency ν will raise the energy of an oscillator of frequency ν by an amount given by $h\nu$; it will have no effect on an oscillator of frequency $\nu' \neq \nu$. Emission of a photon occurs when the oscillator energy drops to the next lower energy; the frequency of the emitted light will correspond to the oscillator frequency.

In what state is an oscillator most likely to be found? If nothing excites it, it is most likely to be found in its lowest energy state or ground state. Hence, at absolute zero one would expect to find all oscillators in the zero energy state according to the above model. This would be true in classical mechanics and can be assumed here without affecting our answer. But we will see later that quantum mechanics predicts a so-called "zero-point motion" at absolute zero instead of the complete cessation of all vibration.[1] At higher temperatures thermal agitation excites some oscillators to higher states so that some sort of distribution of oscillators over all possible states will exist for each temperature.

The distribution function that tells us what fraction of the oscillators will occupy the nth state when the equilibrium temperature is T is known as the Maxwell-Boltzmann distribution function.[2] It is given by

$$N(n) = N_0 e^{-\epsilon_n/kT} \tag{5.22}$$

where $N(n)$ is the number of oscillators in the state n, ϵ_n is the energy of the nth state, k is the Boltzmann constant, T is the absolute temperature, and N_0 is the total number of oscillators. Note that the higher energy states (large ϵ_n) have a smaller probability of being occupied because of the minus sign in the exponent. As the energy increases indefinitely, the number of occupied states becomes vanishingly small. It is evident that this feature eliminates the problem of the ultraviolet catastrophe, because the latter arose as a result of the assumption that oscillators of all energies were excited with equal probability. Instead, a high-energy oscillator has an extremely low probability of excitation. Thus it is clear from this

---

[1] In quantum mechanics, Equation 5.20 is replaced by $\epsilon_n = (n + \frac{1}{2})h\nu$ so that the ground state is $\frac{1}{2}h\nu$ instead of zero.

[2] For a derivation of this, see E. E. Anderson, *Modern Physics and Quantum Mechanics*, W. B. Saunders Co., Philadelphia, 1971, pp. 67–70.

argument that the sharp fall-off in radiated energy observed at short wavelengths (shown in Figure 5–1) can be accounted for.

The discussion of the preceding paragraph is readily confirmed by a calculation of the average energy per oscillator based upon both the Planck hypothesis and the Maxwell-Boltzmann distribution. We define the average energy per oscillator by

$$\langle \epsilon \rangle = \frac{\sum\limits_{n=0}^{\infty} N(n)\epsilon_n}{\sum\limits_{n=0}^{\infty} N(n)}, \tag{5.23}$$

where the sums are taken over the states $n = 0, 1, 2, \ldots$. The numerator of Equation 5.23 represents the total energy of all of the occupied states, while the denominator is simply the total number of occupied states. Substituting Equation 5.22 into Equation 5.23,

$$\langle \epsilon \rangle = \frac{\sum\limits_{n=0}^{\infty} N_0 e^{-nh\nu/kT} \, nh\nu}{\sum\limits_{n=0}^{\infty} N_0 e^{-nh\nu/kT}}$$

$$= \frac{0 + h\nu e^{-h\nu/kT} + 2h\nu e^{-2h\nu/kT} + 3h\nu e^{-3h\nu/kT} + \cdots}{1 + e^{-h\nu/kT} + e^{-2h\nu/kT} + e^{-3h\nu/kT} + \cdots}. \tag{5.24}$$

It is convenient here to make the substitution $x = e^{-h\nu/kT}$. Then Equation 5.24 becomes

$$\langle \epsilon \rangle = h\nu x \left(\frac{1 + 2x + 3x^2 + 4x^3 + \cdots}{1 + x + x^2 + x^3 + \cdots} \right).$$

The numerator of the quantity in parentheses is the binomial expansion of $(1 - x)^{-2}$ while the denominator is the binomial expansion of $(1 - x)^{-1}$. Making these substitutions, we obtain

$$\langle \epsilon \rangle = h\nu x \cdot \frac{(1 - x)^{-2}}{(1 - x)^{-1}} = \frac{h\nu x}{1 - x}$$

$$\langle \epsilon \rangle = \frac{h\nu}{e^{h\nu/kT} - 1}. \tag{5.25}$$

Equation 5.25 is the Planck result for the average energy per oscillator.

Incorporating the result in Equation 5.25 into Equation 5.5, we find that the radiated power per unit area per wavelength (frequency) interval is

$$\left. \begin{aligned} E(\lambda, T) &= \frac{2\pi hc^2}{\lambda^5} \left(e^{hc/\lambda kT} - 1 \right)^{-1} \\[2mm] E(\nu, T) &= \frac{2\pi h\nu^3}{c^2} \left(e^{h\nu/kT} - 1 \right)^{-1} \end{aligned} \right\} \tag{5.26}$$

or

Equation 5.26 is *Planck's radiation law*. An example of how well it describes the radiation curves is illustrated in Figure 5–3, where the circles are experimental

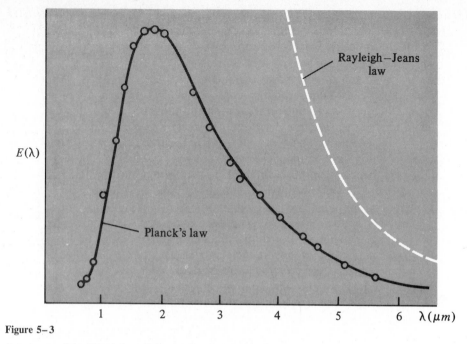

Figure 5–3

Agreement of the Planck law with experimental data at $T = 1600$ K.

points and the solid line is the Planck law curve for a temperature of 1600 K. The problems below will illustrate further how well the Planck law accounted for all that was valid in the older theories as special cases. It thus serves as an excellent example of a conceptual advance that opened exciting new frontiers while still preserving much of the older physics.

It is worth noting before concluding this section that the second factor in Equation 5.26 is the distribution function for photons. This function is now known as the Bose-Einstein distribution function,

$$F_{BE}(h\nu) = \left(\exp \left(\frac{h\nu}{kT} \right) - 1 \right)^{-1}. \qquad (5.27)$$

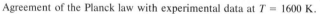

Example 5-4

A photon has a wavelength of 4000 Å. What is its energy? Its momentum?

Solution

In Planck's theory the energy of a photon is given by the energy change of the oscillator which emitted (or absorbed) it. Thus, from Equation 5.19, we have for the photon energy

$$E = h\nu = \frac{hc}{\lambda} = \frac{6.6 \times 10^{-34} \text{J-s} \times 3 \times 10^8 \text{m/s}}{4 \times 10^{-7} \text{m}} = 4.95 \times 10^{-19} \text{J} = 3.1 \text{ eV}.$$

Then

$$p = \frac{E}{c} = \frac{h}{\lambda} = \frac{6.6 \times 10^{-34} \text{J-s}}{4 \times 10^{-7} \text{m}} = 1.65 \times 10^{-27} \text{ kg-m/s}.$$

Example 5-5
Find the wavelength of a photon whose energy is 1 keV.

Solution

$$\lambda = \frac{hc}{E} = \frac{1.98 \times 10^{-25} \text{J-m}}{1.6 \times 10^{-16} \text{J}} = 1.24 \times 10^{-9} \text{m} = 12.4 \text{ Å}.$$

Example 5-6
A radio transmitter operates at a frequency of 97 MHz with a power output of 200 kW. How many photons are emitted per second?

Solution

Each photon has an energy of

$$E = h\nu = 6.6 \times 10^{-34} \text{J-s} \times 9.7 \times 10^{7} \text{s}^{-1} = 6.4 \times 10^{-26} \text{J}.$$

Then the rate of photon emission is

$$N = \frac{2 \times 10^{5} \text{J}}{6.4 \times 10^{-26} \text{J}} = 3.13 \times 10^{30} \text{ quanta per second}.$$

Example 5-7
At absolute zero we would expect the average energy per oscillator to be the ground state energy. At what temperature would the average energy per oscillator be equal to the energy level spacing $h\nu$ above the ground state?

Solution

In Equation 5.25 we note that the condition for $\langle\epsilon\rangle = h\nu$ is that $\frac{h\nu}{kT} = \ln 2$. Therefore,

$$T = \frac{h\nu}{k(\ln 2)} = \frac{h\nu}{6} \times 10^{5},$$

where $h\nu$ is in eV and T is expressed in Kelvin. Thus, if the oscillator levels were separated by 6×10^{-5} eV, a temperature of 1 K would be sufficient to boost the average oscillator energy up to the energy of the first excited state. For a level separation of 6 eV, on the other hand, a temperature of 100,000 K would be required!

Example 5-8
In the first approximation a simple pendulum is a harmonic oscillator. Why were its quantized energy levels not discovered long ago?

Solution

The energy levels of a 1-second pendulum are separated by the energy $h\nu = 6.6 \times 10^{-34}$ joules, which is much too small to be detected. For a pendulum bob having a mass of 100 grams, the quantized levels would correspond to differences in maximum height of the order of 10^{-33} meter! Thus, it is the extremely small value of the Planck constant h which prevents the appearance of quantum phenomena in the everyday world.

5. EINSTEIN'S TRANSITION PROBABILITIES

Thus far we have used the oscillator model to account for the radiation emitted from a hot body and to derive the Planck radiation law. We have stated that each photon or quantum of light that is emitted or absorbed by an oscillator is associated with a change of the energy state of the oscillator. What this generally means in an atomic oscillator is that an electron makes a *transition* from one energy state to another energy state within the atom. If the electron transition results in a *lower* potential energy for the atom, the excess energy is *emitted* as a photon. On the other hand, if the transition *increases* the energy of the atom, then a photon must be *absorbed* in order to provide the additional energy. Such transitions, which are accompanied by the absorption or emission of photons, are known as *radiative transitions*.

Einstein made use of Planck's quantum hypothesis in order to study the radiative transitions for a system of oscillators in equilibrium with a large number of photons having the requisite energy corresponding to the oscillator transition energy. The radiation field may be thought of as a "photon gas" of energy density $I(\nu)$. Let the oscillator system consist of N atoms, with N_1 atoms in the energy state E_1, and N_2 atoms in the excited energy state E_2. That is, $E_2 > E_1$, as shown schematically in Figure 5–4. Photons having energy $h\nu = E_2 - E_1$ may be absorbed by atoms in the E_1 state, thus raising them to the E_2 state. Alternatively, a photon of energy $E_2 - E_1$ can induce a *downward* transition in an atom in the E_2 state, causing it to emit an identical photon of energy $E_2 - E_1$. A photon that is radiated as a result of induced or *stimulated emission* is in phase with the incident photon which brought it into existence. That is to say, the incident and induced photons are *coherent*. An important consequence of that fact will be discussed at length in the next section.

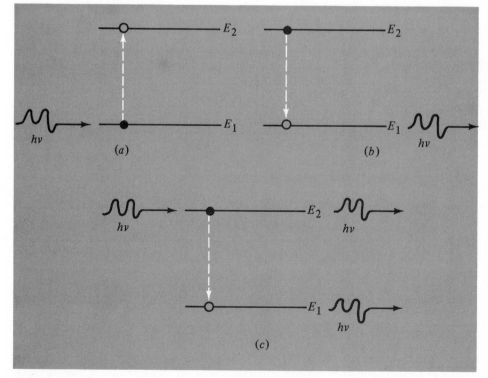

Figure 5–4

Photon interactions in an atom having two energy levels that differ in energy by $E_2 - E_1 = h\nu$. (a) Photon absorption, (b) spontaneous emission, (c) stimulated emission.

When an atom is in an excited state it is generally unstable and it will, in the normal course of events, return to the ground state spontaneously in the absence of any stimulated transitions. It is not possible to predict how long a particular atom will remain in the excited state before emitting a photon, but it is possible to determine an *average lifetime* of the state for a huge collection of identical atoms. A typical lifetime for an atomic state is about 10 nanoseconds, but lifetimes can vary considerably from this, depending upon the particular physical situation. There are some excited states that cannot decay by photon emission (such as the 2s atomic state), so they must reach the ground state through a non-radiative process such as energy transfer by collision. There are other excited states that have long lifetimes because electric dipole transitions, which are normally the predominant mechanism for photon emission, are forbidden for those states. The latter states decay by emitting photons, but they do so at a much lower rate than states for which electric dipole transitions are allowed. All of these long-lived states are called *metastable states*.

Einstein described the processes with which we are concerned here by means of the following three probabilities:

A_{21} is the probability per unit time for a spontaneous transition from E_2 to E_1 with the emission of a photon.

$B_{21} I(\nu)$ is the probability per unit time for an induced transition from E_2 to E_1 with the emission of a photon.

$B_{12} I(\nu)$ is the probability per unit time for an induced transition from E_1 to E_2 with the absorption of a photon.

A and B are constants that are to be determined. The energy density $I(\nu)$ must be included for the induced transitions, since they are dependent upon the presence of the photon gas. The transition rate for photon emission is

$$[A_{21} + B_{21} I(\nu)] \cdot N_2 \tag{5.28}$$

since it is proportional to the product of the population of the excited state and the total probability for a downward transition. The transition rate for photon absorption is simply

$$B_{12} I(\nu) \cdot N_1. \tag{5.29}$$

At equilibrium these two rates must be equal, so we have

$$\frac{N_1}{N_2} = \frac{A_{21} + B_{21} I(\nu)}{B_{12} I(\nu)}. \tag{5.30}$$

Now the equilibrium populations, N_1 and N_2, are dependent upon the temperature of the system; they are given by the distribution function in Equation 5.22. Thus,

$$\frac{N_1}{N_2} = \exp\left(-\frac{E_1}{kT}\right) \exp\left(\frac{E_2}{kT}\right) = \exp\left(\frac{h\nu}{kT}\right), \tag{5.31}$$

where T is the temperature corresponding to the equilibrium state. Then, Equation 5.30 becomes

$$\exp\left(\frac{h\nu}{kT}\right) \cdot B_{12} I(\nu) = A_{21} + B_{21} I(\nu),$$

or
$$I(\nu) = \frac{\dfrac{A_{21}}{B_{12}}}{\exp\left(\dfrac{h\nu}{kT}\right) - \dfrac{B_{21}}{B_{12}}}. \tag{5.32}$$

Equation 5.32 evidently has the same form as the Planck radiation law, Equation 5.26. Multiplying the former by $c/4$ and comparing terms, we see that

and
$$\left.\begin{array}{c} B_{12} = B_{21} \\[2mm] A_{21} = \dfrac{8\pi h\nu^3}{c^3}\, B_{21} \end{array}\right\} \tag{5.33}$$

It is worth pointing out here that the B coefficient can be readily calculated in quantum mechanics by regarding the atom as a dipole oscillator that is driven by the time-dependent electric field intensity of the radiation field.[1] Although the A coefficient cannot be calculated from the dipole model, it can be determined from Equation 5.33 using the calculated value of B.

Example 5-9

Use the Maxwell-Boltzmann distribution, Equation 5.22.

(a) Find the temperature that corresponds to equal populations for two levels such as those in Figure 5–4.

(b) What temperature would be required in order for the population of the upper level to be greater than that of the lower level?

(c) How do you reconcile this use of the term "temperature" with its meaning in thermodynamics?

Solution

(a) Equation 5.31 gives the population ratio in terms of the equilibrium temperature. Setting $N_1 = N_2$ we see that $(h\nu/kT)$ must approach zero in order to satisfy the equation. This means that $T \rightarrow \infty$. In other words, an *infinite temperature* would be required in order to achieve the *equilibrium state* of equal populations.

(b) In order for N_2 to be greater than N_1, we see from Equation 5.31 that the exponent $(h\nu/kT)$ must be negative. It is customary to regard T as the quantity that becomes negative, and this is the origin of the concept of a "negative temperature."

(c) In thermodynamics, temperature is normally used as a state variable to describe a system that is in thermal equilibrium. By means of temperature alone, there is no way ever to achieve inverted populations—or even equal populations—for the two states in Figure 5–4. However, there are other ways of greatly altering the populations of the states from their equilibrium values. When this occurs, the non-equilibrium distribution may be described by a *statistical temperature,* which should be interpreted as follows: "This is the temperature that would be necessary to achieve the distribution by thermal energy alone." It is possible for a material to be at a thermodynamic temperature of 4 K and still have two energy levels with inverted populations corresponding to a negative temperature in the statistical sense.

---

[1] See, for example, E. E. Anderson, *op. cit.,* pp. 359–362.

An artist's conception of the Nova Laser Fusion Facility in Livermore, California. A powerful pulse of laser light from neodymium glass will be directed upon a fusion pellet the size of a pin head situated inside the target chamber. The resulting implosion of the target will produce usable energy in the form of high-speed neutrons. When Nova is completed it will deliver a short pulse lasting 10^{-9} second having a power output of 3×10^8 megawatts. (Courtesy of the Lawrence Livermore National Laboratory operated by the University of California for the U.S. Department of Energy.)

6. AMPLIFICATION THROUGH STIMULATED EMISSION

The extension of the concepts of the previous section to non-equilibrium cases has become the basis for the amplification of signals in laser and maser devices. The term *laser* is an acronym formed from the words "Light Amplification by Stimulated Emission of Radiation." The word *maser* comes from the substitution of "Microwave" for "Light." Lasers normally operate in the visible region of the electromagnetic spectrum, while masers are generally used in the microwave region of the spectrum.

Suppose that the populations of the two levels in Figure 5–4 are somehow altered so that $N_2 > N_1$. This cannot, of course, be achieved by any thermal process, but we will soon see how it can be accomplished. Since the number of downward transitions per second is proportional to N_2 and the number of upward transitions per second is proportional to N_1, there will be a greater number of downward transitions than upward transitions when $N_2 > N_1$. That is, the number of photons emitted will be greater than the number absorbed each second. Since the emitted photons have the same frequency as the incoming photons, this increase in energy density is equivalent to amplification at that frequency.

In order to devise an amplifier for a given frequency, then, one must first find a pair of atomic energy levels whose energy difference corresponds to the photon energy of the signal. Next, a means must exist for producing a population inversion. One approach would be to somehow segregate the excited atoms from those in the lower state. Then the excited atoms could be used for amplification. This approach was followed in the first successful maser device, the ammonia maser, developed by C. H. Townes et al. in 1954.[1] Another approach is to use a process called *pumping* in order to achieve an inverted population. This requires one or more additional energy levels, as illustrated in Figure 5–5.

First consider Figure 5–5(a). The signal to be amplified has photon energy $h\nu = E_2 - E_1$. Energy level E_2 is a metastable state. Consequently, the probabil-

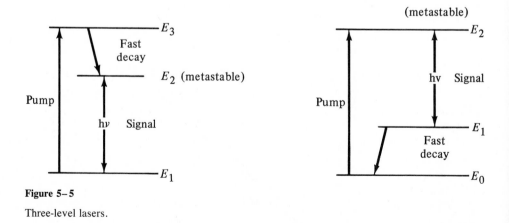

Figure 5–5

Three-level lasers.

[1] J. P. Gordon, H. J. Zeiger, and C. H. Townes, *Phys. Rev. 99,* 1264 (1955).

ity of spontaneous decay from E_2 is so small that the excited state will normally persist until one of the signal photons stimulates or induces its decay to state E_1 with the emission of an additional photon. Now if the atom has a state E_3 with a very short lifetime and a predominant decay mode to E_2, then one can populate E_2 by "pumping" from state E_1 to E_3. The pumping action simultaneously depletes level E_1 and populates level E_2 via the fast decay from E_3 to E_2, thus leading to a population inversion. The pumping frequency, of course, must correspond to the energy difference $E_3 - E_1$ divided by h. It should be a frequency that is readily available and cheap to produce. Pumping energy is sacrificed in order to achieve amplification at the signal frequency, just as the energy expended by the power supply in a hi-fi power amplifier is of little concern provided that we obtain proper amplification of the signal.

A slightly different approach is shown in Figure 5–5(b). Again the signal corresponds to the energy difference $E_2 = E_1$, and we require that $N_2 > N_1$ for amplification. This is achieved by the two following processes: N_2 is increased directly by the pumping action and N_1 is diminished by means of a fast decay from E_1 to E_0.

Sometimes four-level systems are used that incorporate the basic ideas of both of the models discussed above. A hypothetical four-level system is depicted in Figure 5–6.

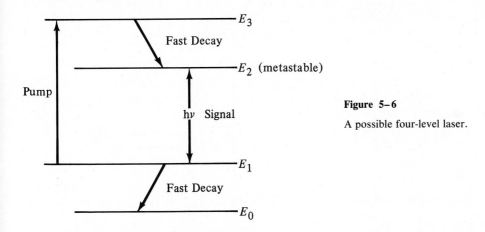

Figure 5–6

A possible four-level laser.

Let us see if there is anything that can be done, once the population inversion has been achieved, to increase the amplification of the signal. A glance at Equation 5.28 tells us that the rate of stimulated emission is directly proportional to the energy density of the photon gas, $I(\nu)$. That is, each photon that is emitted increases the energy density, thus enhancing the emission rate for additional photons. This cascade effect produces an avalanche of photons, provided that the population inversion persists. If, in addition, two reflectors are placed parallel to each other on opposite sides of the laser medium, some photons are focused into a narrow beam that makes multiple transits through the laser material. Under these conditions the region between the two reflectors becomes a resonant cavity for photons having the phase and direction of this beam, and the energy density of the beam increases exponentially. If some of this light is allowed to escape through one of the reflectors, it will be perceived as a narrow beam of *coherent* light. That is, in any cross-section of the beam the phase is the same at every point.

To summarize, the important characteristics of laser light are: it is coherent, it is very nearly monochromatic, it has very little beam spread, and it has a very high energy density. These underlying properties have stimulated hundreds of applications for lasers in all walks of life. Two quite different lasers will be discussed in the next section.

7. THE RUBY AND HELIUM-NEON LASERS

Ruby is aluminum oxide, Al_2O_3, which contains chromium impurities. The red color results from the fact that chromium absorbs a broad band of wavelengths from the visible region, letting only red and blue light through. A pale pink ruby would have about one aluminum atom in 10^4 replaced by a chromium atom. Most natural rubies are unsatisfactory for use in lasers either because they are too small or because they contain too many impurities. However, large single crystals of ruby can be grown in the laboratory; such crystals have the optimum concentration of chromium ions and essentially no other impurities.

The energy levels of interest to us here are shown in Figure 5–7, and a sketch

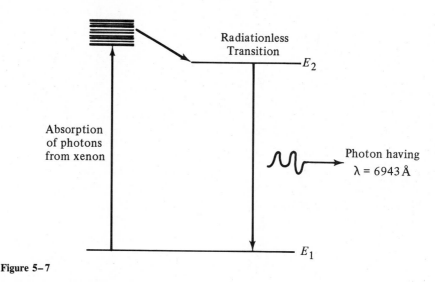

Figure 5–7

Chromium energy levels for the ruby laser.

of a ruby laser is shown in Figure 5–8. The energy difference $E_2 - E_1$ corresponds to the characteristic red light of a ruby laser, which has a wavelength of 6943 Å. Level E_2 is a metastable state with a lifetime of about three milliseconds. E_3 is not a single level but is a whole band of levels corresponding to the broad absorption by chromium in the center of the visible region. When the xenon lamp flashes, it emits light of great intensity in the yellows and greens where it is readily absorbed by the ruby, heavily populating the states labeled E_3. This level has a very short lifetime, decaying to state E_2 without emitting a photon. Hence N_2 quickly becomes greater than N_1 and the stage is set for laser action.

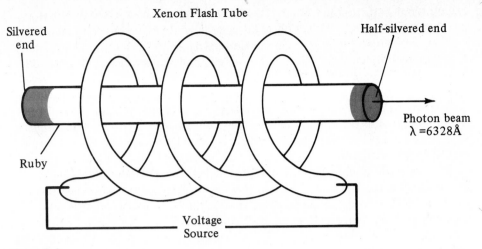

Figure 5–8

Sketch of a ruby laser.

Since some spontaneous transitions always take place, there is a ready supply of photons available to induce additional transitions. The beam intensity builds up rapidly, owing to the combination of multiple reflections and the avalanche of stimulated emission. Ultimately, the beam escapes through the half-silvered end of the rod in a burst lasting about one-half millisecond.

The ruby laser described here is a pulsed system. Each time it fires, it must be cooled and then a new pulse may be built up by refiring the xenon lamp. Short though the pulse is, a carefully focused ruby laser can deliver power equivalent to millions of watts per square centimeter. Modern ruby lasers can deliver an instantaneous optical power of more than 10^9 watts by means of a process called Q-switching.

The helium-neon laser is an example of a continuously operating laser that can be found in most any undergraduate laboratory. It is a low-power device that requires no external cooling. The energy levels involved in the laser action are the neon levels labeled E_1 and E_2 in Figure 5–9. The laser light is again red, though not

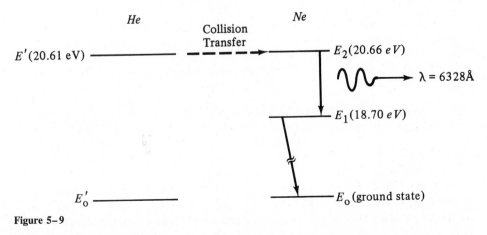

Figure 5–9

Helium-neon laser energy levels.

as deep a red as the output of the ruby laser. An interesting feature of this laser is that the state E_2 is not a metastable state. It decays rapidly by the transition shown and by another that is not shown in the figure. Therefore, it would be impossible to obtain a population inversion between levels E_1 and E_2 without some outside help. This is where the helium gets into the act. Notice that the helium energy level labeled E' is 20.61 eV above its ground state, which nearly equals the energy of the excited state E_2 of neon. An excited helium atom having as little as 0.05 eV of kinetic energy can transfer all of its energy to a neon atom during a collision, leaving the neon atom in the excited state E_2 and the helium atom in its ground state. The reason this process is so significant is that the E' state of helium is a metastable state with an unusually long lifetime. This state cannot decay by emitting a photon, so it must rely on collision transfer with other atoms or the container walls for de-excitation. Consequently, energy can be stored in the E' state to be later funneled through the E_2 state for laser action.

The rapid depletion of the E_1 state also contributes to the laser amplification. Unfortunately, one step in reaching the ground state from state E_1 requires that energy be transferred by collision from the neon atom to the tube wall. This process must be given consideration when choosing the dimensions of the neon tube.

How are helium atoms excited to the E' state? This is achieved simply by means of an electrical discharge. Why are neon atoms not excited by this same discharge? They are, but they decay immediately, causing the gas to glow as in a neon sign. These photons are not emitted in phase, so they cannot set up a standing wave within the laser cavity. On the other hand, when the E_2 state is suddenly highly populated through collision transfer, the photons emitted by stimulated emission are in phase with the stimulating photons. Those having the proper direction set up standing waves as a result of multiple reflections from the mirrors, and they in turn stimulate the emission of additional photons having the same phase and direction. In this manner, a narrow beam of coherent light of high intensity is produced.

There are, of course, numerous possibilities for designing lasers of many different varieties. The interested reader is referred to the references at the end of this chapter.

SUMMARY

The concept of the quantization of electromagnetic energy is the key to Planck's explanation of the spectral distribution of the energy radiated from a thermally excited blackbody. As a consequence of the Planck theory, we now regard the emission or absorption of electromagnetic energy as the exchange of a photon between matter and the electromagnetic field. The energy of a photon is expressed as $E = h\nu$, where ν is the frequency and h is Planck's constant. The approximate value of h is 6.6×10^{-34} J-s. When written as a function of frequency and Kelvin temperature, the Planck radiation law has the following form:

$$E(\nu, T) = \frac{2\pi h \nu^3}{c^2} \cdot \frac{1}{\exp\left(\dfrac{h\nu}{kT}\right) - 1}$$

The Einstein probabilities for spontaneous and induced transitions between electronic energy levels were derived. Such transitions are accompanied by photon emission or absorption, and it was shown that this treatment not only is consistent with the Planck result but provides an alternative way to derive amplification through stimulated emission, or what is commonly called laser action. The prototypes of pulsed and continuous lasers—the ruby and the helium-neon lasers—were discussed in some detail.

Additional Reading

W. E. Kock, *Lasers and Holography,* Doubleday & Co., Garden City, N.Y., 1969.

Lasers and Light, Readings from Scientific American. W. H. Freeman & Co., San Francisco, 1969.

"Lasers in Research." Four articles in *Physics Today, 30,* No. 5 (May 1977).

B. A. Lengyel, *Lasers,* 2nd ed. John Wiley & Sons, Inc., New York, 1971.

Scientific American, special issue on Light, September 1968.

P. A. Tipler, *Modern Physics,* Worth Publishers, Inc., New York, 1978, Chapter 3.

E. H. Wichmann, *Quantum Physics,* Berkeley Physics Course, Vol. 4, McGraw-Hill Book Co., New York, 1971, Chapter 1.

PROBLEMS

5-1. The peak of the radiation curve for a certain blackbody occurs at a wavelength of 10,000 Å. If the temperature is raised so that the total radiated energy is increased 16-fold, at what wavelength will the new intensity maximum be found?

5-2. Assuming that the sun radiates as a blackbody of radius 6.96×10^8 m, estimate its surface temperature from the fact that the solar constant on earth is 1370 W/m^2. (The solar constant is the power per unit area for normal incidence.) The distance from the earth to the sun is 1.49×10^{11} m.

5-3. The fireball of a nuclear device has a radius of 1 m and a surface temperature of 10^7 K. (a) Assuming that it radiates as a blackbody, what is the wavelength λ_p of the peak of the radiation curve? (b) What is the total electromagnetic power radiated? (Use 2.9×10^{-3} m-K for the Wien constant.)

5-4. At the earth's distance from the sun, the average solar power per square meter perpendicular to the sun's rays is 1370 W/m^2. Using 6.37×10^6 m for the radius of the earth, how much solar energy reaches the earth in a 24-hour period?

5-5. Assume that the sun radiates as a blackbody. If $\lambda_p = 5000$ Å for the solar spectrum, what is the surface temperature of the sun? (Use 2.9×10^{-3} m-K for the Wien constant.)

5-6. A blackbody is at an equilibrium temperature of 4000 K. Find (a) the wavelength corresponding to the peak of the radiation curve and (b) the total radiated power per unit area.

5-7. Find the wavelength and frequency of a photon whose energy is (a) 5 eV, (b) 1 MeV.

5-8. Find the wavelength and frequency of a photon whose energy is (a) 1 keV, (b) 1 GeV.

5-9. A photon has a frequency of 1 GHz. Find its energy and wavelength.

5-10. A sodium vapor lamp has a power output of 10 watts. Use 5893 Å as the average wavelength. How many photons are emitted per second?

5-11. An FM radio transmitter has an output of 100 kilowatts at a frequency of 94 MHz. How many quanta are emitted per second by the transmitter?

5-12. A plane wave of frequency 100 MHz has an intensity $E(\nu)$ of 0.6 watt per square meter. (a) Find the energy density $I(\nu)$ in joules per cubic meter. (b) Find the number of photons per cubic meter in the beam.

5-13. A detector located 10 km from a point source of monochromatic radiation records a photon intensity of 3×10^{17} photons per second per square meter normal to the direction of propagation. If the photon frequency is 1 GHz, what is the output power of the source, assuming that it radiates uniformly in all directions?

5-14. What would be the ratio of the populations N_2/N_1 of the states E_2 and E_1 in the helium-neon laser (see Figure 5-9) if they were in thermal equilibrium at 300 K?

5-15. What would be the ratio of the populations N_2/N_1 of the states E_2 and E_1 in the ruby laser (Figure 5-7) if they were in thermal equilibrium at 4 K?

*5-16. By treating the summation in Equation 5.24 as an integration over an energy continuum, show that the average energy per oscillator becomes the classical value, $\langle \epsilon \rangle = kT$.

*5-17. Show that $\langle \epsilon \rangle$ may be written as $\langle \epsilon \rangle = -d/d\alpha \ln \sum\limits_{n=0}^{\infty} x^n = -d/d\alpha \ln (1-x)^{-1}$ where $\alpha = 1/kT$ and $x = e^{-\alpha h\nu}$. Perform the differentiation and obtain Equation 5.25.

*5-18. Use the Planck law as given in Equation 5.26. (a) Show that it reduces to the Rayleigh-Jeans law,

$$E(\lambda, T) = 2\pi c k T/\lambda^4,$$

in the limit as λ gets large. (b) Evaluate the constant in the Wien displacement law, Equation 5.1. (c) Evaluate the constant σ in the Stefan-Boltzmann law, Equation 5.2.

*5-19. Assume that $I(\lambda, T)$ results from N beams, each of energy density $u(\lambda, T)$ and speed c. Derive the result that $E(\lambda, T) = (c/4) \cdot I(\lambda, T)$, where $E(\lambda, T)$ is the radiated power per unit area per wavelength interval.

*5-20. The product of Jeans' number and the Planck distribution function gives the number of photons per unit volume per frequency interval in a blackbody cavity. That is, the number of photons per unit volume with frequency between ν and $\nu + d\nu$ is

$$N(\nu) \cdot d\nu = \frac{8\pi\nu^2 d\nu}{c^3(e^{h\nu/kT} - 1)}.$$

Show that the number of photons per unit volume is $\dfrac{19.2 \, k^3 \pi T^3}{c^3 h^3}$.

*5-21. Show that the Stefan-Boltzmann law can be derived from thermodynamics if one starts with the classical expression, $P = (1/3) u$, where P is the pressure and u the energy density of isotropic radiation in a cavity. Assume that the ideal gas law holds within the cavity.

6 The Particle Nature of Photons

1. INTRODUCTION

Although Isaac Newton was an advocate of a "corpuscular" or particle theory of light, the evidence for the wave nature of light was overwhelming prior to the 20th century. Such phenomena as diffraction, interference, and polarization, which are characteristics of light, were firmly established as properties of waves but not particles. Furthermore, two light beams will pass through each other without any apparent effects, whereas, if they were beams of particle, one would expect them to collide and scatter each other.

It is easy to accept the particle nature of the electron because, as we have already seen, we are able to define its mass, we can accelerate it, and we can determine its kinetic energy by measuring its heat of impact. In other words, we can make electrons behave as we think particles ought to behave. A photon, on the other hand, has no rest mass and cannot be accelerated. It does carry momentum, however, both in the theory of relativity and in classical electromagnetic theory.

We will now discuss two very important experiments whose explanations assume that a single photon interacts directly with an electron. The photon is perceived as if it were a localized particle rather than a wavefront that is extended in space. Thus, the quantum or particle nature of light dominates its wave nature in these experiments. The first is the well-known *photoelectric effect* and the second is the *Compton effect*.

2. THE PHOTOELECTRIC EFFECT

When light is incident upon a clean metal surface it can, under certain conditions, eject electrons from the metal. Such electrons are called photoelectrons. This term does not imply that photoelectrons differ from other electrons, but it merely identifies the source of their excitation.

The valence electrons of a metal are called "free" electrons because they are not localized to a particular atom of the metal. Although they may roam freely within the metal, they are not free to leave its surface because there is an electro- 97 static binding force between each free electron and the positively charged metal ions. There are a number of ways of overcoming this binding force, among which are field emission, thermionic emission, secondary emission, and photoelectric emission. In *field emission* the electron is wrested from the surface by a strong electric field. In *thermionic emission* the electron is thermally excited until it achieves a velocity sufficient to escape from the surface. In *secondary emission* the electron is ejected as a result of an energy transfer from a high-speed particle striking the surface from the outside. In *photoelectric emission* the electron some-how acquires sufficient energy to escape by means of an interaction with incident light. We will be concerned here with just the photoelectric effect. First we will look in more detail at the experimental facts that must be explained by a satisfactory theory, and then we will discuss Einstein's explanation of the interaction.

A typical experimental setup for observing the photoelectric effect is illustrated in Figure 6–1. When the switch is in position 1, the plate of the tube is posi-

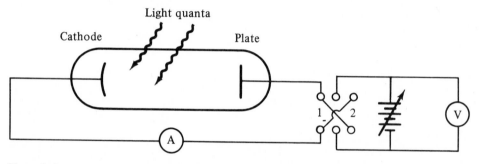

Figure 6–1

Apparatus for determining the photocurrent (switch in position 1) and the maximum kinetic energy of the electrons produced by photoemission (switch in position 2).

tive with respect to the cathode; any photoelectrons that are emitted will strike the plate and contribute to the plate current. A graph of plate current versus plate voltage for two different light intensities is shown in Figure 6–2; the photons in both cases had the same frequency. There are several interesting features about these curves. First, note that they saturate early, meaning that a small voltage is sufficient to collect all of the electrons that are emitted at the light intensity being used. Second, note that the photocurrent is *not* zero when the plate voltage is zero. In fact, a current persists even if the switch is thrown to position 2, making the plate negative with respect to the cathode. What this tells us is that photoelectrons are not just "bubbled off," but that they are ejected with sufficient kinetic energy to overcome an opposing electric field and still reach the plate. The maximum kinetic energy, T_{max}, given to a photoelectron can be expressed as

$$T_{max} = eV_c, \tag{6.1}$$

where V_c is the value of the reversed voltage that reduces the plate current to zero. Third, the same value for T_{max} is obtained, independent of the intensity of the light.

A summary of the experimental facts that must be explained by a satisfactory

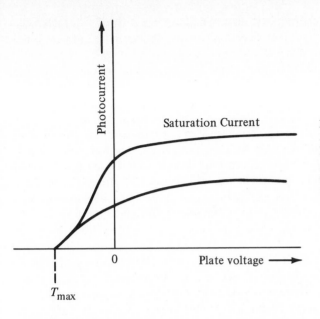

Figure 6–2

Plot of photocurrent *vs.* plate voltage for a given photon frequency. T_{max} is determined from the value of the reversed potential that cuts off the photocurrent.

theory of the photoelectric effect is as follows:

(1) There is a threshold frequency for the emission of photoelectrons. If the light beam contains no photons having frequencies higher than this threshold, then there will be no photocurrent, regardless of the intensity of the light source.

(2) The threshold frequency varies with different metals.

(3) Provided that the threshold frequency is exceeded, photoelectrons are emitted *instantly,* regardless of how low the intensity of the light source is.

(4) When there is a photocurrent, its magnitude is independent of the frequency of the light but it is directly proportional to the intensity of the light.

(5) The maximum kinetic energy of the photoelectrons is independent of the intensity of the light and depends only upon its frequency.

All attempts to explain these phenomena by means of classical electromagnetic theory met with defeat. It was recognized that work must be done in order to remove an electron from the metal, and it seemed reasonable that this *work function* should vary somewhat for different metals. But the idea of a threshold frequency did not seem reasonable. Since an electromagnetic wave has an energy density associated with it, one would expect that the metal could accumulate enough energy, ultimately, to exceed the work function and eject an electron. Light of high frequency and large intensity should do the job more quickly, but there seemed to be no reason why low-frequency illumination, if given enough time, could not produce a photoelectron.

Another related problem for classical theory is that emission is almost instantaneous for frequencies that exceed the threshold frequency. In other words, the emission time is infinite below threshold and nearly zero just above threshold. Yet the energy densities of two waves that satisfy these extreme cases need not vary significantly. Measurements of the time delay between the incidence of light and the appearance of a photocurrent put an upper limit on it of 10^{-9} s.[1] Calculations using the classical model, however, show that delays of minutes or even hours would be expected for a weak source of light.

A third difficulty was provided by the distinct limit on the maximum kinetic energy of the photoelectrons. Classically, photoelectrons would be expected to have a range of kinetic energies depending upon how much energy had been accumulated before the electron escaped. But there was no way to explain a discrete T_{max} which, for a given metal, is a function only of the frequency of the incident light. Classically, T_{max} should increase with the intensity of the light.

The mystery was solved by Einstein in 1905, and he was awarded the Nobel prize for his explanation of the photoelectric effect in 1921. Building upon the quantum concept, he proposed that the photon can deliver all of its energy as a unit to a single electron in the metal. Then, from energy conservation we can write

$$h\nu = W + T_{max} \tag{6.2}$$

where $h\nu$ is the quantum of photon energy, W is the work function of the metal, and T_{max} is the maximum kinetic energy of the photoelectron. The concept of a threshold frequency ν_c is readily understood from Equation 6.2, since the photon energy must exceed W in order for a photocurrent to exist. We define the threshold frequency by

$$\nu_c = \frac{W}{h}, \tag{6.3}$$

where it is assumed that all of the photon energy is required to remove the electron and no energy is left over to provide its kinetic energy. The differences in W from metal to metal explain why ν_c varies. The fact that no significant time interval is required for the electron to accumulate energy is explained by noting that each photon for which $\nu > \nu_c$ carries the requisite amount. A source so weak that it emits a single photon per second could, in principle, produce a photocurrent of one electron per second. One would expect the size of the photocurrent to be directly proportional to the intensity of the light.

[Note: With the advent of lasers capable of emitting coherent radiation at high power levels, two-photon photoemission has been observed in sodium metal.[2] Theory predicts that the double-quantum photocurrent should be proportional to the square of the power of the incident radiation, as opposed to the nearly linear relationship that holds for the single-quantum photoeffect discussed above.[3]]

From Equation 6.2 it is immediately evident that the maximum kinetic energy of the photoelectron is directly proportional to the photon frequency and independent of the light intensity. This functional relationship is shown graphically in Figure 6–3, where the slope of the line is h and its intercepts are W and W/h. Therefore, both h and W can be obtained experimentally by careful measurements of ν and T_{max}. This was first done by Millikan[4] in 1916, for which he was awarded the Nobel prize in 1923.

[1] Lawrence and Beams, *Phys. Rev. 29*, 903 (1927).
[2] M. C. Teich, J. M. Schroer, and G. J. Wolga, *Phys. Rev. Letters 13*, 611 (1964).
[3] R. L. Smith, *Phys. Rev. 128*, 2225 (1962).
[4] R. A. Millikan, *Phys. Rev. 7*, 18, 355 (1916).

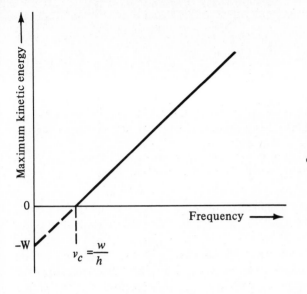

Figure 6–3

Maximum kinetic energy of photo-electrons *vs.* photon frequency.

Example 6-1

A reverse voltage of V_c = 2.5 volts is required to reduce the photocurrent to zero when light of wavelength 4000 Å strikes a certain metal. (a) What is the kinetic energy of the fastest photoelectrons? (b) What is the work function of the metal? (c) What is the threshold frequency for the metal?

Solution

(a) From Equation 6.1,

$$T_{max} = eV_c = 2.5 \text{ eV} = 1.6 \times 10^{-19} \text{ coul} \times 2.5 \text{ volts} = 4.0 \times 10^{-19} \text{ J}$$

(b) The energy of the incident photon is

$$h\nu = \frac{hc}{\lambda} = \frac{2 \times 10^{-25} \text{ J-m}}{4.0 \times 10^{-7} \text{ m}} = 5.0 \times 10^{-19} \text{ J}$$

Then, From Equation 6.2,

$$W = \frac{hc}{\lambda} - T_{max} = 5.0 \times 10^{-19} \text{ J} - 4.0 \times 10^{-19} \text{ J} = 1.0 \times 10^{-19} \text{ J} = 0.63 \text{ eV}$$

(c) The threshold frequency is given by Equation 6.3 as

$$\nu_c = \frac{W}{h} = \frac{1.0 \times 10^{-10} \text{ J}}{6.6 \times 10^{-34} \text{ J-s}} = 1.5 \times 10^{14} \text{ Hz}$$

Example 6-2

What is the threshold wavelength for sodium, whose work function is 3.69 × 10^{-19} J?

Solution

$$\lambda_c = \frac{c}{\nu_c} = \frac{hc}{W} = \frac{2 \times 10^{-25} \text{ J-m}}{3.69 \times 10^{-19} \text{ J}} = 5.42 \times 10^{-7} \text{ m} = 5420 \text{ Å}$$

Example 6-3

It is determined experimentally that the threshold wavelength for the photoemission of electrons from lithium is 5865 Å. What is the work function of lithium?

Solution

$$W = \frac{hc}{\lambda_c} = \frac{2 \times 10^{-25} \text{ J-m}}{5.865 \times 10^{-7} \text{ m}} = 3.41 \times 10^{-19} \text{ J} = 2.13 \text{ eV}$$

3. THE COMPTON EFFECT

It was generally known among the early workers with monochromatic x-rays that a scattered beam always contained a longer wavelength component in addition to the incident wavelength. Compton[1] made a systematic study of the scattering of x-rays from carbon and obtained the spectrum shown in Figure 6–4. The surprising thing is that the wavelength shift is independent of the wavelength of the source and the scattering material, although it varies with the scattering angle. However, Compton succeeded in explaining this peculiar effect by treating the x-ray photon as a particle that undergoes a collision with a rest mass particle such as an electron or an atom as a whole. This is shown schematically in Figure 6–5.

The mathematical analysis of the collision requires only the conservation of energy and momentum. From momentum conservation we have

$$p_0 = p' \cos \theta + p \cos \phi$$

and

$$0 = p' \sin \theta - p \sin \phi,$$

from which we can eliminate ϕ by isolating the term in ϕ on the left side of each equation, squaring, and adding. Then,

$$p^2 = p_0^2 + p'^2 - 2p_0p' \cos \theta. \tag{6.4}$$

The conservation of energy requires that

$$E_0 - E' = T, \tag{6.5}$$

or

$$p_0 - p' = \frac{T}{c}. \tag{6.6}$$

Squaring this,

$$p_0^2 + p'^2 - 2p_0p' = \left(\frac{T}{c}\right)^2. \tag{6.7}$$

Subtracting Equation 6.7 from Equation 6.4,

$$p^2 - \frac{T^2}{c^2} = 2p_0p'(1 - \cos \theta). \tag{6.8}$$

Since the velocity of the particle after the collision could well be relativistic, we

---

[1] A. H. Compton, *Phys. Rev. 22*, 409 (1923).

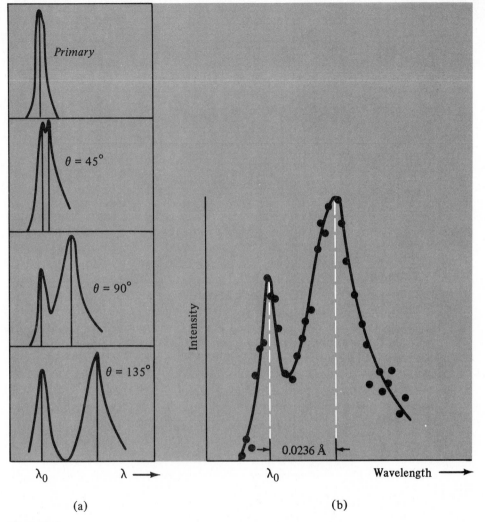

Figure 6–4

Scattering of molybdenum K_α x-radiation from graphite. (a) Variation of the shifted line with scattering angle. The peak at λ_0 is due to the incident x-ray beam. (b) Compton's data for the wavelength shift for 90° scattering. (From A. H. Compton, *Phys. Rev. 22*, 409 (1923), used with permission.)

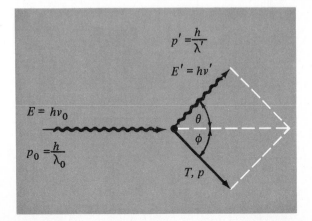

Figure 6–5

Collision of a photon with a free, stationary mass.

must use the relativistic momentum, Equation 4.8,

$$p^2 = \frac{1}{c^2}(T^2 + 2m_0c^2T).$$

Then
$$p^2 - \frac{T^2}{c^2} = 2m_0T = 2m_0c(p_0 - p'), \tag{6.9}$$

where Equation 6.6 has been used to obtain the last expression. Using Equations 6.8 and 6.9,

$$m_0c(p_0 - p') = p_0p'(1 - \cos\theta)$$

$$\frac{1}{p'} - \frac{1}{p_0} = \frac{1}{m_0c}(1 - \cos\theta)$$

or
$$\lambda' - \lambda = \frac{h}{m_0c}(1 - \cos\theta). \tag{6.10}$$

Using the mass of the electron for m_0, we find the quantity $h/m_0c = 0.0243$ angstrom, which is now called the Compton wavelength. Notice that for 90° scattering, Equation 6.10 predicts a new x-ray line just 0.0243 angstrom longer than the primary line. Figure 6–4 shows this line as the large peak measured by Compton. The presence of the primary peak at 90° might at first seem surprising. Compton explained this by considering the scattering of a photon from the atom as a whole. Thus, if one uses the mass of a whole carbon atom instead of the electronic mass in Equation 6.10, the wavelength shift will be reduced by a factor of 20,000 and amounts to roughly one millionth of an angstrom. Therefore, the line scattered by an atom is for all practical purposes unshifted.

Compton's work provided rather convincing evidence that a photon can undergo particle-like collisions with both atoms and unbound electrons. Later studies of the recoil electrons and their energies added further confirmation to the predictions of the theory.[1]

Example 6-4

X-rays having a wavelength of 0.3 Å are Compton scattered. Find the wavelength of a photon scattered at 60° and the energy of the scattered electron.

Solution

From Equation 6.10,

$$\lambda' = \lambda + 0.0243 \text{ Å } (1 - \cos 60°)$$

$$= 0.3 \text{ Å} + 0.0243 \text{ Å } (1 - \tfrac{1}{2})$$

$$= 0.312 \text{ Å}$$

From Equation 6.5,

$$T = E_0 - E' = \frac{hc}{\lambda} - \frac{hc}{\lambda'} = \frac{2 \times 10^{-25} \text{ J-m}}{3.0 \times 10^{-11} \text{ m}} - \frac{2 \times 10^{-25} \text{ J-m}}{3.12 \times 10^{-11} \text{ m}}$$

$$= 6.66 \times 10^{-15} \text{ J} - 6.41 \times 10^{-15} \text{ J}$$

$$= 2.5 \times 10^{-16} \text{ J}$$

$$= 1.56 \text{ keV}$$

[1] C. T. R. Wilson, *Proc. Roy. Soc. (London) 104*, 1 (1923); A. A. Bless, *Phys. Rev. 29*, 918 (1927).

Example 6-5

If the maximum energy given to an electron during Compton scattering is 30 keV, what is the wavelength of the incident photon?

Solution

The maximum energy is given to an electron when the photon is back-scattered, that is, when $\theta = \pi$ in Equation 6.10. Then,

$$\lambda' - \lambda = 2 \times 0.0243 \text{ Å} = 0.0486 \text{ Å}.$$

From Equation 6.5 we have

$$E_0 = E' = 30 \text{ keV} = 30 \times 10^3 \times 1.6 \times 10^{-19} \text{ J} = 4.8 \times 10^{-15} \text{ J}$$

$$\frac{hc}{\lambda} - \frac{hc}{\lambda'} = 4.8 \times 10^{-15} \text{ J}$$

$$\frac{\lambda' - \lambda}{\lambda\lambda'} = \frac{4.8 \times 10^{-15} \text{ J}}{2.0 \times 10^{-25} \text{ J-m}} = 2.4 \times 10^{10} \text{ m}^{-1}$$

$$\frac{0.0486 \text{ Å}}{\lambda(\lambda + 0.0486 \text{ Å})} = 2.4 \text{ (Å)}^{-1}$$

$$\lambda^2 + 0.0486\lambda - 0.02025 = 0$$

$$\lambda = \tfrac{1}{2}(-0.0486 \pm \sqrt{23.62 \times 10^{-4} + 810 \times 10^{-4}})$$

$$= \tfrac{1}{2}(-0.0486 \pm 0.289)$$

$$= 0.120 \text{ Å}$$

4. THE DUAL NATURE OF PHOTONS: THE WAVE PACKET

The particle nature of light, as illustrated by the photoelectric effect and the Compton effect, is no longer viewed as irreconcilable with the overwhelming evidence for its wave-like behavior. Such phenomena as interference and diffraction are peculiar to a wave description wherein the region of interaction is extended over a large portion of the wavefront, in contrast with the localized interactions of particles. Regarding the photon as a *wave packet* consisting of a superposition of many waves imparts to it some of the properties of both waves and material particles. Thus, the photon exhibits its wave nature when it interacts with an object such as a grating, where the details of the instantaneous phase of each of the constituent waves are important. On the other hand, it manifests its particle nature when energy and momentum of the packet as a whole are transferred to another particle. In the latter case the details of the phases of the constituent waves are unimportant. That both the wave and particle aspects of photons are required for the description of light is known historically as *Bohr's principle of complementarity*. That is, the wave and particle aspects complement each other. In order to acquire a better understanding of the nature of a wave packet, let us first consider the superposition of two plane waves.

In classical physics we frequently represent a plane wave traveling in the positive *x*-direction by either the real or the imaginary part of one of the following equivalent expressions:

$$\Psi = Ae^{i(kx-\omega t)} = Ae^{i2\pi(x/\lambda - t/T)} = Ae^{i\omega(kx/\omega - t)}. \tag{6.11}$$

Here A is the wave amplitude, $k = 2\pi/\lambda$ is the propagation constant, $\omega = 2\pi\nu$ is the angular frequency, and T is the period of the harmonic oscillation. The velocity of propagation of the wavefront is the phase velocity, $u = \omega/k = \lambda\nu$.

Suppose that two waves having slightly different frequencies and wavelengths are propagating together through a medium. For simplicity let us take their amplitudes and initial phases to be equal. Then we may represent these two waves by the expression

$$\Psi_1 = A \sin (kx - \omega t) \tag{6.12}$$

and
$$\Psi_2 = A \sin [(k + dk)x - (\omega + d\omega)t], \tag{6.13}$$

where dk and $d\omega$ are infinitesimal quantities. Making use of the trigonometric identity

$$\sin \alpha + \sin \beta = 2 \cos \tfrac{1}{2}(\alpha - \beta) \cdot \sin \tfrac{1}{2}(\alpha + \beta), \tag{6.14}$$

we may express the displacement resulting from the superposition of these two waves by

$$\Psi = \Psi_1 + \Psi_2 = 2A \cos \frac{d\omega}{2} \left(\frac{x}{\frac{d\omega}{dk}} - t \right) \cdot \sin \omega' \left(\frac{x}{\frac{\omega'}{k'}} - t \right). \tag{6.15}$$

Note that $k' = k + dk/2 \sim k$, and $\omega' = \omega + d\omega/2 \sim \omega$, since $dk \ll k$ and $d\omega \ll \omega$. Then the factor containing the sine is essentially the same function as Ψ_1 and may be thought of as a "carrier wave" of phase velocity $u = \omega'/k' \sim \omega/k$. The cosine factor has the effect of modulating the amplitude of the carrier wave, and the modulation envelope moves at the so-called group velocity given by $v = d\omega/dk$. The transmission of energy (that is, a signal) must occur at the group velocity and not with the phase velocity. The reason for this is clear if one understands the Fourier integral theorem (Appendix C), but it may be simply stated here that the transmission of a signal always involves modulation of one kind or another. An infinitely long wave train of a single frequency can never be used to transmit information at its phase velocity. Signaling always involves chopping (keying), amplitude modulation, frequency modulation, or phase modulation. These processes all result in the superposition of many plane waves of different frequencies grouped around some center frequency, that is, the formation of a *wave packet*. Since the envelope of the packet must have a finite spatial extent, the sum of the amplitudes of all of its plane wave components must be zero everywhere except where the packet is localized. The energy carried by the packet at its group velocity is thus analogous to the kinetic energy of a free particle of the same velocity. The packet shown in Figure 6–6, which is made up of only two components, is not zero elsewhere, but its envelope is repeated throughout space. In principle, however, the amplitudes elsewhere can be cancelled if enough component frequencies are used. How this is done analytically will be shown in Chapter 8, where the Fourier integral theorem will be used to construct localized packets to represent material particles.

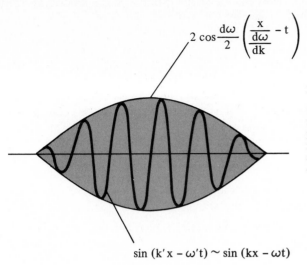

$$2 \cos \frac{d\omega}{2} \left(\frac{x}{\frac{d\omega}{dk}} - t \right)$$

Figure 6–6

One half-wavelength of the modulation envelope formed from the superposition of two plane waves of nearly equal frequencies and equal amplitude. This pattern is repeated continuously throughout space.

$$\sin (k'x - \omega't) \sim \sin (kx - \omega t)$$

If all of the wavelengths of a packet travel through a medium at the same phase velocity, there is said to be no *dispersion*. A physical consequence of this case is that the packet retains its shape as it propagates, regardless of the number of frequencies that it contains. When dispersion occurs, the packet changes shape as it propagates. The dependence of frequency upon the propagation constant is called the *dispersion relation* and may be written

$$\omega = u(k) \cdot k. \tag{6.16}$$

If the phase velocity u is a constant for all wavelengths, there is no dispersion and the group velocity and the phase velocity are equal. That is,

$$v = \frac{d\omega}{dk} = u. \tag{6.17}$$

5. HEISENBERG'S UNCERTAINTY PRINCIPLE

In the previous section it was argued that a localized packet of electromagnetic energy must be composed of a mixture of a large number of plane waves of differing frequencies. Paradoxically, the more nearly monochromatic the packet is, the broader it becomes. Thus, if the frequency spread is nearly zero, then the packet is so broad (its wave train is so long) that its emission time is extremely large. Light radiated by atoms during electronic transitions has a finite wave train which corresponds roughly to the lifetime of the state. In the time domain these bursts, which are of the order of 10^{-8} seconds, result in a spread of frequencies rather than in a single emitted frequency, and produce what is called the *natural linewidth* of the spectral line. The longer the lifetime, the fewer extraneous frequencies in the spectral line.

Denoting the uncertainty in the frequency (the frequency spread) by $\Delta\omega$ and the uncertainty in emission time by Δt, we can write[1]

$$\Delta\omega \sim \frac{2\pi}{\Delta t}. \qquad (6.18)$$

On the other hand, if the degree of monochromaticity is described in terms of the spread in k-values (where $k = \omega/c$), the smaller the value of Δk, the greater the extent of the packet in coordinate space. That is, the uncertainty in the propagation constant and the uncertainty in position are related by

$$\Delta k_x \sim \frac{2\pi}{\Delta x}. \qquad (6.19)$$

Making use of the Planck expression for the energy of the photon and the relativistic value for the photon momentum, the above relations suggest that

$$\left.\begin{aligned} \Delta E \cdot \Delta t &\sim h \\ \Delta p_x \cdot \Delta x &\sim h. \end{aligned}\right\} \qquad (6.20)$$

and

This is an intrinsic limitation to the precision with which certain pairs of physical variables can be simultaneously measured under ideal circumstances in which there are no experimental or instrumental errors. Although these limitations are of great importance in quantum systems, they are completely unnoticeable in macroscopic systems because of the smallness of the Planck constant h. Thus, for a momentum which is known to within 10^{-12} kg-m/s, the uncertainty in x is $\sim 10^{-22}$ m, which is certainly not detectable. The relations in Equation 6.20 are two examples of the well-known *uncertainty principle* of Heisenberg, enunciated in 1927. The same constraints apply to any generalized coordinate and the generalized momentum associated with that coordinate. Such variables are known as *canonically conjugate* variables in classical mechanics. Thus, if q is a generalized coordinate, then p_q is the generalized momentum conjugate to q. Other examples of pairs of conjugate variables are (y, p_y), (z, p_z), (θ, p_θ), and so on. In formal language, the uncertainty principle may be stated as follows: The product of the uncertainties in the measurement of two canonically conjugate variables must be greater than a quantity of the order of $\hbar = h/2\pi$. That is,

$$\Delta q \cdot \Delta p_q \geq \hbar. \qquad (6.21)$$

The uncertainty principle will be discussed quantitatively in Section 4 of Chapter 8 in connection with the behavior of wave packets and the concept of expectation value. Numerous examples of how it can be applied to physical problems will occur throughout the subsequent chapters of this book. The reader who is interested in pursuing this topic further at this point is referred to Heisenberg's own discussion of the subject.[2]

Example 6-6

What is the uncertainty in the location of a photon if its wavelength of 5000 Å is known to an accuracy of one part in 10^7?

---

[1] Equations 6.18 and 6.19 will be derived in Section 4 of Chapter 8.
[2] See Chapter 2 of W. Heisenberg, *The Physical Principles of the Quantum Theory*, Dover Publications, Inc., New York.

[**Top**] Albert A. Michelson, Albert Einstein, and Robert A. Millikan in Pasadena, California in 1931. (Wide World Photos.) [**Bottom**] Albert A. Michelson in his laboratory. (Courtesy of Elmer Taylor and the American Institute of Physics Niels Bohr Library.)

Solution

The momentum of a photon is given by

$$p = \frac{h}{\lambda}$$

Then,

$$|\Delta p| = \left| -\frac{h}{\lambda^2} \Delta\lambda \right|.$$

If this momentum is in the *x*-direction, then the uncertainty in the *x*-position is

$$\Delta x \geq \frac{h}{|\Delta p|} = \frac{\lambda^2}{|\Delta\lambda|}$$

Since $\lambda = 5 \times 10^{-7}$ m and $\Delta\lambda = 5 \times 10^{-4}$ Å $= 5 \times 10^{-14}$ m,

$$\Delta x \geq \frac{(5 \times 10^{-7})^2 \text{ m}^2}{5 \times 10^{-14} \text{ m}} = 5 \text{ m}.$$

Example 6-7

If the lifetime of an electronic excited state is 1×10^{-9} s, what is the uncertainty in the energy of the state?

Solution

From Equation 6.20,

$$\Delta E \geq \frac{h}{\Delta t} = \frac{6.6 \times 10^{-34} \text{ J-s}}{1 \times 10^{-9} \text{ s}} = 6.6 \times 10^{-25} \text{ J} = 4.1 \times 10^{-6} \text{ eV}$$

6. PHOTONS AS EXCHANGE PARTICLES

The Coulomb force between two charged particles has heretofore been treated as action-at-a-distance via the electric field lines connecting the charges. In contemporary theory, however, the force between two charged particles is said to arise from the exchange of photons between the charges. Each charge is surrounded by a cloud of photons, which are constantly being emitted and reabsorbed by the same charge. However, when two charges interact, each will absorb photons that were emitted by the other. This results in an exchange of momentum between them, and they will experience a force proportional to the time rate of the momentum exchange.

A crude but useful model for this idea is to imagine two people playing "catch" with a medicine ball at close quarters. The exchange of momentum tends to push them apart as though a repulsive force were acting. On the other hand, an attractive force would result if one player attempted to pull the ball from the other's hands.

Since we cannot observe these photons, they are called *virtual photons*. If it were possible to detect them, their emission and reabsorption would constitute a violation of the conservation of energy! Then, why should we take the trouble to imagine that they exist? The answer to this question is that *the uncertainty principle permits a violation of the conservation of energy provided that it occurs for such a short time that it cannot be detected.* From Equation 6.20, a virtual photon must be reabsorbed within the time interval given by

$$\frac{h}{\Delta E} = \frac{h}{h\nu} = \frac{1}{\nu},$$

where ν is its frequency.

By means of the same argument, a photon (not a virtual one) may be viewed as continually creating and annihilating virtual electron-positron pairs (see Section 6 of Chapter 4). These processes are shown schematically in Figure 6–7.

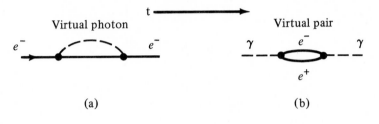

(a) (b)

Figure 6–7

(a) Emission and reabsorption of a virtual photon by an electron. (b) Creation and annihilation of an electron-positron pair by a photon.

SUMMARY

The wave nature of light has been known for several hundred years from studies of such phenomena as interference, diffraction, and polarization. Although some have argued—including Isaac Newton—in favor of a particle theory of light, this point of view had little empirical basis until Planck proposed that light is emitted in discrete units of energy $h\nu$. These units or quanta are called photons and, although they have no rest mass, they do carry momentum as well as energy. Photons appear to behave like particles in some interactions. For example, Einstein explained the photoelectric effect by assuming that a photon can, under the right circumstances, transfer all of its energy to a single electron in a metal, giving it a kinetic energy whose maximum value is equal to the difference between the photon energy and the binding energy of the electron. In equation form this is expressed as

$$h\nu = W + T_{max},$$

where $h\nu$ is the photon energy and W is the work function of the metal.

Another important interaction is the inelastic scattering of x-ray photons by atomic electrons. Compton treated this simply as a two-particle collision and successfully explained the wavelength shift of the scattered radiation. His now famous expression is

$$\lambda' - \lambda_0 = 0.0243 \text{ Å}(1 - \cos\theta),$$

where the scattered wavelength λ' and the incident wavelength λ_0 are in angstroms. The angle θ is the scattering angle.

Thus we are faced with the apparent dilemma that both the wave nature of light and the particle nature of light can be supported by sound experimental evidence. We avoid the dilemma by claiming that the photon is a wave packet that displays both wave and particle characteristics. A packet is a superposition of many waves that are very close in frequency. When such a packet interacts with a diffraction grating or other device where the details of the instantaneous phase of each constituent wave are important, the packet appears to be a wave. On the other hand, the packet manifests its particle nature when the energy and momentum of the packet as a whole are detected. A more simplistic way of saying this is, "A photon is a wave when we look at it with a wave detector and it is a particle when we observe it with a particle detector." We say that the wave and particle aspects of light complement each other, and this idea is known as the principle of complementarity.

There is an intrinsic measurement problem associated with the concept of the wave packet. The longer the packet (wave train), the more precisely the wavelength (or frequency) can be measured. But the longer the packet, the less localized it is; that is, the less we know about its position. According to Heisenberg's uncertainty principle, the product of the uncertainty in momentum and the uncertainty in position must be larger than h, that is,

$$\Delta p_x \cdot \Delta x \gtrsim h.$$

This limitation is of great importance in quantum systems but is completely unnoticeable in macroscopic systems because of the smallness of the Planck constant h.

The virtual exchange of photons between electric charges is now believed to be the origin of the Coulomb electric force. Furthermore, photons with sufficiently great energy are able to produce virtual positron-electron pairs when conditions are right.

Additional Reading

C. H. Blanchard, C. R. Burnett, R. G. Stoner, and R. L. Weber, *Introduction to Modern Physics,* 2nd ed., Prentice-Hall, Inc., Englewood Cliffs, N.J., 1969, Chapter 7.

K. W. Ford, *Classical and Modern Physics,* Vol. 3, Xerox College Publishing, Lexington, Mass., 1974, Chapter 23.

P. A. Tipler, *Modern Physics,* Worth Publishers, Inc., New York, 1978, Chapter 3.

R. T. Weidner and R. L. Sells, *Elementary Modern Physics,* 3rd ed., Allyn and Bacon, Inc., Boston, 1980, Chapter 4.

E. H. Wichmann, *Quantum Physics,* Berkeley Physics Course, Vol. 4, McGraw-Hill Book Co., New York, 1971, Chapters 4 and 6.

PROBLEMS

6-1. If a retarding potential of 5 volts just stops the fastest photoelectrons emitted from cesium, what is the wavelength of the most energetic incident photons? Use 1.8 eV for the work function of cesium.

6-2. If the photoelectric threshold wavelength of sodium is 5420 Å, calculate the maximum velocity of photoelectrons ejected by photons of wavelength 4000 Å.

6-3. (a) Light of wavelength 4000 Å is incident upon lithium. If the work function for lithium is 2.13 eV, find the kinetic energy of the fastest photoelectrons. (b) What would be the maximum wavelength of photons capable of ejecting photoelectrons from lithium at a velocity of 0.95 c?

6-4. If the photocurrent of a photocell is cut off by a retarding potential of 0.92 volts for monochromatic radiation of 2500 Å, what is the work function of the material?

6-5. A monochromatic light source of wavelength λ illuminates a metal surface and ejects photoelectrons having a maximum kinetic energy of 1 eV. The experiment is repeated with the same metal, but using a light source having half the wavelength of the first source. The maximum photoelectron kinetic energy is now found to be 4 eV. What is the work function of the metal?

6-6. Ultraviolet light of wavelength 2500 Å is incident upon potassium. What is the stopping potential for photoelectrons if the work function of potassium is 2.21 eV?

6-7. Photons of wavelength 4500 Å are incident upon an unidentified metal. If the fastest photoelectrons emitted are bent into a circular arc of 20 cm radius by a magnetic field of 2×10^{-5} tesla, what is the work function of the metal?

6-8. The threshold frequency for the emission of photoelectrons from a certain metal is ν_0. What is the maximum kinetic energy of the photoelectrons ejected by photons having frequency $2\nu_0$?

6-9. The threshold wavelength for the emission of photoelectrons from a certain metal is λ_0. What is the maximum kinetic energy of the photoelectrons ejected by photons having wavelength $\lambda_0/2$?

6-10. In Chapter 1 it was shown that atoms in a typical metal surface are about 2 angstroms apart. (a) According to the classical wave theory of light, how much electromagnetic energy is available per atom per second if the intensity of the light incident upon the metal surface is 4.0×10^{-8} watt/m²? (b) Assuming that all of the energy in (a) were absorbed by a single valence electron, how much time would be required in order for the electron to accumulate 1 eV of energy?

6-11. If a light pulse delivers 10^{10} photons having a wavelength of 6328 Å in 10^{-6} second, what are the energy and momentum of the pulse? What is the average power output of the pulse?

6-12. A ruby laser delivers a pulse of 1 megawatt average power in a time interval of 1 nanosecond (10^{-9} s). If all of the photons in the pulse are of wavelength 6943 Å, how many photons are contained in the pulse?

6-13. A photocell having a work function of 2 eV is given a reversed bias so that no photocurrent will flow unless photons having wavelengths less then 3500 Å strike its cathode. What is the value of the bias potential?

6-14. A spaceship having a mass of 5000 kg is accelerated by photon propulsion. (a) If the photon rocket has an output power of 500 megawatts, what acceleration can be achieved? (b) If the acceleration remains constant at the maximum value, how long would it take to achieve a speed of 1 km/s from rest?

6-15. Show that a photon cannot transfer all of its energy to a free electron. What is the maximum recoil kinetic energy that can be given to an electron by a photon of energy E_0? (See Problem 4-15.)

6-16. What is the recoil kinetic energy of a free electron, initially at rest, after a Compton scattering event in which a 1 MeV gamma photon is scattered at 90°? Find the recoil angle of the electron.

6-17. A 2 MeV gamma photon is scattered through an angle of 180° by an electron. What is the recoil kinetic energy of the electron?

*6-18. Photons of energy 0.1 MeV undergo Compton scattering. Find the energy of a photon scattered at 60°, the recoil angle of the electron, and the recoil kinetic energy of the electron.

6-19. Gamma rays of energy 1.02 MeV are scattered from electrons which are initially at rest. Find the angle for symmetric scattering at this energy (that is, the angles θ and ϕ are equal). What is the energy of the scattered photon for this case?

*6-20. In a Compton scattering event the scattered photon was found to have an energy of 120 keV, and the recoil electron was given a kinetic energy of 40 keV. (a) What was the wavelength of the incident photon? (b) What is the scattering angle θ for the 120 keV photon? (c) What is the angle ϕ at which the electron is scattered?

6-21. An electron's position along the x-axis is known to within 10 Å. What is the minimum uncertainty in its momentum?

6-22. A particle of mass m is confined to a one-dimensional box. (a) Can it be at rest in the box? (b) Use the uncertainty principle to estimate the smallest kinetic energy the particle can have.

6-23. What is the minimum uncertainty in the energy of an excited state of a system if the average lifetime of the energy state is 1×10^{-12} s?

7 The Quantum Theory of the Atom

1. THE ATOMIC MODELS OF THOMSON AND RUTHERFORD

It had long been suspected that atomic matter is held together by electrical forces, and the work of 19th century scientists added more and more evidence for that concept. The discovery of the electron by J. J. Thomson[1] in 1897 and the determination of the electron charge[2] and mass confirmed the identity of at least one of the atomic building blocks. These experiments were discussed in Chapter 2. The assumption that electrons are constituents of all atoms was a reasonable inference from a variety of experiments; these range from the simple act of charging an object by friction to more sophisticated experiments such as the ionization of gases in discharge tubes, x-ray bombardment, the photoelectric effect, and beta emission during radioactive decay. A difficulty still existed, however, when one attempted to design a stable atom that could contain both electrons and the positive charge necessary to make the atom electrically neutral.

Prout had in 1815 proposed that the elements were all composites of hydrogen atoms. However, the large discrepancies that exist between the atomic number and the mass number of most elements provided a stumbling block to that approach. It was later revived somewhat as a result of J. J. Thomson's experiments with beams of positive ions.[3] Thomson was able to determine the ratio of charge to mass (e/m) of the particles in the beam, and he found that the mass of a particle having a positive charge equal to the magnitude of the electronic charge corresponds to the mass of an ionized hydrogen atom. It would not be unreasonable to suggest that all elements consist of an appropriate number of hydrogen ions and electrons so as to account for both the mass number and the atomic number.

As early as 1898, Thomson had proposed an atomic model in which the electrons were embedded in successive rings within a massive matrix of positive

---

[1] J. J. Thomson, *Phil. Mag. 44,* 293, 310, 318 (1897).
[2] R. A. Millikan, *Phys. Rev. 32,* 349 (1911); *2,* 109 (1913).
[3] J. J. Thomson, *Phil. Mag. 13,* 561 (1907).

charge, which was thought to occupy a volume of roughly one atomic diameter. It has long been known that the diameters of atoms must be between 1 and 10 angstrom units, a fact inferred from the sizes of molar volumes and Avogadro's number. This model of Thomson's has been called the "plum pudding" model because the electrons could be imagined to be arranged like the raisins in a pudding. A definitive test of this model of the atom arose within little more than a decade.

Since there is no direct way to find out just how an atom is constructed, an indirect method must be used. An absurd example of an indirect method would be the determination of the positions of the trees in a strip of woodland by carefully noting the emerging paths of thousands of golf balls fired systematically and uniformly into the forest. That is, one could perform a scattering experiment on the forest, assuming that the only interaction present is an elastic collision when a ball hits a tree. In the atomic scattering experiment it is the location of a charge that concerns us, so it is evident that the "golf ball" should be a charged particle itself. The interaction here would be the Coulomb force between the charge of the projectile and the charge of the atom. The discovery of the alpha particle by Rutherford and his collaborators[1,2] provided the "golf ball." An alpha particle is a helium nucleus; that is, it has a mass number of 4 and a charge of $+2$.

At the suggestion of Rutherford, a number of experiments were performed by scattering alpha particles of known energy from thin foils of gold.[3] The deflection of an alpha particle that would be caused by a Thomson atom can be estimated from the following argument. Suppose that an alpha particle of mass M and velocity v strikes an atom. How much of its momentum can it transfer to an electron of mass m_0 in a head-on collision? Simply from the conservation of energy and linear momentum, the maximum momentum transfer to the electron would be $2m_0v$. Then the maximum possible angle of deflection of the alpha particle due to a collision with an electron would be given by

$$\Delta\phi_{max} = \frac{2m_0v}{Mv} \approx \frac{2m_0}{4(1836\,m_0)} \approx 10^{-4} \text{ radian,} \qquad (7.1)$$

since the momentum given to the electron represents the change in momentum of the alpha particle. Similarly, one can estimate the deflection due to the continuous charge within the atom by calculating the total impulse given the alpha particle by the Coulomb force during its transit through the positive "pudding." Thus,

$$\Delta\phi_{max} = \frac{(Ft)_{max}}{Mv}. \qquad (7.2)$$

Now, if R is the radius of the atom and Z is its atomic number, $(Ft)_{max}$ would be the impulse given to an alpha particle that enters the charge along a diameter. Then,

$$(Ft)_{max} = \int_{-R}^{R} F\,dt = 2k \int_{0}^{R} \frac{2eq}{r^2} \cdot \frac{dr}{v} = 4k \int_{0}^{R} \frac{Ze^2}{r^2} \left(\frac{r}{R}\right)^3 \frac{dr}{v}$$

$$= \frac{4Ze^2k}{R^3v} \int_{0}^{R} r\,dr = \frac{2Ze^2k}{Rv}.$$

[1] E. Rutherford and H. Geiger, *Proc. Roy. Soc. (London)* **81**, 141 (1908).
[2] E. Rutherford and T. Royds, *Phil. Mag.* **17**, 281 (1909).
[3] H. Geiger and E. Marsden, *Proc. Roy. Soc. (London)* **82**, 495 (1909); *Phil. Mag.* **25**, 605 (1913).

Then,
$$\Delta\phi_{max} = \frac{2Ze^2k}{MRv^2}. \tag{7.3}$$

Taking $R \approx 1$ Å, $v = 2 \times 10^7$ m/s, $M \approx 6 \times 10^{-27}$ kg, $Z \approx 10^2$, and $k = 9 \times 10^9$ N-m^2/coul^2, we obtain

$$\Delta\phi_{max} \approx 10^{-4} \text{ radian.}$$

Since the probability of scattering from more than one electron within a single atom is quite small, the scattering per atom would be the sum of Equations 7.1 and 7.3, that is, approximately 1×10^{-4} radian. Multiple scatterings from many atoms can occur, of course, but since they are completely random they obey the relation

$$\overline{\Delta\phi} = \sqrt{N} \cdot \Delta\phi_{max}, \tag{7.4}$$

where N is the number of atoms contributing to the scattering and $\overline{\Delta\phi}$ is the average total deflection of the alpha particle. Geiger[1] found that the average deflection of alpha particles passing through a 0.5 micron gold foil is about 1°. This is quite reasonable in light of the above discussion, since a one micron film would be about 10^4 atoms thick and the average deflection for random scattering from Equation 7.4 would be

$$\overline{\Delta\phi} = \sqrt{10^4} \cdot 10^{-4} \text{ rad} \approx 10^{-2} \text{ rad} \approx 1°.$$

So far, the Thomson model is in agreement with experiment, but let us now examine its predictions for large-angle scattering. It turns out that the probability for large-angle scattering goes as $\exp[-(\Delta\phi/\overline{\Delta\phi})^2]$. In particular, only one particle in $e^{8100} = 10^{3518}$ would be scattered through an angle of 90°! Here the Thomson model was in serious trouble, because Geiger and Marsden[2] had shown that nearly one alpha particle in 10^4 was scattered 90° or more! Though this might seem like a small number, it is a huge number in comparison with any prediction based on the Thomson model of the atom.

In order to account for the unexpected large-angle scattering of alpha particles, Rutherford[3] proposed a model of the nuclear atom, in which most of the atomic mass and all of the positive charge are concentrated in a nucleus so small that it may be regarded as a point mass as a first approximation. By considering only the Coulomb interaction between the incident alpha particle and the target nucleus, Rutherford calculated the scattering cross section and obtained results that agree remarkably with the experimental results. His calculation will be shown after a brief discussion of the rudiments of scattering theory.

2. CLASSICAL SCATTERING CROSS-SECTIONS

A schematic diagram of a simple scattering experiment is shown in Figure 7–1. The incoming particles are collimated so that they constitute a uniform beam of monoenergetic projectiles. We define the beam intensity I as the number of par-

[1] H. Geiger, *Proc. Roy. Soc. (London)* *83*, 492 (1910).
[2] H. Geiger and E. Marsden, *Proc. Roy. Soc. (London)* *82*, 495 (1909).
[3] E. Rutherford, *Phil. Mag. 21*, 669 (1911).

ticles per unit area normal to the beam per second. Consider a particle path that would miss the scattering center by the distance s if it were undeflected. This "miss distance" is known as the *impact parameter*. Note that the interaction between the projectile and the target scatters the particle through an angle θ. Since there is cylindrical symmetry about the beam axis, all of the particles having impact parameter s will be scattered along the conical surface defined by the angle θ. All of the incident particles whose impact parameters are less than s will be outside this cone. Any real detector must have an angular width $d\theta$, so that it actually counts the particles that enter through the conical wedge bounded by θ and $\theta + d\theta$. These particles have come from the beams whose impact parameters lie between s and $s - ds$. That is, all of the particles that pass through the washer-like area defined by $d\sigma = 2\pi s \, ds$ are scattered into the conical wedge bounded by θ and $\theta + d\theta$. The area $d\sigma$ is called the *scattering cross-section,* and it is related to $dN(\theta)$, the number of particles per second scattered into the conical wedge defined by θ and $\theta + d\theta$, by

$$dN(\theta) = I \, d\sigma = 2\pi I s \, ds. \tag{7.5}$$

From Equation 7.5 we see that the scattering cross-section can be determined experimentally by simply dividing the detector count per second by the beam intensity. But $dN(\theta)$ evidently depends upon the angular size of the detector, so different detectors would give different values for the cross-section. In order to avoid such ambiguity, we prefer to normalize all counts by dividing both sides of Equation 7.5 by the solid angle subtended by the detector, namely, $d\Omega = \sin \theta \, d\theta \, d\phi$. Then,

$$\frac{d\sigma}{d\Omega} = \frac{1}{I} \frac{dN(\theta, \phi)}{d\Omega}, \tag{7.6}$$

where we define $\dfrac{d\sigma}{d\Omega}$ as the *differential scattering cross-section.*

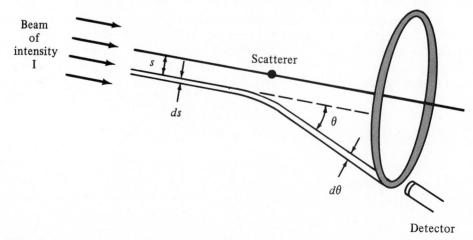

Figure 7–1

Schematic diagram of a scattering experiment.

Using the right-hand part of Equation 7.5 we may write

$$\frac{d\sigma}{d\Omega} = \frac{2\pi s\, ds}{2\pi \sin \theta\, d\theta} = \frac{s}{\sin \theta} \cdot \frac{ds}{d\theta}. \tag{7.7}$$

What Equation 7.7 tells us is that we can calculate the differential scattering cross-section for a specific kind of interaction between the projectile and the target, provided that we can obtain a relationship between the impact parameter s and the scattering angle θ. Then, a comparison of the values obtained from Equations 7.6 and 7.7 furnishes us with a means of checking experiment against theory.

Example 7-1

Consider the elastic scattering of a hard sphere of radius a from a stationary sphere of radius b. The incident sphere will rebound from the tangent plane at an angle equal to its angle of incidence as shown in Figure 7–2. Show that $d\sigma/d\Omega = \frac{1}{4}(a + b)^2$ and find σ. The small sphere is incident from the left and the scattering angle is 2α.

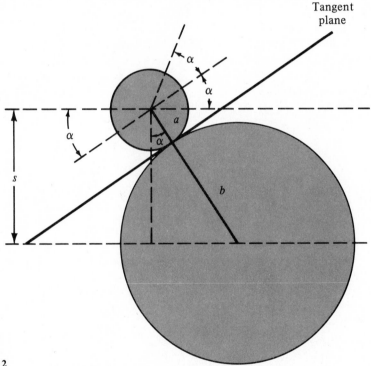

Figure 7–2

Classical scattering of hard spheres.

Solution

$$s = (a + b) \cos \frac{\theta}{2}$$

$$\left|\frac{ds}{d\theta}\right| = \tfrac{1}{2}(a + b) \sin \frac{\theta}{2}$$

$$\frac{d\sigma}{d\Omega} = \frac{s}{\sin\theta}\left|\frac{ds}{d\theta}\right| = \frac{(a+b)^2 \sin\frac{\theta}{2} \cdot \cos\frac{\theta}{2}}{2\sin\theta} = \tfrac{1}{4}(a+b)^2$$

$$\sigma = \int_\Omega \frac{d\sigma}{d\Omega}\, d\Omega = \int_0^{4\pi} \tfrac{1}{4}(a+b)^2 \cdot d\Omega = \pi(a+b)^2$$

In Example 7-1 we considered the case of scattering when the only interaction between the particles is the repulsive force that develops as the masses make contact. Now suppose that both particles are electrically charged so that they interact through the very long-ranged Coulomb force. In this case scattering occurs whether or not the masses ever make contact. In fact, the scattering begins as soon as each particle "feels" the electric field of the other, and it continues until the distance between the particles is so great that the Coulomb interaction is negligible. In particular, consider two positively charged particles, a light projectile of mass m and charge ze and a stationary (massive) scattering center of charge Ze. Since z and Z are both positive, the Coulomb force is repulsive and the scattered particle will have a hyperbolic trajectory as shown in Figures 7–3 and 7–4. The important relationships between impact parameters and scattering angle for this scattering interaction are derived in Appendix B. They are

$$s = \frac{D}{2}\cot\frac{\theta}{2} \tag{7.8}$$

and

$$\frac{ds}{d\theta} = \frac{D}{4}\csc^2\frac{\theta}{2}, \tag{7.9}$$

where D is given by

$$D = \frac{kzZe^2}{\tfrac{1}{2}mv_0^2} = \frac{kzZe^2}{T_0}. \tag{7.10}$$

Here T_0 is simply the classical kinetic energy of the incident particle. The quantity D is known as the *collision diameter*. It is the closest distance between the two particles in the case of a head-on collision. That is, D is the distance from the target at which all of the kinetic energy of the projectile is converted to electrostatic potential energy. This is analogous to the problem of finding the maximum height of a baseball by setting its initial kinetic energy equal to the maximum gravitational potential energy of the ball.

In order to find the differential scattering cross-section for Coulomb scattering, we substitute Equations 7.8 and 7.9 into Equation 7.7 to obtain

$$\frac{d\sigma}{d\Omega} = \frac{1}{\sin\theta} \cdot \frac{D}{2}\cot\frac{\theta}{2} \cdot \frac{D}{4}\csc^2\frac{\theta}{2}$$

$$= \frac{D^2}{16} \cdot \frac{1}{\sin\frac{\theta}{2}\cdot\cos\frac{\theta}{2}} \cdot \frac{\cos\frac{\theta}{2}}{\sin\frac{\theta}{2}} \cdot \frac{1}{\sin^2\frac{\theta}{2}}$$

$$= \frac{D^2}{16} \cdot \frac{1}{\sin^4\frac{\theta}{2}}. \tag{7.11}$$

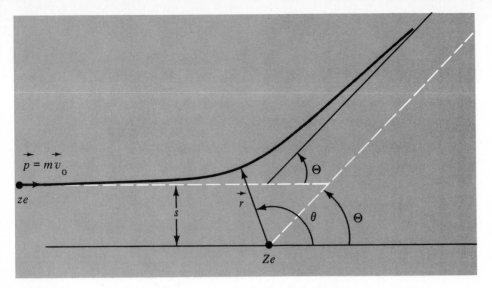

Figure 7-3

Coulomb scattering of a particle of charge ze by a nucleus of charge Ze.

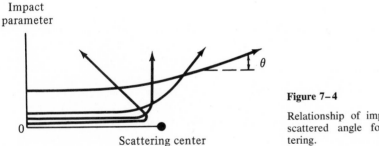

Figure 7-4

Relationship of impact parameter to scattered angle for Coulomb scattering.

Then the fraction of the incident particle flux I that will be scattered at a given angle θ by a single scattering center is, from Equation 7.6,

$$\frac{dN(\theta)}{I} = \frac{d\sigma}{d\Omega}\,d\Omega = \frac{D^2}{16} \cdot \frac{2\pi \sin\theta\, d\theta}{\sin^4 \dfrac{\theta}{2}}. \tag{7.12}$$

Here we have written $d\Omega$ as $2\pi \sin\theta\, d\theta$, since we assume cylindrical symmetry about the axis of the beam. If there are C scattering centers per unit area of target, then we write

$$\frac{dN(\theta)}{I} = \frac{\pi C D^2}{8} \cdot \frac{\sin\theta\, d\theta}{\sin^4 \dfrac{\theta}{2}}. \tag{7.13}$$

Since the experiments that were used to test Equation 7.13 employed metallic foils as targets, let us first calculate C, the number of atoms per unit area of foil. If we assume that the foil is thin enough to prevent the obscuring of atoms within the foil by those nearer the surface, then C is simply the product of the number of atoms per unit volume and the thickness of the foil. That is,

$$C = nt. \tag{7.14}$$

The number of atoms per unit volume is found by dividing the number of atoms per mole (Avogadro's number) by the molar volume (atomic weight divided by the density). Thus,

$$n = \frac{\rho N_0}{A}, \tag{7.15}$$

where ρ is the density of the metal, A is its atomic weight, and N_0 is Avogadro's number. Then,

$$C = \frac{\rho N_0 t}{A}. \tag{7.16}$$

Example 7-2

Find the number of atoms per unit volume for (a) gold, (b) silver, and (c) platinum.

Solution

(a) The density of gold is 19.3 g/cm³ and its atomic weight is 197. Then, for gold,

$$n = \frac{19.3 \times 6.02 \times 10^{23}}{197} = 5.90 \times 10^{22} \text{ atoms/cm}^3$$

(b) Silver has a density of 10.5 g/cm³ and atomic weight 107.9. Therefore,

$$n = \frac{10.5 \times 6.02 \times 10^{23}}{107.9} = 5.86 \times 10^{22} \text{ atoms/cm}^3.$$

(c) Platinum has a density of 21.45 g/cm³ and atomic weight 195.1. Hence,

$$n = \frac{21.45 \times 6.02 \times 10^{23}}{195.1} = 6.62 \times 10^{22} \text{ atoms/cm}^3.$$

Example 7-3

What is the collision diameter when alpha particles having 5 MeV of kinetic energy are scattered from gold?

Solution

For 5 MeV alpha particles on gold, $z = 2$ and $Z = 79$. Then,

$$D = \frac{kzZe^2}{T_0}$$

$$= \frac{9 \times 10^9 \text{ J-m-coul}^{-2} \times 2 \times 79 \times (1.6)^2 \times 10^{-38} \text{ coul}^2}{5 \times 1.6 \times 10^{-13} \text{ J}} = 4.55 \times 10^{-12} \text{ cm}.$$

Example 7-4

Alpha particles having 5 MeV of kinetic energy are scattered from a gold foil one micron thick. What fraction of the particles will be scattered into the conical wedge bounded by $\theta = 90°$ and $\theta = 90° + d\theta$?

Solution

Using the value for n obtained in Example 7-2 and a foil thickness of $t = 1 \times 10^{-4}$ cm, we find that

$$C = nt = 5.90 \times 10^{18} \text{ atoms/cm}^2.$$

Using this and the value of D for 5 MeV alpha particles from Example 7-3, we get

$$\frac{\pi CD^2}{8} = \frac{\pi}{8} \times 5.90 \times 10^{18} \text{ cm}^{-2} \times (4.55 \times 10^{-12})^2 \text{ cm}^2 = 4.80 \times 10^{-5}.$$

Then,

$$\frac{dN(90°)}{I} = 4.8 \times 10^{-5} \times \frac{\sin \frac{\pi}{2}}{\sin^4 \frac{\pi}{4}} d\theta = 1.9 \times 10^{-4} \, d\theta.$$

Example 7-5

What fraction of the alpha particles in Example 7-4 will be scattered through angles of 60° or more?

Solution

$$\frac{dN(\theta > 60°)}{I} = 4.8 \times 10^{-5} \int_{\pi/3}^{\pi} \frac{\sin \theta \, d\theta}{\sin^4 \frac{\theta}{2}}$$

The integral is easily evaluated as follows:

$$\int_{\pi/3}^{\pi} \frac{\sin \theta \, d\theta}{\sin^4 \frac{\theta}{2}} = \int_{\pi/3}^{\pi} \frac{2 \sin \frac{\theta}{2} \cos \frac{\theta}{2} \, d\theta}{\sin^4 \frac{\theta}{2}}$$

$$= -2 \int_{\pi/3}^{\pi} \sin^{-3} \frac{\theta}{2} \cdot \cos \frac{\theta}{2} \, d\theta$$

$$= \left[\frac{-2}{\sin^2 \frac{\theta}{2}} \right]_{\pi/3}^{\pi} = 6.$$

Then,

$$\frac{dN(\theta > 60°)}{I} = 4.8 \times 10^{-5} \times 6 = 2.9 \times 10^{-4}.$$

The following alternative solution does not require an integration. The value of the impact parameter s that corresponds to 60° scattering is given by Equation 7.8 as

$$s = \frac{4.6 \times 10^{-12}}{2} \cot 30° = 4.0 \times 10^{-12} \text{ cm}.$$

All impact parameters *less than* this value will result in scattering at angles *greater* than 60°. That is, the fraction of the particle flux that will be scattered more than 60° by a single scatterer is simply the fraction that will pass through a disc of radius s. Then,

$$\frac{dN(\theta > 60°)}{I} = \pi Cs^2 = \pi \times 5.9 \times 10^{18} \times (4.0 \times 10^{-12})^2 = 2.9 \times 10^{-4}.$$

There were two important results of the scattering experiment of Rutherford and his collaborators. First, the evidence was overwhelming for an atomic model in which all of the positive charge is concentrated in a small core or nucleus. Second, an upper limit of about 10^{-12} centimeters was obtained for the size of the nucleus. It was found that for impact parameters that were less than this value, the scattering of alpha particles from light elements showed anomalies that could *not* be accounted for on the basis of the Coulomb force alone. Therefore, it was proposed that a new force, the *nuclear force,* becomes important when nuclei get within about 10^{-13} cm of each other. This force is a very strong, attractive, short-ranged force which will be mentioned further in a later chapter.

3. BOHR'S THEORY OF ATOMIC SPECTRA

The nuclear atom proposed by Rutherford settled the problem associated with the scattering of alpha particles, but did not explain the stability of the atom. Since it is impossible to have a stable configuration of charges subject to electrostatic forces only, a dynamical system was proposed, analogous to a planetary system. Such a system could account for the fact that the nucleus is only of the order of 1×10^{-12} cm while the atom as a whole has an effective diameter of the order of 1×10^{-8} cm. However, a serious problem arose in connection with electromagnetic theory, namely, that a charge undergoing continuous centripetal acceleration should radiate continuously. If this were the case, the energy of the dynamic system would decrease continuously, and the planetary charge would spiral into the nucleus after a nominal lifetime of about 10^{-8} second. That this does not occur is borne out by the infinite lifetimes of most elementary atoms and by the nature of their radiation spectra. Atoms do not radiate unless excited; and when radiation does occur, its spectrum consists of discrete frequencies rather than the continuum of frequencies required by the classical theory of radiation.

Much was known about the optical spectrum of hydrogen, for it had been studied intensively by many investigators in the nineteenth century. For example, it was well established that hydrogen when properly excited emits strong spectral lines in the ultraviolet as well as a multitude of lines in the visible region.

Further, it was found that certain groups of lines had wavelengths that were related to each other by a simple mathematical expression. Each such group became known as a *spectral series,* which later bore the name of its discoverer. Thus, for hydrogen the Lyman series, the Balmer series, and the Paschen series were known prior to 1913, while the Brackett series and Pfund series were found later. The Balmer series, consisting of visible wavelengths, was the first to be discovered (see Figure 7–5); it was later shown by J. R. Rydberg that it could be expressed mathematically as follows:

$$\frac{1}{\lambda} = \text{Ry} \left(\frac{1}{2^2} - \frac{1}{n^2} \right), \tag{7.17}$$

where $n = 3, 4, 5, \ldots$, and where Ry is a constant that was determined experimentally. The best experimental value for the Rydberg constant today is $109{,}677.576 \pm 0.012$ cm^{-1}.

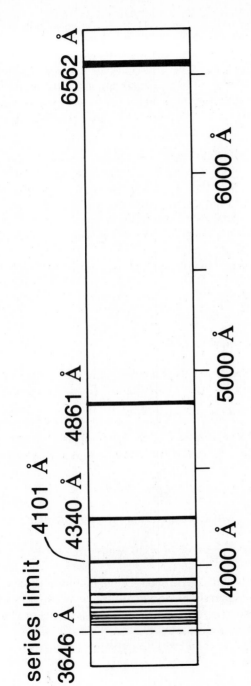

Figure 7–5

The Balmer series of lines in the spectrum of hydrogen.

In 1913, Bohr[1] proposed a theory that was successful in explaining the radiation spectra of one-electron atoms, although it is in direct disagreement with the classical theory of radiation. Bohr's postulates may be summarized as follows:

(1) The Coulomb force on a planetary electron provides the centripetal acceleration required for a dynamically stable circular orbit.

(2) The only permissible orbits are those in the discrete set for which the angular momentum of the electron equals an integer times $\hbar$, where $\hbar = h/2\pi$.

(3) An electron moving in one of these stable orbits does not radiate.

(4) Emission or absorption of radiation occurs only when an electron makes a transition from one orbit to another.

From the second postulate, note that we now have angular momentum (as well as charge and energy) quantized in atomic systems. The third postulate rejects the troublesome claim that an accelerated charge must radiate in atomic systems, in spite of its validity in the macroscopic world. The fourth postulate provides the link with Planck's theory of radiation, since the frequency of the photon emitted or absorbed is given by the energy difference of the two states divided by h.

We can readily obtain Bohr's result for hydrogen by using the first postulate to write

$$\frac{mv^2}{r} = \frac{ke^2}{r^2}, \tag{7.18}$$

where $k = \dfrac{1}{4\pi\epsilon_0}$ and has the approximate value of 9×10^9 N-m²-coul⁻². Here we have neglected any motion of the nucleus. From the second postulate we have the relationship

$$mvr = n\hbar, \tag{7.19}$$

where the integer n is called the *orbital quantum number*. Combining Equations 7.18 and 7.19, we derive the expression for the radius of the nth orbit,

$$r_n = \frac{n^2\hbar^2}{mke^2} = n^2 a_0. \tag{7.20}$$

The quantity a_0 is often simply referred to as *the Bohr radius*. Its value is

$$a_0 = \frac{\hbar^2}{mke^2} = 5.29 \times 10^{-11} \text{ m} \approx 0.53 \text{ Å}. \tag{7.21}$$

It is the radius of the first orbit of hydrogen, calculated for a rigid nucleus, and it is a quantity that appears frequently in atomic physics. Note that Equation 7.20 specifies that the orbital radii of hydrogen are indeed discrete and that they increase as the square of the integers.

Using Equation 7.18, we may write the kinetic energy of the electron as

$$T = \frac{1}{2} mv^2 = \frac{ke^2}{2r}. \tag{7.22}$$

---

[1] N. Bohr, *Phil. Mag.* 26, 1 (1913).

Since the potential energy of the electron in the Coulomb field of the nucleus is

$$V = -\frac{ke^2}{r},\tag{7.23}$$

then the total energy is

$$E = T + V = -\frac{ke^2}{2r}.\tag{7.24}$$

Using the values of r that correspond to the discrete set of allowed orbits, we find that the orbital energies are

$$E_n = -\frac{ke^2}{2n^2 a_0} = -\frac{mk^2 e^4}{2\hbar^2 n^2} = -\frac{w_0}{n^2},\tag{7.25}$$

where w_0 is given by

$$w_0 = \frac{mk^2 e^4}{2\hbar^2} = 13.6 \text{ eV}.\tag{7.26}$$

Equation 7.25 expresses the important result that the energies of atoms are quantized. That is, the energy of a given hydrogen atom is restricted to the set of values that corresponds to the allowed radial states for the electron. Can an electron change its radial state? Yes, if it gains enough energy to take it precisely to a higher energy state, then it can make a *transition* to that state. It can acquire the necessary energy from heat (collision), from an electric or magnetic field, or from photon absorption as postulated by Planck. Likewise, a transition to a lower energy state can occur if the excess energy is removed by photon emission or some other process.

Suppose that a transition occurs from the nth level to the kth level, where $n > k$. The frequency of the emitted photon is given by

$$h\nu = E_n - E_k = w_0 \left(\frac{1}{k^2} - \frac{1}{n^2}\right).\tag{7.27}$$

Since optical spectroscopists generally use *wave numbers* (reciprocal wavelengths) rather than frequencies, let us convert Equation 7.27 to this unit, which is denoted by $\bar{\nu}$. Then,

$$h\nu = hc \left(\frac{1}{\lambda}\right) = hc\bar{\nu},\tag{7.28}$$

and

$$\bar{\nu} = \frac{w_0}{hc} \left(\frac{1}{k^2} - \frac{1}{n^2}\right) = \text{Ry}_\infty \left(\frac{1}{k^2} - \frac{1}{n^2}\right).\tag{7.29}$$

The symbol Ry_∞ represents the Rydberg constant for a fixed nucleus. Its value is

$$\text{Ry}_\infty = \frac{w_0}{hc} = 109{,}737.31 \text{ cm}^{-1}.\tag{7.30}$$

Note that Equation 7.29 is identical with the Rydberg equation (7.17) for $k = 2$. It turns out that all of the known spectral series for hydrogen can be satisfactorily explained by Equation 7.29. Thus, the Lyman series is obtained if we set

$k = 1$ and allow n to take on the values $n = 2, 3, 4, \ldots$; the Balmer series corresponds to $k = 2$, $n = 3, 4, 5, \ldots$; the Paschen, Brackett, and Pfund series correspond to $k = 3, 4,$ and 5, respectively, with n running through the integers greater than k. These series are illustrated pictorially in Figure 7–6.

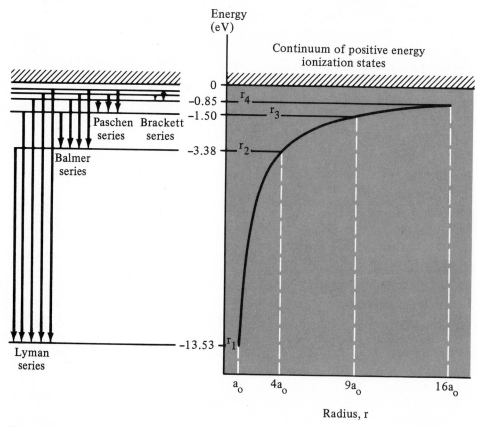

Figure 7–6

Bohr radii and allowed energies of hydrogen. The first few lines of several of the spectral series of hydrogen are shown at the left.

Equation 7.27 can be used also to calculate the ionization energy for hydrogen, which is known from experiment to be about 13.6 eV. Since ionization occurs when the electron escapes from the nucleus, this process is equivalent to letting $n \rightarrow \infty$. Then, for $k = 1$, we see that the Bohr theory predicts an ionization energy of w_0, whose value was calculated to be 13.6 eV in Equation 7.26.

This excellent agreement between the Bohr theory and the experimental values for the Rydberg constant, the ionization potential, and the emission spectrum of hydrogen was a great triumph for the theory. The agreement is even better if a correction is made for the motion of the nucleus instead of simply assuming that it is fixed. We know from elementary mechanics that *both* the nucleus and the electron will revolve about their common center of mass. Therefore, the kinetic energy that was attributed to the electron in Equation 7.22 is too large, since a part of it is associated with the nuclear motion. It is easy to show[1] that the correct electron kinetic energy relative to the moving nucleus can be obtained by using, instead of the electron mass, the *reduced mass*

$$\mu = \frac{mM}{M + m}, \tag{7.31}$$

where m is the electron mass and M is the nuclear mass.

Example 7-6
    Using Equation 7.17, find the wavelength of the Balmer line for $n = 3$.

Solution

$$1/\lambda = 109{,}677 \text{ cm}^{-1}(1/2^2 - 1/3^2) = 15{,}233 \text{ cm}^{-1}$$

$$\lambda = 6.56 \times 10^{-5} \text{ cm} = 6560 \text{ Å}.$$

Example 7-7
    (a) What is the rotational frequency of an electron in the nth Bohr orbit of hydrogen?
    (b) Compare this result with the frequency emitted by hydrogen during a transition from the nth to the $(n\text{-}1)$th state when $n \gg 1$.

Solution
    (a) In terms of orbital speed and radius, the frequency is

$$\nu = \omega/2\pi = v/2\pi r_n.$$

Using v from Equation 7.19 and r_n from Equation 7.20,

$$\nu = n\hbar/2\pi mr_n^2 = (n\hbar/2\pi m) \cdot (mke^2/n^2\hbar^2)^2 = mk^2e^4/2\pi n^3\hbar^3.$$

    (b) From Equations 7.26 and 7.27,

$$\nu = (w_0/h)[1/(n - 1)^2 - 1/n^2] = \left(\frac{mk^2e^4}{4\pi\hbar^3}\right) \cdot \left[\frac{2n - 1}{n^2(n - 1)^2}\right].$$

For $n \gg 1$,

$$(2n - 1)/n^2(n - 1)^2 \approx 2n/n^2n^2 = 2/n^3.$$

Then,

$$\nu \approx mk^2e^4/2\pi n^3\hbar^3.$$

Example 7-8
    Use the Bohr theory to obtain the allowed radii and energies of the electron in singly ionized helium.

Solution
    In the previous derivation it is understood that $Z = 1$. In order to include Z explicitly, we must write Equation 7.18 as

$$mv^2/r = kZe^2/r^2.$$

---

[1] See problem 7-15.

It follows from this that

$$r_n = n^2\hbar^2/mkZe^2 = n^2a_0/Z,$$

and
$$E_n = -mk^2Z^2e^4/2n^2\hbar^2 = -Z^2w_0/n^2.$$

In singly ionized helium $Z = 2$, so the allowed energies and radii (neglecting nuclear motion) are:

$$r_n = \frac{1}{2}n^2a_0$$

and

$$E_n = -4w_0/n^2.$$

Example 7-9

Use the Bohr theory to calculate the allowed circular orbits and energies for muonium. Muonium is a hydrogen-like atom consisting of a proton of mass $1836m$ and a muon whose charge is $-e$ and whose mass is $207m$, where m is the electron mass.

Solution

The reduced mass of the muon is

$$\mu = (1836m)(207m)/(1836 + 207)m = 186m.$$

Then,
$$r_n = n^2\hbar^2/186mkZe^2 = n^2a_0/186,$$

and
$$E_n = 186mk^2Z^2e^4/2n^2\hbar^2 = -186w_0/n^2.$$

4.   THE FRANCK-HERTZ EXPERIMENT

The experiment of Franck and Hertz in 1914 provided additional confirmation of the discrete energy states predicted by the Bohr theory. In this experiment mercury vapor was bombarded with electrons of known kinetic energy. When this kinetic energy is less than the energy of the first excited state of atomic mercury, the only energy that the colliding electron loses is the small amount of kinetic energy (about one part in 10^5) that it can transfer to the massive mercury atom by an elastic collision. However, when the kinetic energy of the incident electron just exceeds the energy of the first excited state of mercury, then the electron gives up virtually all of its kinetic energy to the mercury atom in an inelastic collision. Thus, by comparing the kinetic energy of the electrons before and after the collision one can determine how much energy is transferred to the target atoms.

Franck and Hertz used a simple apparatus consisting of a mercury tube containing a cathode, a plate, and an accelerating grid which was located physically near the plate (see Figure 7–7). Both the plate and the grid were maintained at positive potentials with respect to the cathode, but the plate potential was slightly lower than the grid potential. This small retarding potential on the plate had the effect of preventing contributions to the plate current from electrons having negligible kinetic energy. Hence, if electrons were to lose most of their kinetic energy in inelastic collisions with mercury atoms, the retarding potential would prevent them from reaching the plate and a drop in plate current would result. Such a drop in plate current was found to occur at 4.9 volts, as shown in Figure 7–8. Further-

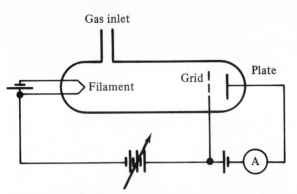

Figure 7–7

Apparatus for Franck-Hertz experiment.

more, Franck and Hertz noted that at this voltage the 2536 Å spectral line of mercury appeared in the emission spectrum of the vapor. A simple calculation shows that the photon energy of the 2536 Å line corresponds to 4.86 electron-volts! At slightly higher voltages a large drop in plate current occurs and new lines appear in the emission spectrum of the mercury vapor. This behavior is repeated at multiples of 4.9 volts, as shown in the figure. We must conclude from this that the Bohr concept of discrete energy states is qualitatively correct.

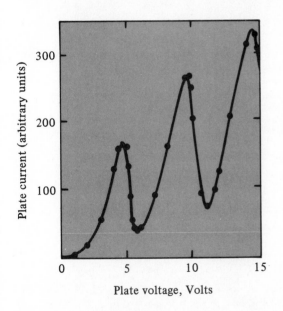

Figure 7–8

Current *vs.* voltage in the Franck-Hertz experiment.

5. X-RAY SPECTRA AND THE BOHR THEORY

When high-energy electrons are used to bombard a target made of a heavy metal, the electrons are decelerated and they will radiate. The resulting radiation produces a continuous spectrum called *Bremsstrahlung,* which is characterized by a sharp cutoff at the short wavelength end. A typical spectrum is shown in Figure 7–9. This continuous spectrum can be accounted for classically by the theory of electromagnetic radiation[1] or quantum mechanically by invoking an inverse

---

[1] F. K. Richtmyer, E. H. Kennard, and T. Lauritsen, *Introduction to Modern Physics,* 6th ed., McGraw-Hill Book Co., New York, 1969, p. 349.

Enrico Fermi (left) and Niels Bohr. (Photo taken by S. A. Goudsmit. Courtesy of the American Institute of Physics Niels Bohr Library.)

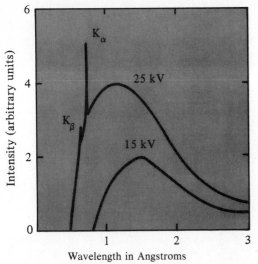

Figure 7–9

X-ray spectrum of molybdenum. The continuous spectrum is the Bremsstrahlung radiation.

photoelectric effect. That is, the change in kinetic energy provides an upper limit on the photon frequency, whereas in the photoelectric effect the photon frequency provides an upper limit on the change in kinetic energy. Since the initial kinetic energy of an electron is determined by the voltage on the x-ray tube, the short wavelength cutoff is given by

$$\lambda_c = \frac{hc}{eV} = \frac{12,400}{V}, \qquad (7.32)$$

where V is in volts and λ is in angstroms. Thus, a 12 kilovolt x-ray tube has a cut-off wavelength of about 1 angstrom. Not many photons having wavelength λ_c are emitted, because the majority of electrons perform mechanical work on the target when they stop and a large fraction of their kinetic energy is converted into heat. Equation 7.32 provides an accurate method[1] for determining the ratio e/h, since a plot of ν_c vs. V is a straight line whose slope is e/h.

Superimposed upon the continuous spectrum are a few discrete peaks whose wavelengths are characteristic of the target material. A few of these peaks are also shown in the figure. The characteristic x-ray spectra of many elements were studied by Moseley,[2] who showed that the frequency of a given line varies from element to element as the square of the atomic number. A plot of some of Moseley's data is shown in Figure 7–10. This is the result one would expect from the Bohr theory if the atomic number Z were included in the Coulomb force term in Equation 7.18. (See problem 7-13.)

The Bohr theory accounts for the discreteness and the order of magnitude of the photon energies of the characteristic spectrum by assuming that these photons are emitted as a result of transitions involving inner electron shells. If, for example, a high-energy electron knocks out an electron from the first shell of a target atom, the vacancy may be filled by a transition from the second shell, or the third, or higher shells. Since the electrons of the first shell have been traditionally called

[1] J. DuMond and V. L. Bollman, *Phys. Rev. 51*, 400 (1937).
[2] H. G. J. Moseley, *Phil. Mag. 26*, 1024 (1913); *27*, 703 (1914).

K-electrons, the photon emitted by a transition from the second shell to the K shell is called K_α radiation, that due to a transition from the third shell to the K shell is called K_β radiation, and so on. The second shell is designated as the L shell, the third M, and so forth, so that $L_\alpha, L_\beta, \ldots, M_\alpha, M_\beta, \ldots$ radiation is defined in a similar fashion.

Since Figure 7–10 shows that the photon energy is proportional to Z^2, a tran-

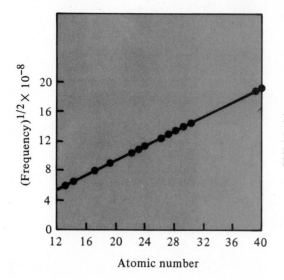

Figure 7–10

Moseley's results for the effect of atomic number on the frequency of a given x-ray line.

sition from the second to the first shell for copper ($Z = 29$) would result in the emission of a photon having an energy about 841 times the energy of the longest line of the Lyman series of hydrogen. This approximate calculation gives 1.44 Å for the K_α line of copper, whereas the correct value is 1.54 Å. The agreement is not nearly so good for transitions involving shells of higher order because of the shortcomings of the Bohr theory and the important screening effect of the nuclear charge by electrons in the innermost shells.

A process which generally competes with x-ray emission is the internal photoelectric effect or *Auger effect*. Here, the x-ray photon does not actually appear, but an equivalent amount of kinetic energy is given to an outer electron, which is in turn ejected from the atom.

Example 7-10

What is the shortest wavelength that can be produced by a 30,000-volt x-ray tube?

Solution

The shortest or cut-off wavelength in angstroms is given by Equation 7.32, namely, $\lambda_c = 12,400/V = 12,400/30,000 = 0.41$ Å.

Example 7-11

Show that the Bohr theory predicts Moseley's result that the frequency of a characteristic x-ray line is proportional to the square of the atomic number.

Solution

It was shown in Example 7-8 that if Z is included as a parameter in the Bohr

Five Nobel Laureates at a meeting in Berlin in 1928. Left to right: W. Nernst, A. Einstein, M. Planck, R. A. Millikan and M. von Laue. (Courtesy of the American Institute of Physics Niels Bohr Library.)

theory, then Equation 7.27 becomes

$$h\nu = Z^2 w_0[1/k^2 - 1/n^2],$$

where $n > k$. Therefore,

$$\nu^{1/2} \sim Z,$$

as shown in Figure 7–10.

SUMMARY

The work of Rutherford and his associates culminated in a model of the atom in which all of the positive charge and most of the mass of the atom are concentrated in a nucleus whose diameter is about 10^{-5} times the atomic diameter. Charge neutrality is achieved by placing as many electrons outside the nucleus as there are protons within the nucleus, and these electrons must be moving rapidly in order for the atom to be stable. Bohr proposed that each "planetary" electron moves in a circular orbit in which the centripetal force is provided by the Coulomb attraction between the nucleus and the electron. Perhaps the most startling of Bohr's postulates is that the orbital angular momentum associated with an electron is quantized in units of $\hbar = h/2\pi$, where h is Planck's quantum constant. This restriction on the possible values of the angular momentum leads to the quantization of the energy of the atom. That is, the discrete set of angular momenta restricts the orbital radii and energies to discrete values.

The Bohr theory predicts the energy levels of hydrogen and hydrogen-like atoms reasonably well, although serious discrepancies arise for non-hydrogenic atoms.

An additional postulate of Bohr, namely, that the absorption or emission of a photon by an atom occurs when an electron makes a transition between two of the atomic energy levels, was an important link between the quantum theory of radiation and the quantum theory of matter.

Additional Reading

C. H. Blanchard, C. R. Burnett, R. G. Stoner, and R. L. Weber, *Introduction to Modern Physics,* 2nd ed., Prentice-Hall, Inc., Englewood Cliffs, N.J., 1969, Chapter 6.

K. W. Ford, *Classical and Modern Physics,* Vol. 3, Xerox College Publishing, Lennox, Mass., 1974, Chapter 23.

G. Gamow, *Mister Tomkins in Paperback,* Cambridge University Press, New York, 1967.

G. Gamow, *Thirty Years That Shook Physics: The Story of the Quantum Theory,* Doubleday & Co., Inc., Garden City, N.Y., 1965.

P. A. Tipler, *Modern Physics,* Worth Publishers, Inc., New York, 1978, Chapter 4.

R. T. Weidner and R. L. Sells, *Elementary Modern Physics,* 3rd ed., Allyn & Bacon, Inc., Boston, 1980, Chapter 6.

E. H. Wichmann, *Quantum Physics,* Berkeley Physics Course, Vol. 4, McGraw-Hill Book Co., New York, 1971, Chapters 1 and 3.

PROBLEMS

7-1. What fraction of incident alpha particles having 2.5 MeV of kinetic energy will be scattered through an angle of 60° or more by a gold foil 2 microns thick?

7-2. For alpha particles of energy 7.68 MeV scattered from a platinum foil two microns thick, find: (a) the collision diameter, (b) the differential scattering cross section, and (c) the fractional number of particles scattered through angles of 90° or more.

*7-3. An accelerator produces a current of 2 nano-amperes consisting of a very narrow beam of 2.5 MeV alpha particles. The alpha particles are then scat-

PROBLEMS 137

tered from a gold foil having a thickness of 0.5 micron. A detector having a cross-sectional area of 2 cm² is located at a distance of 1 meter from the foil. How many alpha particles per second will be detected at an angle of 45° from the incident beam direction?

7-4. Using the simple Bohr theory for circular orbits, calculate the second ionization potential of helium.

7-5. Using the simple Bohr theory, calculate the third ionization potential of lithium.

7-6. What is the radius of the first Bohr orbit in singly ionized helium? Doubly ionized lithium? Triply ionized beryllium?

7-7. Determine the shortest and longest wavelengths in the Lyman series for hydrogen.

7-8. Determine the shortest and longest wavelengths in the Balmer series for hydrogen.

7-9. What value of n is associated with the Lyman series line in hydrogen whose wavelength is 1026 Å?

7-10. Find the wavelength and frequency of the photon that could excite a hydrogen atom from the ground state to the $n = 2$ state.

7-11. What are the frequency and wavelength of the photon emitted when an excited hydrogen atom decays from the $n = 3$ state to the $n = 2$ state?

7-12. What is the wavelength of the photon required to raise the electron in singly-ionized helium to the first excited state?

7-13. Include the atomic number Z in the Coulomb force term in Equation 7.18 and derive the equations for the Bohr radii and energies for any value of Z. These equations should reduce to Equations 7.20 and 7.25 for $Z = 1$.

7-14. If hydrogen gas in a discharge tube is excited by a potential difference of 11 volts, which, if any, spectral lines corresponding to the Lyman series of hydrogen will be excited?

7-15. Derive the expression for the reduced mass given in Equation 7.31.

7-16. Find the reduced masses (Equation 7.31) for the electrons in hydrogen, deuterium, and tritium.

7-17. Determine the effect of nuclear mass on the first line of the Balmer series of hydrogen by calculating its wavelength shift in deuterium and tritium.

7-18. Positronium is a hydrogen-like atom consisting of a positive and a negative electron revolving about one another. Each particle has a mass m and a charge of magnitude e. Use the Bohr theory and the reduced mass to obtain the allowed radii and energies.

*7-19. Muonic lead is formed when Pb-208 captures a muon of charge $-e$ and mass equal to 207 electron masses in lieu of one of the orbital electrons. Find the radius and energy of the ground state. (Hint: What assumption can be made about the nuclear charge that will act upon the muon?)

*7-20. Estimate the wavelengths of the K_α and K_β lines of the characteristic spectrum of vanadium by using the elementary Bohr theory. Compare these estimates with the accepted values.

*7-21. Use the Bohr theory to estimate the kinetic energy of an Auger electron resulting from a radiationless transition from the L to the K shell of chromium.

*7-22. A negative muon, having a mass of 207 electron masses and a charge $-e$, is captured by a deuteron to form a muonic atom. The deuteron has a rest mass of 3670 electron masses. (a) Find the energies of the ground state and the first excited state. (b) What is the wavelength of the photon emitted when the atom decays from the first excited state to the ground state?

*7-23. In classical physics an orbiting electron would lose energy continuously through radiation. Estimate the lifetime of the hydrogen atom, assuming that the power radiated is given by the expression

$$P = \frac{2ke^2 \ddot{r}^2}{3c^3},$$

where $\ddot{r}$ is the acceleration of the electron.

*7-24. (a) If the Bohr atom is interpreted as a classical oscillator whose frequency is $\nu_0 = v/2\pi r$, show that the classical oscillator frequency can be expressed as

$$\nu_0 = \frac{2w_0}{hn^3},$$

where w_0 is given in Equation 7.26. (b) Show that in the limit as n gets very large, the Bohr frequency relation reduces to the expression in (a). This is an example of the *correspondence principle*, which states that a quantum theory result should agree with the equivalent classical solution in the limit of large quantum numbers.

8 The Wave Nature of Particles

1. THE DE BROGLIE WAVELENGTH

In Chapter 5 we discovered that light, which had heretofore been assumed to be a wave, can behave like a particle. It is then only natural to ask, "Can a particle ever act like a wave?" This question was given serious thought by Louis de Broglie, who proposed in his doctoral thesis[1] that an electron has both a wave and a particle nature. His work was partially motivated by the mystery of the Bohr orbits that we discussed in the preceding chapter. He attempted to explain them by regarding the electron as a standing wave around the circumference of an orbit, as shown in Figure 8–1. Thus, de Broglie required that

$$n\lambda = 2\pi r, \tag{8.1}$$

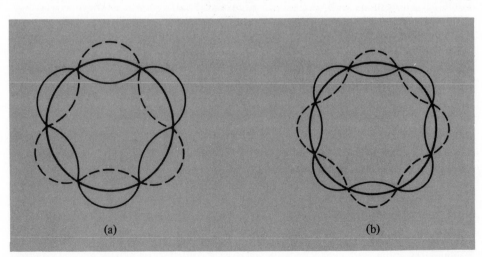

(a)   (b)

Figure 8–1

Standing waves around the circumference of a circle suggest a way of accounting for the quantized Bohr orbits by means of de Broglie waves. In (a) the circumference contains exactly three wavelengths, and in (b) there are four wavelengths.

---

[1] L. de Broglie, *Ann. Phys. (Paris) 3*, 22 (1925).

where λ is the wavelength associated with the electron in the nth orbit and r is the radius of the orbit. Combining this with the quantization of angular momentum proposed by Bohr,

$$mvr = n\hbar, \tag{8.2}$$

we immediately obtain the result that

$$\lambda = \frac{2\pi r}{n} = \frac{h}{mv} = \frac{h}{p}. \tag{8.3}$$

Thus, a material particle having a momentum p is accompanied by a wave whose wavelength is given by Equation 8.3. This wavelength is now called the *de Broglie wavelength*. It should be noted that this result is in exact agreement with the expression for the momentum of a photon given in Equation 6.21. Assuming complete symmetry between particles and photons, de Broglie proposed that any material particle of total energy E and momentum p must be accompanied by a phase wave whose wavelength is given by $\lambda = h/p$ and whose frequency is given by the Planck formula, $\nu = E/h$. The Planck and de Broglie relations may be expressed in the useful forms

and

$$\left.\begin{array}{l} E = \hbar\omega \\[6pt] p = \hbar k, \end{array}\right\} \tag{8.4}$$

where $\hbar = h/2\pi$, $k = 2\pi/\lambda$, and $\omega = 2\pi\nu$.

The concept of the de Broglie wavelength is one of the cornerstones of modern quantum theory, and the simple relationship

$$\lambda = \frac{h}{p}$$

holds for photons as well as for both relativistic and nonrelativistic material particles, provided that the appropriate expression for p is used.

The physical nature of a particle wave was not clearly described by de Broglie. Unlike a classical wave, the energy E of the particle wave is not thought of as spread out over the extent of the wave, but is regarded as localized with the particle. However, the accompanying wave is essential in order to account for the phenomenon of particle diffraction, which will be discussed in the next section.

Example 8-1

Find the de Broglie wavelength associated with a bullet with a mass of 10 g, which has a velocity of 500 m/s.

Solution

$$\lambda = \frac{h}{p} = \frac{6.63 \times 10^{-34} \text{ J-s}}{10 \times 10^{-3} \text{ kg} \times 5 \times 10^2 \text{ m/s}} = 1.33 \times 10^{-34} \text{ m}.$$

Example 8-2.

What is the de Broglie wavelength of an electron that is accelerated from rest by a potential difference of 150 volts?

Solution

The kinetic energy given to the electron is 150 eV = $150 \times 1.6 \times 10^{-19}$ J = 2.40×10^{-17} J. Since this is small compared to the rest mass energy, the classical expression for the kinetic energy may be used. That is,

$$T = \frac{p^2}{2m},$$

or

$$p = \sqrt{2mT} = \sqrt{2 \times 9.1 \times 10^{-31} \text{ kg} \times 2.40 \times 10^{-17} \text{ J}}$$

$$= 10^{-24} \sqrt{43.7} \text{ kg-m-s}^{-1}$$

$$\lambda = \frac{h}{p} = \frac{6.63 \times 10^{-34}}{6.62 \times 10^{-24}} m = 1 \text{ Å}$$

2. THE DIFFRACTION OF WAVES FROM CRYSTALS

Whether or not a particle of a given momentum will exhibit its wave properties will be determined by the relative magnitude of its de Broglie wavelength in comparison with the physical dimensions of the environment in which it is found. We know from our study of light waves, sound waves, and water waves that, for wavelengths that are much smaller than the dimensions of apertures and obstacles, diffraction and other wave effects are not ordinarily observed. In such cases we can assume rectilinear propagation, and problems can be treated by means of ray diagrams. Thus, we do not expect to see around doors, trees, or other obstacles in our daily lives because the dimensions of such objects are about 10^6 or 10^7 times as large as the wavelength of visible light. On the other hand, for wavelengths that approximate or exceed the dimensions of objects, diffraction effects become quite important and ray diagrams become meaningless. For example, our speech produces wavelengths of the order of meters or tens of meters, and we are quite accustomed to the diffraction of these waves around obstacles. Our world would seem quite strange indeed if audible sound propagation were to obey ray diagrams!

In the case of particle waves, it is the extremely small size of Planck's constant that has prevented their detection until this century. The bullet in Example 8-1 has a de Broglie wavelength that is hopelessly beyond any reasonable limit of detection. The same holds true for any object in our everyday world. It is only when we consider the elementary particles themselves that we can obtain moving particles which have momenta sufficiently small compared to the value of h to produce a detectable wavelength. Thus, we note from Example 8-2 that electrons having energies of the order of 100 eV have de Broglie wavelengths that correspond to the atomic spacings of most crystalline solids.[1] It was therefore conjectured that electrons of these energies ought to be diffracted by crystals in the same manner as x-rays if de Broglie waves truly exist.

X-ray diffraction from crystals was first reported by researchers in Germany, but it is commonly called *Bragg reflection* or *Bragg diffraction* in honor of W. L. and W. H. Bragg, who thoroughly developed the method.[2] The results are analyzed by first assuming that each plane of atoms acts like a mirror, in that the angle of reflection equals the angle of incidence. Next, the reflected waves from succes-

---

[1] See Section 2 of Chapter 1.
[2] W. H. Bragg and W. L. Bragg, *Proc. Roy. Soc.* (London) **88**, 428 (1913).

sive layers of atoms are allowed to interfere. If the reflected waves from adjacent planes are in phase, the interference will be constructive and the reflection will be intense. On the other hand, reflections that are out of phase will tend to cancel each other. Figure 8–2 shows two parallel rays incident on two successive atomic

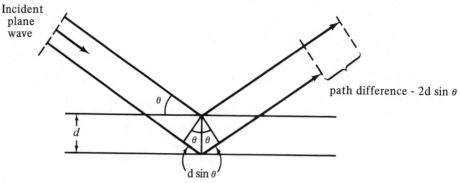

Figure 8–2

X-ray scattering from two parallel atomic planes.

planes of a cubic crystal. Note that the path difference between the two rays is $2d$ sin θ. If the path difference is an integral number of wavelengths, the rays are in phase and the reflection will be strong. However, if the path difference is equal to an odd number of half wavelengths, the reflection will be weak. Thus, we have the Bragg equation for strong reflections,

$$n\lambda = 2d \sin \theta, \qquad (8.5)$$

where the integer n is called the *order* of the maximum. The first order maximum is given by $n = 1$, the second order maximum by $n = 2$, etc. Since the path difference varies with the angle θ, it is evident that angular positions of the reflection maxima will depend upon both the wavelength and the angle of incidence of the radiation.

Many different sets of parallel planes can be drawn through the atoms in a real crystal. Figure 8–2 considers reflections from only the planes parallel to the surface of the crystal, whereas a somewhat more realistic case is illustrated in Figure 8–3. In the latter figure note that the interplanar distance is related to the atomic spacing a by the relation

$$d = a \sin \phi.$$

Furthermore, since sin θ = cos ϕ, Equation 8.5 may be expressed in the form

$$n\lambda = 2a \sin \phi \cdot \cos \phi = a \sin 2\phi. \qquad (8.6)$$

The diffraction of particles was first confirmed experimentally in 1927 by the work of Clinton Davisson[1] and Lester Germer in the United States and by George P. Thomson[2] in England. In the Davisson-Germer experiment, low-energy elec-

[1] C. Davisson and L. H. Germer, *Phys. Rev. 30,* 705 (1927).
[2] G. P. Thomson, *Proc. Roy. Soc. A 117,* 600 (1927); *Nature 120,* 802 (1927).

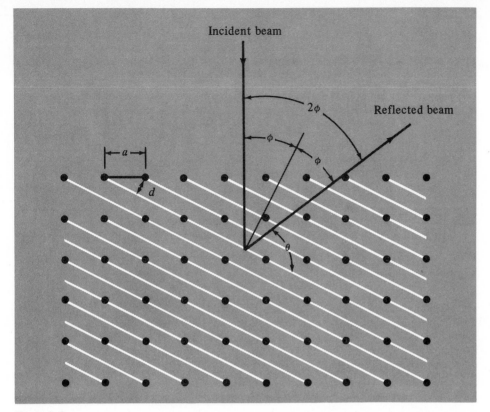

Figure 8–3

Bragg reflections of electron waves or x-rays for normal incidence on a crystal.

trons were scattered from a single crystal of nickel. Although some electrons were scattered at nearly all angles, it was noted that a distinct peak in the intensity occurred at an angle of 50°. Further, the size of this peak varied with the energy of the incident electrons, and was found to reach a maximum of intensity for 54-volt electrons. These experimental results are shown graphically in Figure 8–4. The following examples will help the reader to understand the significance of this experiment.

Example 8-3

Calculate the de Broglie wavelength for 54-volt electrons.

Solution

$$\lambda = \frac{h}{p} = \frac{h}{\sqrt{2mE}}$$

$$= \frac{6.63 \times 10^{-34} \text{ J-s}}{(2 \times 9.1 \times 10^{-31} \text{ kg} \times 54 \times 1.6 \times 10^{-19} \text{ J})^{1/2}}$$

$$= 1.67 \times 10^{-10} \text{ m} = 1.67 \text{ Å}$$

Example 8-4

If we assume an interatomic spacing of 2.15 Å for the nickel crystal, what wave-

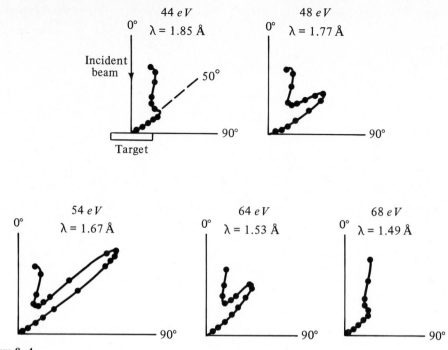

Figure 8–4

Angular plots of scattered intensity for low-energy electrons incident upon a nickel single crystal. The energies and de Broglie wavelengths of the electrons are given for each plot. (From C. Davisson and L. H. Germer, *Phys. Rev. 30,* 705 (1927). Used with permission.)

length corresponds to a first-order maximum in the diffraction pattern for a scattering angle of 50°?

Solution

Since the angle between the incident and the scattered beams is $2\phi = 50°$, the distance $a = 2.15$ Å, and $n = 1$, Equation 8.6 becomes

$$\lambda = 2.15 \sin 50° = 1.65 \text{ Å}.$$

The excellent agreement between the calculated de Broglie wavelength for 54-volt electrons and the Bragg condition illustrated by the preceding examples was immediately accepted as dramatic evidence for the wave nature of electrons. Actually, the agreement turned out to be even better than these examples show. Further experimentation confirmed that particle waves are refracted by crystals in the same way that light waves are refracted by glass. The index of refraction for nickel was found to be about 1.02. This means that the wavelength within the nickel crystal is shorter than that obtained in Example 8-3 by approximately 2 percent, which brings it very close to the value obtained in Example 8-4.

Thus far we have discussed the study of diffraction by back-reflection as depicted schematically in Figure 8–3. Diffraction is also observed when waves of appropriate wavelengths are transmitted through crystals or thin films of metal, as shown in Figure 8–5. If a monochromatic beam is incident upon a pure, single crystal, a given set of parallel atomic planes will give rise to a single spot on the

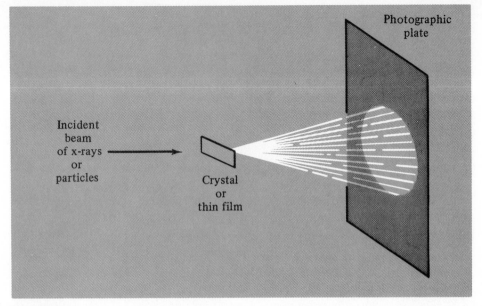

Figure 8–5

Laue diffraction.

film for each order (n value) of the diffraction that is intense enough to produce an image. The array of spots produced by all sets of planes is called a *Laue diffraction pattern* after Max von Laue, who first used this method for x-rays. Two examples of the Laue diffraction of particles by single crystals are shown in Figures 8–6 and 8–7.

If a polycrystalline specimen is used instead of a single crystal, each spot in the Laue diffraction pattern becomes a ring about the scattering axis. The reason for this is that a polycrystal contains a large number of minicrystals, which are randomly oriented with respect to the beam. Their collective effect would be equivalent to rotating the crystal in Figure 8–3, for example, about the incident

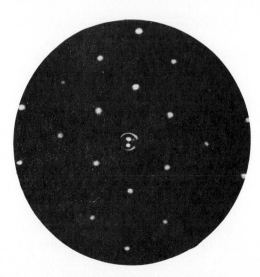

Figure 8–6

Diffraction of 300 volt electrons from a clean (110) surface of a tungsten single crystal. (Photograph kindly furnished by Dr. L. H. Germer. After L. H. Germer and J. W. May, *Surface Science 4*. 452 (1966). Used with permission of North-Holland Publishing Co., Amsterdam.)

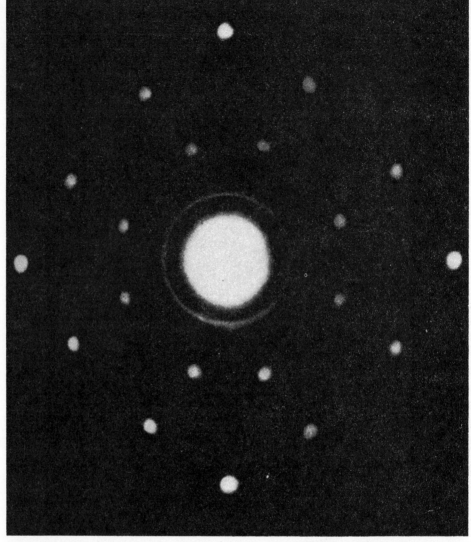

Figure 8–7

Neutron Laue photograph of NaCl. (Photograph kindly furnished by Dr. E. O. Wollan.)

beam. Figure 8–8 shows a Laue photograph of the diffraction pattern for electrons scattered from a polycrystalline film of an alloy of copper and gold.

3. THE WAVE FUNCTION AND PROBABILITY

As we saw in the previous section, a beam of mono-energetic particles behaves very much like a beam of photons for the proper choices of energies and physical dimensions. Diffraction patterns similar to those shown have now been obtained for many different particles as well as for atoms and molecules. Although

Figure 8–8

Diffraction of 50 kV electrons from a disordered film of Cu_3Au alloy. The alloy film was 400 Å thick. (Photograph kindly furnished by Dr. L. H. Germer.)

we will be using electrons as our example in what follows, it should be understood that all that is said will be true for any elementary or atomic particle.

Consider an experimental arrangement similar to Young's double-slit experiment in optics.[1] Let a mono-energetic beam of electrons be incident on two slits whose dimensions and spacing are chosen to be of the same order of magnitude as the de Broglie wavelength of the electrons, so that diffraction effects can be detected. A sketch of the experiment is shown in Figure 8–9(a). The detector may be regarded as an array of microscopic counters or as a fluorescent screen that can be photographed. If either slit were blocked off, the pattern observed on the screen would look like that shown in Figure 8–9(b). However, if both slits are open, the pattern resembles Figure 8–9(c), where the interference effects are strikingly evident. If electrons were to behave like classical particles, there would be no interference effects and the pattern on the screen would approximate that shown in Figure 8–9(d). By repeating the experiment with lower beam intensities (fewer electrons per second), the same pattern would be observed after a sufficient time. Surprisingly enough, even if only one electron at a time were fired at the slits, the interference pattern of Figure 8–9(c) would be produced after a sufficiently large

---

[1] See, for example, the review of this subject in Hugh D. Young, *Fundamentals of Optics and Modern Physics,* McGraw-Hill Book Co., New York, 1968.

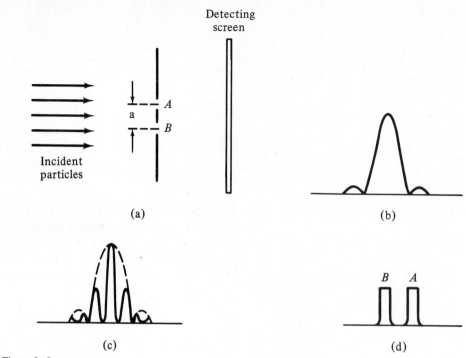

Figure 8–9

(a) Double slit experiment with particles. (b) Distribution of particles recorded on the screen due to diffraction from either slit A or B. (c) Distribution of particles recorded on the screen due to diffraction with both slits A and B open. (d) Hypothetical distribution of particles recorded on the screen if wave effects are neglected.

number of electrons hit the screen. The computer-simulated growth of the two-slit pattern is shown in Figure 8–10.

   This experiment forces us to conclude that *each electron interacts with both slits at once* even though our classically trained intuition tells us that an electron can pass through only one of the slits. Thus, from the point of view of modern physics, it is meaningless to ask which slit the electron goes through. Any attempt to determine experimentally *which* slit an electron goes through will destroy the interference pattern just as effectively as if the other slit were blocked off! It follows, then, that it is impossible to predict at which point a given electron will strike the detecting screen. However, we can relate the relative height of the particle distribution curve at position x to the relative *probability* that an electron will strike the screen at position x. Drawing upon the optical analogy again, if monochromatic photons had been incident upon the slits of Figure 8–9(a), the curve in (c) would be a plot of relative light intensity versus position on the screen. The light intensity at each point is interpreted as the square of the amplitude of a wave which can be represented by either the real or the imaginary part of the function[1]

$$\Psi(x,\ t) = Ae^{i(kx-\omega t)}.\tag{8.7}$$

---

[1] See Section 4 of Chapter 6.

(a) After 28 electrons

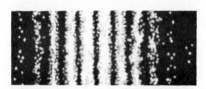

(b) After 1000 electrons

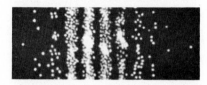

(c) After 10,000 electrons

(d) Two slit electron pattern

Figure 8–10

(a), (b), and (c). Computer-simulated growth of a two-slit interference pattern for electrons. (d) An actual photograph of a two-slit pattern produced by electrons. [Parts (a), (b), and (c) from E. R. Huggins, *Physics I,* W. A. Benjamin, Inc., New York, 1968. Part (d) is from C. Jönsson, *Zeitschrift für Physik, 161,* 454 (1961). Used with permission.]

In like manner, we define a *wave function* for an electron, $\Psi(x, t)$, such that the square of its magnitude is proportional to the *probability* of finding the electron at position x at time t. We write the wave function and its complex conjugate Ψ^* as

$$\left.\begin{array}{l} \Psi(x, t) = \Psi_0 e^{i(kx-\omega t)} = \Psi_0 e^{i(xp_x - Et)/\hbar}, \\[2mm] \Psi^*(x, t) = \Psi_0^* e^{-i(kx-\omega t)} = \Psi_0^* e^{-i(xp_x - Et)/\hbar} \end{array}\right\} \tag{8.8}$$

where both Ψ and Ψ_0 are complex numbers instead of vectors. The de Broglie relations have been incorporated in the second expressions of Equation 8.8 in order to replace the frequency and wave number by the energy and the momentum of the electron. Although this complex wave function is not directly observable (that is, measurable), its physical significance rests on the assumption that the quantity

$$|\Psi(x_1, t_1)|^2 dx = \Psi^*(x_1, t_1) \cdot \Psi(x_1, t_1) dx \tag{8.9}$$

is proportional to the probability of finding the electron in the element dx centered at x_1 at time t_1. Then the total probability for finding the electron *anywhere* in the one-dimensional space defined by $-\infty < x < \infty$ is proportional to the integral of

$|\Psi(x, t)|^2$ over all of space. Thus, the total probability of finding the electron somewhere along the x-axis is

$$P = \int_{-\infty}^{\infty} |\Psi(x, t)|^2 dx. \tag{8.10}$$

This total probability should be unity if there is a single electron in the space. It will, of course, be zero if there is no electron in the space, since in that case the wave amplitude Ψ_0 will be zero. If the integration in Equation 8.10 results in some finite number N that is not equal to unity, then the amplitude of Ψ should be multiplied by the factor $1/\sqrt{N}$ in order to insure that the total probability is indeed unity. The factor $1/\sqrt{N}$ is known as the *normalization factor,* and the process of correcting the amplitude of Ψ so that $P = 1$ is simply called *normalizing* the wave function.

In the event that the wave function cannot be normalized because the integral in Equation 8.10 diverges (becomes infinite), then it must be examined to see whether it is suitable for representing an electron. If the wave function itself diverges, it is said to be "not well-behaved" and it is rejected as a possible wave function.

The explanation of the interference effects illustrated in Figure 8–9 hinges upon the assumption that *the principle of superposition* is valid. When we apply this principle to coherent light in optics, we add the amplitudes at a point on the screen vectorially. Then the square of the resultant amplitude is the intensity at that point. Thus,

$$I \sim (\vec{A}_1 + \vec{A}_2)^2 = A_1{}^2 + A_2{}^2 + 2\vec{A}_1 \cdot \vec{A}_2 \sim I_1 + I_2 + 2\vec{A}_1 \cdot \vec{A}_2, \tag{8.11}$$

where I represents the time average of the intensity over a full cycle. On the other hand, when light is incoherent the relative phases of the different sources are washed out and the resultant intensity just goes as the sum

$$I \sim I_1 + I_2. \tag{8.12}$$

That is, the interference term $2\vec{A}_1 \cdot \vec{A}_2$ vanishes so that there are no minima or secondary maxima in the intensity. Applying the principle of superposition to the experiment of Figure 8–9, we write

$$|\Psi(x, t)|^2 = |\Psi_A(x, t) \pm \Psi_B(x, t)|^2 = |\Psi_A|^2 + |\Psi_B|^2 \pm |\Psi_A^*\Psi_B| \pm |\Psi_B^*\Psi_A|. \tag{8.13}$$

Here, the last two terms are the interference terms, which depend upon the relative *phases* of the two waves. The phase factors in the wave functions Ψ_A and Ψ_B play a role analogous to that of vector addition in the above example from optics, thus indicating the importance of choosing complex wave functions.

By way of a summary, we have seen that the experimental confirmation of de Broglie's hypothesis led to the representation of an electron (or any other atomic particle) by a wave function. Any well-behaved complex function can be considered as a possible wave function. "Good behavior" requires that the function be finite, continuous, and single-valued throughout all of space in order that it may be normalized as discussed above. The latter requirement is necessary in order to insure that the square of the wave function (the modulus squared) is indeed the probability for finding the particle in a given locality. We also accept the validity of the superposition principle. That is, we can study the interference of two or more wave functions by taking into account the instantaneous amplitude and phase of each when finding their combined effect at a point in space. The superposition principle also allows us to combine two or more wave functions to obtain another function. It can be shown that such a *sum* of well-behaved functions is itself a well-behaved function. Thus, the superposition principle provides us with a very important idea, namely, that a more general and useful representation of a particle may be obtained by the combination of a large number of particular wave functions. We will now use this idea to develop the concept of the wave packet.

4. REPRESENTING PARTICLES BY WAVE PACKETS

In Section 4 of Chapter 6 we learned how the concept of the wave packet resolved the dilemma of the wave-particle duality in the case of photons. It is certainly reasonable to extend this concept to the atomic and elementary particles, which also manifest both wave and particle properties. Experiments that depend upon the particle-like properties can be regarded as observations of the behavior of the wave packet as a whole, whereas experiments that depend upon interference and diffraction are observations of individual wavelength components of the packet. Instead of using just two wave functions that differ slightly in phase, as was done in deriving Equation 6.15, we will sum an infinite number of waves having infinitesimal phase differences. These waves are so closely spaced that we can assume that their frequencies and k-values form a continuum, so that the summation can be replaced by an integration. The integration is accomplished by means of the Fourier integral theorem, which is discussed in Appendix C. This has the advantage of producing a truly localized packet having zero amplitude except in one region of space, as shown schematically in Figure 8–11. By way of contrast, recall that any finite sum of waves whose frequencies differ by discrete amounts can produce a modulation envelope that resembles a packet (see Figure 6–6), but that this wave form repeats itself regularly throughout space.

In order to avoid unnecessary mathematical complications, we will consider just two elementary but important wave packets. The essential physical ideas can be extracted from the following examples without a detailed understanding of the Fourier integral theorem.

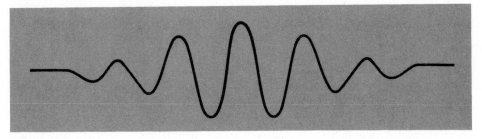

Figure 8–11

A hypothetical wave packet that is localized in space.

Example 8-5

The output of an oscillator that produces a pure sine wave of frequency ω_0 is switched on and off (that is, keyed or chopped) such that it is "on" for $2T$ seconds. What is the distribution of frequencies in the output pulses?

Solution

If we take the zero for time to be at the center of the pulse, then we can express the harmonic behavior in time as

$$f(t) = e^{i\omega_0 t}, \text{ for } -T \le t \le T,$$

$$f(t) = 0, \text{ for } |t| > T.$$

The chopping (or keying) process will introduce many new frequencies in varying amounts, given by the Fourier transform, Equation C.5 in Appendix C,

$$g(\omega) = \frac{1}{\sqrt{2\pi}} \int_{-\infty}^{\infty} dt\, f(t) e^{-i\omega t}$$

$$= \frac{1}{\sqrt{2\pi}} \int_{-T}^{T} e^{i(\omega_0 - \omega)t}\, dt$$

$$= \sqrt{\frac{2}{\pi}} \cdot T \cdot \frac{\sin(\omega_0 - \omega)T}{(\omega_0 - \omega)T}. \tag{8.14}$$

This function, which is called the *spectral distribution function*, is plotted in Figure 8–12(a). The continuous sine wave has only one frequency, ω_0, but the keying process

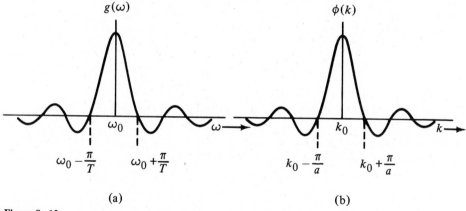

(a)   (b)

Figure 8–12

The spectral distribution resulting from chopping a pure, infinite sine wave in the time domain (a) and in the spatial domain (b).

results in the mixing of an infinite number of frequencies. The principal contribution is still at the frequency ω_0. Nature provides an example of this type of chopping in the emission of photons during electronic transitions in atoms.

The spatial counterpart of the above example would be the chopping of a plane wave of light having wave vector $k_0 = \omega_0/c$ by means of a shutter, such that the extent of the packet in the x-direction is $2a$. Here $2a = 2cT$, where c is the speed of light and $2T$ is the time interval that the shutter is open. We can represent the plane wave mathematically by the function

$$\psi(x) = e^{ik_0x} \text{ for } -a \leq x \leq a,$$

$$\psi(x) = 0, \text{ for } |x| > a.$$

Then, from Equation C.6 of Appendix C,

$$\phi(k) = \frac{1}{\sqrt{2\pi}} \int_{-\infty}^{\infty} \psi(x)e^{-ikx} \, dx$$

$$= \frac{1}{\sqrt{2\pi}} \int_{-a}^{a} e^{i(k_0-k)x} \, dx$$

$$= \sqrt{\frac{2}{\pi}} \cdot a \cdot \frac{\sin(k_0 - k)a}{(k_0 - k)a}. \tag{8.15}$$

This function is shown in Figure 8–12(b). Note that the spatial chopping of the plane wave introduces a continuum of k-values centered around k_0. The resemblance to the optical diffraction pattern produced by a single slit is not coincidental. In diffraction we simply regard k_0 as a constant and allow its *direction* to take on a range of values. This is equivalent to treating the slit as a source of Huygen's wavelets.

Notice that the central peak in Figure 8–12(a) has a half width given by

$$\Delta\omega = 2\pi\Delta\nu = \frac{\pi}{T},$$

from which we see that

$$2T \cdot \Delta\nu = 1. \tag{8.16}$$

Multiplying by h and replacing T by $\Delta t/2$, we have the statement that the minimum uncertainty product for this type of packet is

$$\Delta t \cdot \Delta E = h. \tag{8.17}$$

Thus we may express Heisenberg's uncertainty principle for simultaneous measurements of energy and time for such a packet as

$$\Delta t \cdot \Delta E \geq h. \tag{8.18}$$

Similarly, the central peak of Figure 8–12(b) has a half width of

$$\Delta k = \frac{\Delta p_x}{\hbar} = \frac{\pi}{a},$$

from which we obtain the result

$$2a \cdot \Delta p_x = h.$$

Since $\Delta x = 2a$, in this case we may write the uncertainty principle for simultaneous measurements of position and momentum for the packet as

$$\Delta x \cdot \Delta p_x \geq h. \tag{8.19}$$

For the packet shape shown in Figure 8–11, the minimum value of the uncertainty product is h at time $t = 0$. However, as time progresses, the packet will broaden in coordinate space (x-space) so that, for a given value of Δp, the uncertainty product increases with time. It can be shown that the more massive the particle, the more slowly its packet spreads with time. This is just what we want in order to satisfy the correspondence principle enunciated by Bohr. Thus, for an electron that is localized to 1 Å the wave packet retains its form for only about 10^{-16} second, whereas for a .22 calibre bullet localized to 0.1 mm, the packet would remain unchanged for about 10^{24} seconds. It is evident that the concept of a classical trajectory, though meaningless for the electron, is quite appropriate for a macroscopic particle.

Example 8-6

Now consider a packet whose spectrum of k-values is the Gaussian function (see Figure 8-13),

$$\phi(k) = Ae^{-k^2/2\sigma^2},$$

where σ is the standard deviation of the distribution. That is, σ is a measure of the spread of the packet in k-space. Find the shape of the packet in x-space.

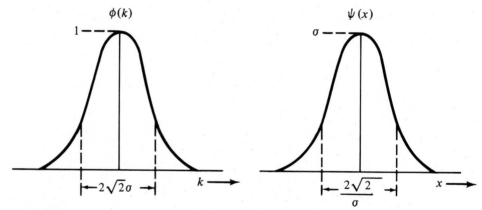

Figure 8-13

Gaussian packets. Each packet is the Fourier transform of the other. The breadth of the packet is arbitrarily taken to be its standard deviation, as shown.

Solution

We use the Fourier transform given by Equation C.6 to write

$$\psi(x) = \frac{A}{\sqrt{2\pi}} \int_{-\infty}^{\infty} dk\, e^{-k^2/2\sigma^2}\, e^{ikx}.$$

This may be readily integrated by first completing the square in the exponential. In order to do this we add and subtract the quantity $x^2\sigma^2/2$ as follows:

$$\psi(x) = \frac{A}{\sqrt{2\pi}} \int_{-\infty}^{\infty} dk\, \exp\left[-\frac{k^2}{2\sigma^2} + ikx + \frac{x^2\sigma^2}{2} - \frac{x^2\sigma^2}{2}\right]$$

$$= \frac{A}{\sqrt{2\pi}} e^{-x^2\sigma^2/2} \int_{-\infty}^{\infty} dk\, \exp\left[-\frac{1}{2\sigma^2}(k^2 - 2ikx\sigma^2 - x^2\sigma^4)\right]$$

$$= \frac{A}{\sqrt{2\pi}} e^{-x^2\sigma^2/2} \int_{-\infty}^{\infty} dk\, \exp\left[-\frac{1}{2\sigma^2}(k - ix\sigma^2)^2\right]$$

$$= \frac{\sqrt{2}A\sigma}{\sqrt{2\pi}} e^{-x^2\sigma^2/2} \int_{-\infty}^{\infty} e^{-u^2} \, du$$

$$= A\sigma e^{-x^2\sigma^2/2}. \tag{8.20}$$

We note that Equation 8.20 is also a Gaussian curve, and that the spread of the packet in x-space is given by $1/\sigma$, the reciprocal of the spread in k-space. This illustrates the reciprocal nature of Δk and Δx as required by the uncertainty principle.

Before we conclude our brief introduction to wave packets, it should be pointed out that in the classical limit the particle velocity is associated with the group velocity of the packet. This is another example of the *correspondence principle*, which states that quantum mechanics must give the same result as classical mechanics in the appropriate classical limit. For particles, the classical limit occurs for very short de Broglie wavelengths.

The group velocity for a wave packet was given in Equation 6.17 as

$$v_{\text{group}} = \frac{d\omega}{dk}.$$

Using the de Broglie relations we can write

$$v_{\text{group}} = \frac{d(\hbar\omega)}{d(\hbar k)} = \frac{dE}{dp}. \tag{8.21}$$

Since the energy of a free particle in the classical limit is simply its kinetic energy, $p^2/2m$, we find that

$$v_{\text{group}} = \frac{dE}{dp} = \frac{d}{dp}\left(\frac{p^2}{2m}\right) = \frac{p}{m} = v, \tag{8.22}$$

where v is the particle velocity.

The speed of any particular component wave is called its *phase velocity*,

$$u = \lambda\nu = \frac{\omega}{k}. \tag{8.23}$$

Phase velocities may be greater than or less than the group velocity. That is, a component wave crest might appear to be running off the forward end of the packet or receding off the rear end. A crude but instructive way to visualize this is to imagine a spinning barber pole moving through space. The speed of the barber pole would be the analog of the group velocity, while the apparent speed of the stripe could be likened to the phase velocity. Although the stripe might appear to be moving faster than the pole, all of the energy is carried by the pole. In wave packets representing photons or high-speed particles, it can happen that phase velocities exceed the speed of light. However, since no energy (that is, no signal or information) is transmitted at the phase velocity, the fact that u is greater than c constitutes no violation of the postulates of special relativity.

SUMMARY

De Broglie's postulate that a particle has a wavelength was later followed by the experimental observation of particle diffraction.

We do not observe the wave properties of massive objects in our everyday world because their wavelengths are too short to be detected. If Planck's constant h were a much larger number, then such activities as ball games, billiards, and target shooting would be characterized by diffraction effects rather than by the precise trajectories with which we are familiar.

In order to account for the wave nature of a particle, we define a wave function to represent it mathematically. The square of the absolute value of the wave function is interpreted as the relative probability for finding the particle at a given point in space and time. It turns out that we can define physical measurements that correspond to the *square* of the wave function, although the wave amplitude itself is *not* an observable quantity.

The concept of the wave packet resolves the classical wave-particle dualism through the de Broglie relations, which connect the energy and momentum of the particle with the frequency and propagation constant of the wave. One of the fundamental assumptions of quantum mechanics is that a wave packet composed of a superposition of complex wave functions can be used to represent a partially localized particle. As required by the uncertainty principle, the more narrowly localized the packet in coordinate space, the greater the number of k-values required to form the packet. The Fourier integral theorem provides the mathematical tool necessary for constructing packets. A completely unlocalized particle may be represented by a single plane wave having one frequency and one k-value; that is to say, the Fourier transform of an infinite plane wave is a delta function in k-space.

Finally, we note that the group velocity of a packet corresponds to the particle velocity, whereas the phase velocity of the packet has no physical significance.

Additional Reading

Louis de Broglie, *Matter and Light: The New Physics,* Dover Publications, Inc., New York, 1939.

K. W. Ford, *Classical and Modern Physics,* Vol. 3, Xerox College Publishing, Lexington, Mass., Chapter 23.

G. Gamow, *Mister Tomkins in Paperback,* Cambridge University Press, New York, 1967.

L. H. Germer, "The Structure of Crystal Surfaces," *Sci. Amer. 212* (3), 32 (March 1965)

P. Tipler, *Modern Physics,* Worth Publishers, Inc., New York, 1978, Chapter 5.

R. T. Weidner and R. L. Sells, *Elementary Modern Physics,* 3rd ed., Allyn & Bacon, Inc., Boston, 1980, Chapter 5.

E. H. Wichmann, *Quantum Physics,* Berkeley Physics Course, Vol. 4, McGraw-Hill Book Co., New York, 1971, Chapter 5.

PROBLEMS

8-1. At what kinetic energy will an electron have a de Broglie wavelength equal to the diameter of the hydrogen atom? How does this energy compare with the ground state energy of hydrogen?

8-2. What is the de Broglie wavelength of a 30 keV electron?

8-3. Calculate the de Broglie wavelength of a 1 MeV alpha particle.

8-4. Find the de Broglie wavelength of a neutron having a kinetic energy of 1/40 eV. Such a neutron is called a "thermal neutron" because it is in thermal equilibrium with gas molecules at room temperature.

8-5. A beam of 25-volt electrons is scattered by silver, which has an atomic spacing of 4.08 Å. At what angle is the first-order Bragg peak?

8-6. A beam of 0.025 eV neutrons is incident upon a crystal whose lattice constant is 3.14 Å. At what angle is the first-order Bragg peak?

8-7. A 0.082 eV neutron beam is directed perpendicular to a face of a cubic crystal, and a strong Bragg reflection peak occurs at an angular separation of 44° between the incident and reflected beams. (a) What is the atomic spacing (lattice constant) of the crystal? (b) What is the distance between adjacent reflecting planes causing the peak?

8-8. A crystal having an interatomic spacing of 1.25 Å is being used to select only those gamma rays having an energy of 50 keV from an incident beam containing a distribution of energies. For normal incidence (as in Figure 8–3), at what angle would the desired beam occur?

8-9. In order for an electron to be confined to a nucleus, its de Broglie wavelength would need to be less than about 1×10^{-14} m. What kinetic energy would such an electron have?

8-10. The theoretical maximum resolving power of a microscope is limited to the wavelength of the "light." In order to observe an object 5 Å in diameter, what is the minimum energy that photons must have? What is the minimum energy required if electrons are used?

8-11. Show that the ratio of the Compton wavelength to the de Broglie wavelength for a relativistic electron is given by

$$\frac{\lambda_c}{\lambda} = \beta\gamma = \left[\left(\frac{E}{m_0 c^2} \right)^2 - 1 \right]^{1/2}.$$

8-12. Show that the wave packet for a particle having relativistic momentum $p = mv$ has a group velocity given by βc and a phase velocity given by $u = c^2/v$.

8-13. Prove that the probability defined by Equation 8.9 or 8.10 must always be a real quantity even though the wave function is complex.

8-14. Perform a thought experiment to distinguish between adding the amplitudes or adding the intensities from two sources, as in Equations 8.11 and 8.12. Why are such effects not observed from two lights in the classroom?

*8-15. What is the de Broglie wavelength of an oxygen molecule having the rms speed characteristic of the Maxwell distribution at 27°C?

8-16. Find the phase velocities of 1 MeV protons and 1 MeV electrons.

8-17. A system is to be designed to handle up to 1 million electric pulses per second. Use the uncertainty principle to estimate the frequency bandwidth that will be required in the receiving equipment.

8-18. What must be the frequency bandwidth of the detecting and amplifying stages of a radar system operating at pulse widths of 0.1 μsec? If the radar is used for ranging (distance measurements), what is the uncertainty in the range?

8-19. Use the uncertainty principle to obtain the uncertainty in the momentum of a particle of mass m constrained to the volume of a cubical box of side a. What is the uncertainty in the kinetic energy of the particle?

8-20. For a free particle, show that the wave packet in coordinate space broadens with time, corresponding to the increasing uncertainty in the position of the particle.

8-21. What are the phase and group velocities of an electron whose de Broglie wavelength is 0.01 Å? What is the kinetic energy of the electron?

8-22. A particle is defined in the space $-\infty < x < \infty$ by the wave function

$$\Psi(x, t) = Ae^{-x^2}e^{i(kx - \omega t)}.$$

(a) Find the normalization constant A. (b) What is the probability of finding the particle in the interval bounded by x and $x + dx$ at time t? (c) What is the total probability of finding the particle somewhere between $-\infty$ and $+\infty$?

8-23. A particle may be represented in the space $-a \leq x \leq a$ by either of the wave functions (a) $\Psi(x) = A \cos \pi x/2a$ or (b) $\Psi(x) = B \sin \pi x/a$. Find the normalization constants A and B.

*8-24. A particle may be described in the space $-a \leq x \leq a$ by a superposition of the two wave functions of the preceding problem, namely,

$$\Psi(x) = A \cos \frac{\pi x}{2a} + B \sin \frac{\pi x}{a}.$$

(a) Normalize this new function. (b) Sketch the probability density as a function of x.

*8-25. (a) Find the normalization constant N for the Gaussian wave packet

$$\psi(x) = Ne^{-(x-x_0)^2/2K^2}.$$

(b) Obtain the Fourier transform of $\psi(x)$ and verify that it is normalized.

9 The Schrödinger Wave Equation

1. CLASSICAL WAVES

We know from our previous experience in physics that mechanical waves will propagate in elastic media, and electromagnetic waves can be transmitted vast distances through free space. The periodic variations in the amplitude of the wave in both space and time can be predicted easily if one knows the initial conditions of the wave and the *wave equation* that it must satisfy. Thus, if Ψ is the amplitude of a wave being propagated in the x-direction with speed v, then the wave equation,

$$\frac{\partial^2 \Psi}{\partial x^2} = \frac{1}{v^2} \frac{\partial^2 \Psi}{\partial t^2}, \tag{9.1}$$

describes the wave for all x and t. Partial derivatives are used here because Ψ is a function of both x and t. For a wave traveling along a string or wire, Ψ is the transverse displacement from the equilibrium position and v depends upon the tension of the string and its mass per unit length. For a sound wave, Ψ is the compression or rarefaction and v is calculated from the density and the bulk elastic modulus. For an electromagnetic wave, Ψ may be either the electric or magnetic field intensity and v is then the speed of light.

Equation 9.1 is easily generalized to more than one dimension by writing it in the form

$$\frac{\partial^2 \Psi}{\partial x^2} + \frac{\partial^2 \Psi}{\partial y^2} + \frac{\partial^2 \Psi}{\partial z^2} = \frac{1}{v^2} \frac{\partial^2 \Psi}{\partial t^2}. \tag{9.2}$$

This expression, in turn, may be simplified by defining the Laplacian operator,

$$\nabla^2 = \frac{\partial^2}{\partial x^2} + \frac{\partial^2}{\partial y^2} + \frac{\partial^2}{\partial z^2}. \tag{9.3}$$

The wave equation then becomes

$$\nabla^2 \Psi = \frac{1}{v^2} \frac{\partial^2 \Psi}{\partial t^2}, \tag{9.4}$$

for any number of dimensions.

Perhaps the most important class of waves in physics is that in which the periodic variations of the source may be described by simple harmonic motion. It is therefore customary to assume that the functional dependence on both the time and space variables is sinusoidal. There are a number of good reasons for this, among them the following. (1) For small displacements, many physical problems involving periodic vibrations may be approximated by simple harmonic oscillators (e.g., the simple pendulum). (2) The solutions of the harmonic oscillator equation are the sine and cosine functions, which are relatively easy to handle mathematically. (3) More complex periodic motions can, by Fourier analysis, be reduced to a set of harmonic motions. Conversely, almost any complex periodic wave may be constructed by the *superposition* of an appropriate number of sine or cosine functions having different frequencies and amplitudes.

The *principle of superposition* simply says that two or more waves traveling in the same medium act independently. Hence, the displacement of a particle at a given instant in time is the vector sum of the amplitudes of all of the waves acting on that particle. Our everyday experience with light and sound confirms the principle of superposition, except for extreme situations that produce shock waves or turbulence. In mathematical terms we say that superposition is valid if the wave equation is linear. Squaring any term in Equation 9.1 would therefore invalidate the superposition principle.

Waves may be classified not only according to their functional dependence on space and time, but also according to the type of wavefront that exists during propagation. A wavefront is the surface formed by joining all of the comparable disturbances having the same phase at that instant. Thus, a pulsating sphere would produce a *spherical* wavefront, since the locus of successive sets of crests and troughs would be a set of concentric spheres. Similarly, a group of parallel light rays having the same initial phase is said to be a *plane wave*, since the wavefront is planar.

The frequency and wavelength are simply related to the phase velocity of a wave by the expression

$$v = \lambda \nu = \frac{\omega}{k}, \tag{9.5}$$

where $\omega = 2\pi\nu$ and $k = 2\pi/\lambda$. Here ω is known as the angular frequency and k is called the propagation constant.

A harmonic plane wave traveling in the positive x-direction may be represented mathematically by either

or
$$\left.\begin{array}{l} \Psi = A\cos(kx - \omega t) \\ \Psi = A\sin(kx - \omega t). \end{array}\right\} \tag{9.6}$$

However, it is more convenient to use the exponential form,

$$\Psi = Ae^{i(kx-\omega t)}, \tag{9.7}$$

which is equivalent to

$$\Psi = A\cos(kx - \omega t) + iA\sin(kx - \omega t). \tag{9.8}$$

Note that either the real part *or* the imaginary part of Equation 9.8 should be used to represent a wave since Equation 9.8 corresponds to both functions in Equations 9.6.

2. THE SCHRÖDINGER WAVE EQUATION

Now consider a one-dimensional system in which a particle of mass m and momentum p is subject to an interaction that can be represented by a potential V, which is assumed to be independent of time. We now postulate that the wave function $\Psi(x, t)$ that represents this particle must satisfy the following *wave equation:*

$$-\frac{\hbar^2}{2m}\frac{\partial^2 \Psi}{\partial x^2} + V\Psi = i\hbar\frac{\partial \Psi}{\partial t}. \tag{9.9}$$

As in the case of Equation 9.4, the operator ∇^2 may be used to express Equation 9.9 in the form

$$-\frac{\hbar^2}{2m}\nabla^2\Psi + V\Psi = i\hbar\frac{\partial \Psi}{\partial t}, \tag{9.10}$$

which is valid for any number of dimensions. Either Equation 9.9 or 9.10 is called the *Schrödinger wave equation.*

Comparing Equations 9.4 and 9.10 (or 9.1 and 9.9), note that both wave equations require second derivatives with respect to the spatial variables. However, the classical equation requires the second derivative with respect to time, whereas the Schrödinger equation contains only the first time derivative. Other differences are: (1) the classical equation explicitly contains the wave velocity; (2) the Schrödinger equation includes the interaction potential; (3) the classical equation is real, whereas the right side of the Schrödinger equation is imaginary.

In order to get a feeling for the plausibility of the Schrödinger equation, consider a particle whose wave function is the plane wave

$$\Psi = Ae^{i(kx-\omega t)},$$

in a region where the potential has the constant value V_0 (that is, the force is zero). Differentiating Ψ with respect to x and t, we obtain:

$$\frac{\partial \Psi}{\partial x} = ik\Psi \qquad \frac{\partial \Psi}{\partial t} = -i\omega\Psi \qquad \frac{\partial^2 \Psi}{\partial x^2} = -k^2\Psi$$

Substituting these results into Equation 9.9:

$$-\frac{\hbar^2}{2m}(-k^2\Psi) + V_0\Psi = i\hbar(-i\omega\Psi)$$

$$\frac{\hbar^2 k^2}{2m}\Psi + V_0\Psi = \hbar\omega\Psi$$

or

$$\frac{\hbar^2 k^2}{2m} + V_0 = \hbar\omega. \tag{9.11}$$

From the de Broglie relations, recall that

$$\hbar k = \frac{h}{2\pi}\cdot\frac{2\pi}{\lambda} = \frac{h}{\lambda} = p$$

and

$$\hbar\omega = \frac{h}{2\pi}\cdot 2\pi\nu = h\nu = E,$$

so that Equation 9.11 may be written as

$$\frac{p^2}{2m} + V_0 = E. \tag{9.12}$$

Since the first term is simply the kinetic energy of the particle, it is evident that Equation 9.12 is a reaffirmation of the conservation of energy. This is in no sense a derivation or a proof of the Schrödinger equation. It does show, however, that the Schrödinger equation is consistent with the de Broglie relations and energy conservation for harmonic wave functions.

*3. SEPARATION OF THE SPACE AND TIME VARIABLES

The plane wave given by Equation 9.7 could just as easily be written as the product of a function of x only times a function of t only. Thus,

$$\Psi(x, t) = Ae^{i(kx-\omega t)} = Ae^{ikx} \cdot e^{-i\omega t}.$$

Whenever the wave function can be written in such a product form, and provided that the potential V is *not* a function of time, the Schrödinger equation can be separated into two differential equations. In order to see this, let us define a general wave function by

$$\Psi(x, t) = \psi(x) \cdot f(t). \tag{9.13}$$

Substituting this into Equation 9.9, we obtain:

$$-\frac{\hbar^2}{2m} \frac{\partial^2 \psi}{\partial x^2} \cdot f(t) + V(x)\psi(x) \cdot f(t) = i\hbar \, \psi(x) \cdot \frac{\partial f}{\partial t}.$$

Dividing each term by $\psi(x) \cdot f(t)$,

$$-\frac{\hbar^2}{2m} \frac{\frac{\partial^2 \psi}{\partial x^2}}{\psi(x)} + V(x) = i\hbar \frac{\frac{\partial f}{\partial t}}{f(t)}.$$

The left-hand side is a function of x only, and the right-hand side is a function of t only. That is, the differential equation is *separated* and each side of the equation may be set equal to the same constant, called the *separation constant*. If we denote the separation constant by E, we then obtain two ordinary differential equations:

$$\frac{df}{dt} = -\frac{iE}{\hbar} f(t) \tag{9.14}$$

and

$$-\frac{\hbar^2}{2m} \frac{d^2 \psi}{dx^2} + V(x) \cdot \psi(x) = E \, \psi(x). \tag{9.15}$$

We chose the symbol E for the separation constant because it does turn out to be the total energy of the system. Equation 9.15 is the statement of the conservation of energy as shown in Equation 9.12.

Equation 9.14 may be integrated at once to obtain

$$f(t) = Ce^{-i\omega t}. \tag{9.16}$$

Equation 9.15 cannot be solved so readily. First, let us rewrite it as follows:

$$\left(-\frac{\hbar^2}{2m}\frac{d^2}{dx^2} + V\right)\psi = E\psi \tag{9.17}$$

or

$$Q\psi = E\psi, \tag{9.18}$$

where we have defined the operator

$$Q = -\frac{\hbar^2}{2m}\frac{d^2}{dx^2} + V. \tag{9.19}$$

We understand Equation 9.18 to mean that when Q operates on the wave function ψ, the result is simply the same wave function multiplied by the energy of the system (a scalar). Equation 9.18 is known as the *energy eigenequation*. Such equations arise frequently in physics, and their properties will be briefly discussed in the next section.

4. THE EIGENVALUE PROBLEM

Let Q represent an operator that performs a mathematical operation on a function f. This is indicated by the product Qf. Suppose further that the operation Qf results in reproducing the *same function* f multiplied by a constant. That is,

$$Qf = \pm cf, \tag{9.20}$$

where c is a positive real constant. When this occurs, f is called an *eigenfunction* of the operator Q, and $\pm c$ is called the *eigenvalue* associated with f. For instance, for the function $f = \sin 2x$ and the operator $Q = d^2/dx^2$, the eigenequation is

$$Qf = -4 \sin 2x = -4f.$$

For this operator, the eigenfunction $f = \sin 2x$ has the eigenvalue -4.

Example 9-1

Which of the following are eigenfunctions of the operator d^2/dx^2? Give the eigenvalue when appropriate.

(a) $\sin x$	(d) e^{2x}
(b) $\cos x$	(e) e^{ix}
(c) $\sin^2 x$	(f) $\sin nx$

Solution

(a) $Qf = \dfrac{d^2}{dx^2} \sin x = -\sin x = -f$; the eigenvalue is -1.

(b) $Qf = \dfrac{d^2}{dx^2} \cos x = -\cos x = -f$; the eigenvalue is -1.

(c) $Qf = \dfrac{d^2}{dx^2} \sin^2 x = 2 - 4\sin^2 x$. Not an eigenfunction.

(d) $Qf = \dfrac{d^2}{dx^2} e^{2x} = 4f$; the eigenvalue is $+4$.

(e) $Qf = \dfrac{d^2}{dx^2} e^{ix} = -f$; the eigenvalue is -1.

(f) $Qf = \dfrac{d^2}{dx^2} \sin nx = -n^2 f$; the eigenvalue is $-n^2$.

Not all functions are suitable as eigenfunctions, as the above example shows. For some functions there is no operator, (other than the identity operator) such that an eigenvalue exists.

The eigenvalue problem, Equation 9.20, is one of the most important problems in mathematical physics. One surprising feature of the problem is that we usually want to find *both the function f and the eigenvalue c when only the operator is known.* This is made possible by the combination of the properties of the operator and the constraints imposed by the boundary conditions in the problem. By way of illustration, the determination of the standing waves on a string is an eigenvalue problem. Here the eigenvalue is related to the number of antinodes (loops) of the standing wave pattern and it is determined by the length of the string as the following example shows.

Example 9-2
Find the standing wave solutions for a string of length L that is fastened at both ends.

Solution
If we assume that the amplitude function can be written as a product function, as in Equation 9.13, and that the time-dependent part has the form of Equation 9.16, then

$$\Psi(x, t) = f(x) \cdot e^{-i\omega t}. \tag{9.21}$$

Let us now substitute this function into the wave equation for classical waves, Equation 9.1. First, we obtain the derivatives:

$$\frac{\partial^2 \Psi}{\partial x^2} = e^{-i\omega t} \frac{d^2 f}{dx^2}$$

$$\frac{\partial^2 \Psi}{\partial t^2} = -\omega^2 f e^{-i\omega t}$$

Then, we substitute into Equation 9.1,

$$\frac{d^2 f}{dx^2} = -\frac{\omega^2}{v^2} f = -k^2 f, \tag{9.22}$$

where the wave relationship of Equation 9.5 has been used. Equation 9.22 is an eigenequation having eigenvalues $-k^2$. Let us now see what restrictions, if any, are imposed upon the values of k by the boundary conditions.

Equation 9.22 should be recognized as the classical harmonic oscillator equation. It has solutions of the form

$$f(x) = A \sin kx + B \cos kx.$$

Since the string is clamped at its ends, $f = 0$ for $x = 0$ and for $x = L$. That is,

$$0 = A \sin (0) + B \cos (0) \tag{9.23}$$

$$0 = A \sin kL + B \cos kL. \tag{9.24}$$

From Equation 9.23 we see that $B = 0$, and from Equation 9.24 we obtain the condition that

$$kL = n\pi$$

or

$$k = \frac{n\pi}{L}. \tag{9.25}$$

Thus the eigenvalues are

$$-\left(\frac{\pi}{L}\right)^2, \ -\left(\frac{2\pi}{L}\right)^2, \ -\left(\frac{3\pi}{L}\right)^2, \ \ldots\ldots, \ -\left(\frac{n\pi}{L}\right)^2,$$

and the eigenfunctions are

$$f_n(x) = A_n \sin \frac{n\pi x}{L}. \tag{9.26}$$

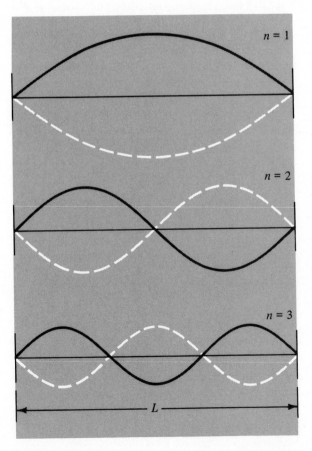

n = 1

n = 2

n = 3

L

Figure 9–1

The standing waves for $n = 1$, 2, and 3 for a string of length L fixed at both ends. The nodes (points of zero amplitude) remain fixed. The antinodes (loops) vary from one extreme position to the other sinusoidally in time.

In the classical example just discussed, the wave amplitude and its variation in time can be observed directly. In contrast to this, the wave function in quantum mechanics is not a physical observable. However, we are aided in selecting an appropriate solution of the eigenequation through the requirement that a wave function must be well-behaved. As you will recall from Section 4 of Chapter 8, this means that the eigenfunction must be finite, continuous, and single-valued throughout all of space. The next example will illustrate how the concept of "good behavior" is used.

Example 9-3

Examine the solutions of $Qf = \pm cf$, where $Q = \dfrac{d^2}{dx^2}$, and find the eigenfunctions for the space $-\infty < x < \infty$. (c is a positive constant.)

Solution

First, consider the plus sign. The solution of

$$\frac{d^2f}{dx^2} = cf$$

is

$$f = Ae^{\sqrt{c}\cdot x} + Be^{-\sqrt{c}\cdot x}.$$

This function diverges as $x \to \pm\infty$, so it is not a suitable eigenfunction. However, if A were zero, the function

$$f = Be^{-\sqrt{c}\cdot x}$$

would be well behaved in the positive half-space ($x > 0$). For $B = 0$, the function

$$f = Ae^{\sqrt{c}\cdot x}$$

would be well behaved in the negative half-space ($x < 0$).

To obtain well-behaved eigenfunctions for all of space, we must consider eigenvalues $-c$. Then,

$$\frac{d^2f}{dx^2} = -cf$$

and the solution of this equation is

$$f = D \sin \sqrt{c} \cdot x + E \cos \sqrt{c} \cdot x.$$

This function is well-behaved for all x, and an eigenfunction of Q exists for *all* eigenvalues $-c$. For this situation we obtain a continuum of possible eigenvalues, whereas in many physical problems the eigenvalues form a discrete set of numbers.

Example 9-4

Suppose that the "space" of Example 9-3 is reduced to $-L < x < L$ instead of $-\infty < x < \infty$. That is, the boundary conditions are such that $f = 0$ for $x \le -L$ and for $x \ge L$. What are now the allowed eigenfunctions and eigenvalues?

Solution

Since $f = D \sin \sqrt{c} \cdot x + E \cos \sqrt{c} \cdot x$ must be zero for $x = \pm L$, it can no longer be written as a sum of the sine and cosine of the same angle. Thus, either D or E must now be zero. Letting $E = 0$, we have at $x = L$,

$$0 = D \sin \sqrt{c} \, L.$$

For non-zero D this is satisfied only under the condition that

$$\sqrt{c}\, L = n\pi,$$

where n is an integer. Hence, the continuum of eigenvalues in Example 9-3 is now reduced to a discrete set of values given by

$$c_n = \left(\frac{n\pi}{L}\right)^2,$$

and a discrete set of wave functions given by

$$f_n = D \sin \frac{n\pi x}{L}. \tag{9.27}$$

It should be noted that this result is identical to that obtained for the classical case in Example 9-2.

5. ENERGY STATES IN QUANTUM SYSTEMS

We are now ready to apply what we have learned to simple quantum systems. Given a particle of mass m and the forces acting upon it, one question of primary importance is, "What are the possible energy states of the particle?" If the particle is an elementary particle such as an electron, its wave function must first satisfy the Schrödinger equation. We have seen, however, that if steady-state solutions exist then we need not use the entire Schrödinger equation, which includes the time derivative; instead, we can focus our attention on the separated energy eigenequation, Equation 9.17 of Section 3. Hence, in the one-dimensional case when no forces act upon the particle, the eigenequation becomes

$$\frac{d^2\psi}{dx^2} - \frac{2mE}{\hbar^2}\,\psi = 0.$$

In Example 9-3 it was shown that the solutions of this are the plane waves

$$\psi = A e^{ikx},$$

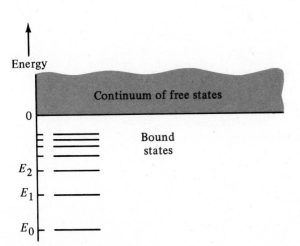

Figure 9–2

Energies of a hypothetical quantum system. The lowest energy state in a bound-particle system is called the ground state.

where $k^2 = 2mE/\hbar^2$. It is especially important to note that *all* positive values for the energy E are allowed. That is, for a free particle the allowed energies form a continuum of values, as contrasted with the discrete energy levels of bound systems. This illustrates a very important property of quantum systems, namely, that *the energy states for a bound particle are quantized*. The reader should recall that discrete energy levels were obtained in Section 3 of Chapter 7 for the allowed states of an electron bound to a proton in the Bohr theory of the hydrogen atom.

6. EXPECTATION VALUES

We have discovered that the uncertainty principle does not permit us to measure trajectories, positions, velocities, and so forth in the world of the fundamental particles as is customary in classical mechanics. Instead, we must represent an elementary particle by a wave function, where it is understood that the wave function contains all of the information that can be known about the particle. Even then, this information is available to us only in the form of *probabilities*. How do we obtain this information, and how is it related to physical measurements on the system?

In Equation 8.9 we defined the probability that a particle in one dimension would be found within the element dx as $|\psi|^2 dx$. It follows that the *average value* of x would be

$$\bar{x} = \frac{\int_{-\infty}^{\infty} x\,|\psi|^2 dx}{\int_{-\infty}^{\infty} |\psi|^2 dx}, \qquad (9.28)$$

where the probability is the weighting factor for each value of x. If the wave function ψ is normalized (as described in Section 3 of Chapter 8), the denominator of Equation 9.28 is simply unity and may be dropped. In a classical problem Equation 9.28 would be correct as it is; in quantum mechanics, however, the x in the integrand must be regarded as an operator, and the numerator is written according to the following convention:

$$\int_{-\infty}^{\infty} \psi^* x_{op}\psi dx.$$

Thus the average value of x in quantum mechanics, which is called the *expectation value*, is

$$\langle x \rangle = \frac{\int_{-\infty}^{\infty} \psi^* x_{op}\,\psi dx}{\int_{-\infty}^{\infty} |\psi|^2 dx}. \qquad (9.29)$$

No prediction can be made about what value will be obtained from a single measurement. Furthermore, we must anticipate that successive measurements of x will in general give different values. However, the average of a large number of measurements should be a value very close to the expectation value given by Equation 9.29.

In quantum mechanics, every physical observable can be represented by an appropriate mathematical operator which can, in turn, be used to predict the results of measurements on the system. For the position variable x there is a position operator x_{op}, the momentum p has its operator p_{op}, and so on. Some of the operators are quite simple, while others are differential operators. A few common operators for wave functions in coordinate space are given in Table 9-1.

The importance of these concepts will become clearer when they are applied to examples in later chapters.

Table 9–1
Operators for $\psi(x,t)$

Variable	Operator
x	x
x^2	x^2
p_x	$-i\hbar\dfrac{\partial}{\partial x}$
$(p_x)^2$	$-\hbar^2\dfrac{\partial^2}{\partial x^2}$
E	$i\hbar\dfrac{\partial}{\partial t}$

SUMMARY

Since elementary particles can be represented by wave functions or wave packets in the quantum world, it is important to have a wave equation in order to describe their motions as a function of both space and time. The equation that does this is called the Schrödinger wave equation, and it has the following form:

$$\left(-\frac{\hbar^2}{2m}\nabla^2 + V\right)\Psi = i\hbar\frac{\partial\Psi}{\partial t},$$

where Ψ is a function of the position variables and time. When the potential V is independent of the time, Schrödinger's equation can be separated into two equations, one a function of time only and the other a function of position coordinates only. The equation in time is readily integrated and has the solution

$$f(t) = Ce^{-iEt/\hbar}.$$

The spatial equation is known as the *energy eigenequation* and has the form

$$\left(-\frac{\hbar^2}{2m}\nabla^2 + V\right)\psi = E\psi,$$

where ψ is independent of the time. For a given solution ψ_n of this equation, a solution of the Schrödinger equation can be obtained by forming the product

$$\Psi_n(x, y, z, t) = \psi_n(x, y, z)\,f_n(t) = \psi_n(x, y, z)e^{-iEt/\hbar}$$

In one dimension the differential operator ∇^2 becomes simply d^2/dx^2 and the potential V is a function of x only. Applications to one-dimensional problems will be considered in the next chapter.

Additional Reading

P. A. Tipler, *Modern Physics,* Worth Publishers, Inc., New York, 1978, Chapter 6.

F. W. Van Name, Jr., *Modern Physics,* 2nd ed., Prentice-Hall, Inc., Englewood Cliffs, N.J., 1962, Chapter 5.

E. H. Wichmann, *Quantum Physics,* Berkeley Physics Course, Vol. 4, McGraw-Hill Book Co., New York, 1971, Chapter 7.

PROBLEMS

9-1. Show that the functions given in Equation 9.6 are solutions of the classical wave equation, by differentiating them and substituting into Equation 9.1.

9-2. Show that Equation 9.7 is a solution of the classical wave equation, Equation 9.1.

9-3. (a) Show that the function $\Psi = A \exp[i(kx - \omega t)]$ satisfies the Schrödinger equation, Equation 9.9, for a particle of mass m in a region of constant potential V. (b) Does the function $\Psi = A \sin(kx - \omega t)$ satisfy Equation 9.9?

9-4. A particle of mass m is confined to a harmonic oscillator potential given by $V = \frac{1}{2}\omega^2 m x^2$, where $\omega^2 = K/m$ and K is the force constant. The particle is in a state described by the following wave function:

$$\Psi = A \exp\left(-\frac{\omega m x^2}{2\hbar} - \frac{i\omega t}{2}\right).$$

Verify that this wave function is a solution of Schrödinger's equation.

9-5. Normalize the wave function in problem 9-4. That is, determine A so that

$$\int_{-\infty}^{\infty} \Psi^*\Psi \, dx = 1.$$

9-6. Using the normalized wave function from problems 9-4 and 9-5, find the expectation values of the operators x and x^2.

9-7. Using the normalized wave function from problems 9-4 and 9-5, find the expectation values of the operators p_x and p_x^2.

9-8. Using the normalized wave function from problems 9-4 and 9-5, find the expectation value of the energy operator.

9-9. Use the results of problems 9-4 through 9-8 to verify that

$$\langle E \rangle = \langle T \rangle + \langle V \rangle,$$

where $\langle E \rangle$ is the expectation value of the total energy,

$$\langle T \rangle = \frac{1}{2m}\langle p_x^2 \rangle$$

is the expectation value of the kinetic energy, and

$$\langle V \rangle = \frac{1}{2}\omega^2 m \langle x^2 \rangle$$

is the expectation value of the potential energy.

9-10. Determine whether or not the function $f(x) = Ax \exp(-x^2/2)$ is an eigenfunction of the operator $Q = d^2/dx^2 - x^2$. If it is, what is its eigenvalue?

9-11. For what values of the constant C will the function $f(x) = Ae^{-\alpha x}$ be an eigenfunction of the operator

$$Q = \frac{d^2}{dx^2} + \frac{2}{x} \cdot \frac{d}{dx} + \frac{C}{x}?$$

What is the corresponding eigenvalue?

9-12. Normalize the function $\psi(x) = Ax \exp(-x^2/2)$ over the space defined by $-\infty \le x \le \infty$.

*9-13. If the normalized function $\psi(x) = (2/\sqrt{\pi})^{1/2} x \exp(-x^2/2)$ represents a particle of mass m in the space defined by $-\infty \le x \le \infty$, find the expectation values $\langle x \rangle$, $\langle x^2 \rangle$, $\langle p_x \rangle$, and $\langle p_x^2 \rangle$.

9-14. The wave function for a particle in the region bounded by $-L \le x \le L$ is given by

$$\psi = A(L^2 - x^2)e^{ikx}.$$

(a) What is the probability of finding the particle in the interval between x and $x + dx$? (b) Find the value of A that will normalize the wave function.

9-15. Given the following wave function for a Gaussian packet in x-space at $t = 0$:

$$\psi(x) = \left(\frac{1}{\sqrt{\pi}\sigma}\right)^{1/2} \exp\left(ik_0 x - \frac{x^2}{2\sigma^2}\right).$$

Find the probability density.

9-16. Find the expectation value of the momentum for the wave function given in problem 9-15.

*9-17. For the wave packet in problem 9-15, find the expectation value $\langle p_x^2 \rangle$ and the kinetic energy, $\dfrac{1}{2m}\langle p_x^2 \rangle$.

*9-18. Show that the wave function

$$\Psi(x, y, z, t) = A \sin k_x x \cdot \sin k_y y \cdot \sin k_z z \cdot \exp\left(-\frac{i\hbar k^2 t}{2m}\right)$$

satisfies the Schrödinger equation, Equation 9.10, where $V = 0$ and $k^2 = k_x^2 + k_y^2 + k_z^2$.

*9-19. Show that the ground state wave function for the two-dimensional harmonic oscillator,

$$\Psi(x, y, t) = \left(\frac{\alpha}{\pi}\right)^{1/4} \exp\left(-\frac{\alpha}{2}x^2 - \frac{\alpha}{2}y^2 - \frac{iEt}{\hbar}\right),$$

satisfies the Schrödinger wave equation, Equation 9.10, when

$$V = \tfrac{1}{2}m\omega^2 \cdot (x^2 + y^2)$$

and

$$\alpha = \frac{\omega m}{\hbar}.$$

What is the energy of this state?

*9-20. A three-dimensional harmonic oscillator is in a state given by the wave function

$$\Psi(x, y, z, t) = \left(\frac{\alpha}{\pi}\right)^{1/4} \sqrt{\alpha}\, x \exp\left[-\frac{\alpha}{2}(x^2 + y^2 + z^2) - \frac{iEt}{\hbar}\right],$$

where $\alpha = \omega m/\hbar$. Find the energy of this state by substituting into the Schrödinger equation, Equation 9.10, with the potential energy given by

$$V = \tfrac{1}{2}m\omega^2(x^2 + y^2 + z^2).$$

10 Applications to One-Dimensional Quantum Systems

1. PARTICLE IN A BOX OR AN INFINITE POTENTIAL WELL

Imagine a particle of mass m in a one-dimensional box that has perfectly rigid walls at $x = 0$ and $x = L$, confining the particle to the region defined by $0 < x < L$. This is equivalent to placing a charged particle in an infinitely deep potential well, as shown in Figure 10–1. The terms "perfectly rigid" and "infinitely deep"

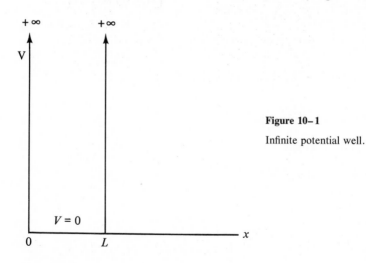

Figure 10–1

Infinite potential well.

are used to insure that there is no penetration of the walls. Therefore, the wave function for the particle must be zero at $x = 0$ and at $x = L$. Since $V = 0$ inside the well, Equation 9.17 becomes

$$-\frac{\hbar^2}{2m}\frac{d^2\psi}{dx^2} = E\psi,$$

172 or

$$\frac{d^2\psi}{dx^2} + k^2\psi = 0$$

where $k^2 = 2mE/\hbar^2$. As we have seen already in Example 9-2, this equation has solutions of the form $\psi = A \sin kx + B \cos kx$. Using the boundary condition that $\psi = 0$ at $x = 0$, we see that B must be zero. Hence, the solutions are the sine functions. From the condition that $\psi = 0$ when $x = L$ we find that k must satisfy the relation

$$kL = n\pi.$$

Thus, we find that

$$E_n = \frac{\hbar^2 k^2}{2m} = \frac{n^2 \pi^2 \hbar^2}{2mL^2}, \tag{10.1}$$

and the eigenfunctions are

$$\psi_n = A \sin k_n x. \tag{10.2}$$

Sketches of the three eigenfunctions of lowest energy and their probability densities are shown in Figure 10–2. The number of solutions is infinite for the infinite potential well.

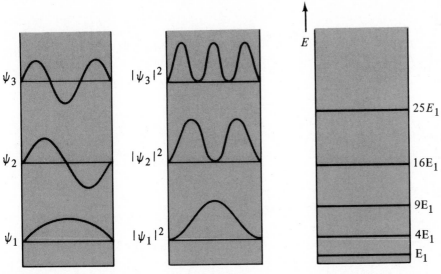

Figure 10–2

Some eigenfunctions, probability densities, and energy levels in the infinite well.

Note that the only energies available to the particle are the discrete values $E_1 = \pi^2\hbar^2/2mL^2$, $E_2 = 4E_1$, $E_3 = 9E_1$, ... , $E_n = n^2E_1$. One might well ask why the lowest state would not be $E = 0$ instead of $E_1 = \pi^2\hbar^2/2mL^2$. One way to answer this question is to say that if $n = 0$, then $k = 0$ and $\psi = 0$ everywhere within the well. But if ψ is zero in the well, then $|\psi|^2$ is zero and there is no particle anywhere within the well.

Example 10-1
A classical particle can rest indefinitely in a well. That is, it can have both zero kinetic energy and zero potential energy relative to the bottom of the well. However, the uncertainty principle prohibits this for a quantum particle. Show that the kinetic

energy required by the uncertainty principle provides a reasonable estimate of the ground state energy.

Solution

The minimum uncertainty product $\Delta x \cdot \Delta p$ is a number of the order of h.

Let
$$\Delta x \cdot \Delta p = \frac{h}{2}.$$

Then, for $\Delta x = L$,
$$\Delta p = \frac{h}{2L}.$$

In order for the momentum to have an uncertainty of $h/2L$, it must be at least as large as $h/2L$. Then, the minimum kinetic energy of the particle must be

$$E_{min} = \frac{p^2}{2m} = \frac{h^2}{8mL^2} = \frac{4\pi^2\hbar^2}{8mL^2} = \frac{\pi^2\hbar^2}{2mL^2},$$

which is in agreement with the result obtained above.

Example 10-2

Normalize the wave functions for the infinite well given in Equation 10.2.

Solution

In order to normalize a wave function, we require that the integral of its square over all of space be unity. Since the wave function is zero everywhere but inside the well, we need integrate only over the space $0 < x < L$. Then,

$$1 = \int_0^L |\psi_n|^2 dx = |A|^2 \int_0^L \sin^2 k_n x \, dx = |A|^2 \cdot \frac{L}{2}.$$

$$\therefore A = \sqrt{\frac{2}{L}} \text{ and } \psi_n = \sqrt{\frac{2}{L}} \sin k_n x.$$

2. FREE PARTICLE AND A FINITE POTENTIAL WELL

A potential well of finite depth is shown in Figure 10–3. The bottom of the well is at zero potential and the top of the well is at potential V_0. Whenever a particle acquires sufficient kinetic energy to carry it to the top of the well, the particle will escape and become a free particle. As we saw earlier, the wave function for a free particle is a plane wave, which can be depicted as a continuous sine or cosine function (Equation 9.6). Accordingly, for a particle with total energy E that is greater than V_0, the wave function will be a continuous sinusoid, but it will undergo changes in wavelength, velocity, and amplitude at the edges of the well. In order to understand this, let us consider each region of the space in detail.

In regions 1 and 3, that is for $x < -a$ and $x > a$, the energy eigenequation is

$$\frac{d^2\psi}{dx^2} + \frac{2m}{\hbar^2} (E - V_0) \psi = 0.$$

A general solution for region 1 is

$$\psi_1 = Ae^{ik_1x} + Be^{-ik_1x} \text{ (for } x < -a), \tag{10.3}$$

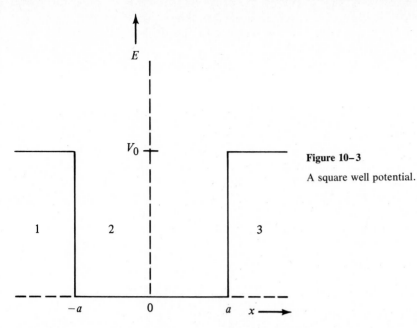

Figure 10–3

A square well potential.

where $\hbar k_1 = \sqrt{2m \, (E - V_0)}$. The first term is a harmonic wave traveling to the right, and the second term is a wave traveling to the left. That is, A represents the amplitude of an *incident* particle coming from the left. B represents the *net* amplitude from the right due to the combination of incident particles from the right and to reflections from the boundaries at $x = -a$ and $x = a$.

In region 2, the potential energy of the particle is zero and the kinetic energy equals the total energy E. Then the energy eigenequation becomes

$$\frac{d^2\psi}{dx^2} + \frac{2mE}{\hbar^2} \, \psi = 0,$$

which has the general solution

$$\psi_2 = Ce^{ik_2x} + De^{-ik_2x} \text{ (for } -a < x < a), \tag{10.4}$$

where $\hbar k_2 = \sqrt{2mE}$. Here C is the net amplitude of transmitted and reflected waves traveling to the right; D is the net amplitude to the left. It is evident that $k_2 > k_1$, from which it follows that $\lambda_1 > \lambda_2$. Thus the wavelength in the region within the well boundaries is shorter than the wavelength outside the well.

In region 3, where the k-value is the same as in region 1, a solution of the energy eigenequation is

$$\psi_3 = Fe^{ik_1x} + Ge^{-ik_1x} \text{ (for } x > a). \tag{10.5}$$

Here F takes into account both incident and reflected particles moving from left to right, while G is due only to the amplitude of particles incident from the right.

The change in amplitude that occurs at the boundaries of the regions may be justified as follows. The particle is less apt to be "found" in a region where it is traveling faster. Hence, the probability amplitude should be smaller in the region of higher speed, that is, where the wavelength is shorter.

The above discussion helps us to understand the qualitative features of the particle wave function. However, in order to obtain a well-behaved function to represent a particle, we require that the function be single-valued and that it and its derivatives be continuous at the boundaries of each region. The mathematical statements of these boundary conditions are as follows:

$$\psi_1(-a) = \psi_2(-a) \qquad \text{and} \qquad \psi_2(a) = \psi_3(a).$$

Also,

$$\left.\left(\frac{d\psi_1(x)}{dx}\right)\right|_{x=-a} = \left.\left(\frac{d\psi_2(x)}{dx}\right)\right|_{x=-a} \quad \text{and} \quad \left.\left(\frac{d\psi_2(x)}{dx}\right)\right|_{x=a} = \left.\left(\frac{d\psi_3(x)}{dx}\right)\right|_{x=a}. \tag{10.6}$$

By means of these relations, four of the coefficients in Equations 10.3 to 10.5 may be expressed in terms of one of the others. Even further simplification is possible by utilizing the initial conditions or assumptions. For example, if no particles are incident from the right and all particles come only from the left, then $A \neq 0$ but $G = 0$. Then all of the coefficients can be expressed in terms of the amplitude of the incident particle flux A.

A wave function for a free particle that interacts with a finite well is shown by the wave labeled $E > V_0$ in Figure 10–4.

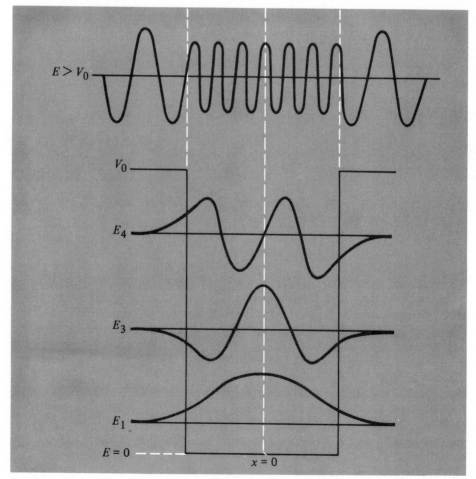

Figure 10–4

Eigenfunctions for the square well potential.

3. PARTICLE BOUND TO A FINITE POTENTIAL WELL

We have seen that a quantum particle, like its classical counterpart, will escape from a potential well when its kinetic energy is sufficiently great. There is another aspect of the behavior of quantum particles, however, that is in striking contrast to the common-sense world of classical particles. This phenomenon is known as "barrier penetration." What happens is that the wave function for a bound particle "leaks" into the wall on each side of the well. The penetration distance is not very large, since the wave function in the classically forbidden region decays exponentially. Nevertheless, it occurs and it has some important consequences, which will soon be evident.

Let us look briefly at the wave functions in the three regions as we did earlier. In regions 1 and 3 there can be no traveling waves as before, since the total particle energy is less than V_0. Next we note that k_1 in Equation 10.3 is now an imaginary quantity, since $V_0 > E$. If we define the quantity K_1 such that

$$K_1 = ik_1 = \frac{1}{\hbar} \sqrt{2m(V_0 - E)}, \qquad (10.7)$$

then Equation 10.3 becomes

$$\psi_1 = Ae^{K_1 x} + Be^{-K_1 x},$$

where K_1 is real. Now for negative values of x the first term decays exponentially, whereas the second term increases exponentially. The latter cannot be allowed for a well-behaved wave function, so only the first term can be used. In region 2, ψ_2 has the same form as before, i.e., Equation 10.4. In region 3, we see that ψ_3 will diverge for positive x unless F in Equation 10.5 is set equal to zero. Our three solutions then become:

$$\left. \begin{array}{llll} \psi_1 = Ae^{K_1 x} & \text{for} & x < -a \\[4pt] \psi_2 = Ce^{ik_2 x} + De^{-ik_2 x} & \text{for} & -a < x < a \\[4pt] \psi_3 = Ge^{-K_1 x} & \text{for} & x > a \end{array} \right\} \qquad (10.8)$$

A mathematical analysis using the boundary conditions given in Equation 10.6 yields the result that $C = \pm D$.[1] For $C = D$ we also obtain $A = G$, and ψ_2 becomes a cosine function. The solution is then symmetric about the center of the well, as illustrated by E_1 and E_3 in Figure 10–4. For $C = -D$ we obtain $A = -G$, and ψ_2 becomes a sine function that is anti-symmetric about the center of the well, as illustrated by E_4 in the same figure.

Note that penetration of the walls increases with increasing energy. In general, the deeper and wider the well, the less the wave function leaks out. A measure of the "strength" of a well in confining a particle goes roughly as the product of its depth and breadth. In a "strong" well the wave functions and the energies of the bound states can be approximated by those of the infinite well given by Equations 10.1 and 10.2.

---

[1] E. E. Anderson, *Modern Physics and Quantum Mechanics*, W. B. Saunders, Philadelphia, 1971, p. 167.

4. TUNNELING THROUGH BARRIERS

Suppose that the two identical potential wells of Figure 10–5 are separated by a narrow barrier, as shown schematically in Figure 10–6. The term "narrow" is somewhat ambiguous since it depends also upon the "strength" of the well. Let

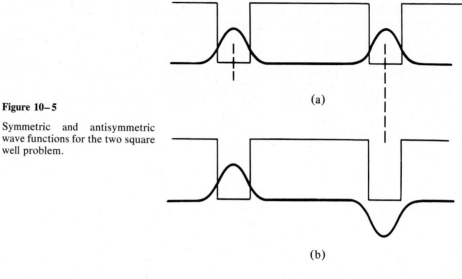

(a)

Figure 10–5

Symmetric and antisymmetric wave functions for the two square well problem.

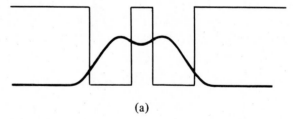

(b)

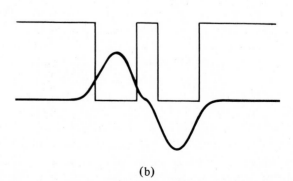

(a)

Figure 10–6

Two identical wells separated by a narrow barrier.

(b)

us be more specific, however, and say that a narrow barrier is one whose thickness is less than the penetration depth of a given wave function. When this occurs, the wave function can leak through the barrier and the particle is said to have *tunneled* its way to the other well. Tunneling is strictly a quantum mechanical phenomenon, which has no classical analog.

Under conditions that are favorable for tunneling, it is impossible to know which well contains the particle after sufficient time has elapsed for tunneling to occur. Then there is equal probability of finding the particle in either well, and the wave function for the two wells must be a linear combination of the two single-well ground state functions. The two possibilities are shown schematically in Figure 10–5 for two widely separated wells and in Figure 10–6 for two wells having a narrow barrier. In (a) the total wave function is symmetric in the two wells, and in (b) it is antisymmetric. The energies of these two states are not the same, and the energy splitting increases as the width of the barrier decreases. This is shown schematically in Figure 10–7. The following qualitative argument is helpful in

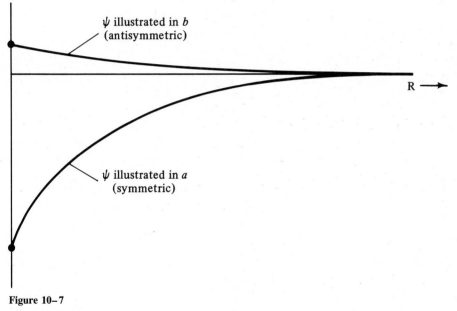

ψ illustrated in *b* (antisymmetric)

R →

ψ illustrated in *a* (symmetric)

Figure 10–7

Energy versus barrier width *R*.

understanding this energy splitting. In the limit as the barrier vanishes, the symmetric wave function corresponds to the ground state of a well having twice the width of the original wells; hence the energy of the symmetric state *decreases*. On the other hand, as the wells coalesce the antisymmetric function corresponds to the first excited state of the well of double width. Thus, the energy decrease due to the increased width of the well is more than compensated by excitation of the particle to the first state above the ground state.

The most important feature of this model is that the interaction of two wells, which have the same energy levels at large separations, splits each of the original energy levels into two levels.

5. THE HARMONIC OSCILLATOR

When a displacement of a mass from its equilibrium position results in a restoring force that is proportional to the displacement x, we may express this mathematically as

$$F_x = -Kx. \tag{10.9}$$

The constant K is known as the force constant or spring constant. The potential energy stored in the spring is readily obtained by calculating the work done in achieving the displacement x. That is,

$$V = -W = -\int_0^x F_x \, dx = -\int_0^x -Kx \, dx = \tfrac{1}{2} Kx^2. \tag{10.10}$$

If the mass is released, the system will oscillate in simple harmonic motion (for small displacements) at an angular frequency given by

$$\omega = \sqrt{\frac{K}{m}}. \tag{10.11}$$

From Equation 10.11 we may express K as $K = \omega^2 m$, and the potential energy V may then be written as

$$V = \tfrac{1}{2} \omega^2 m x^2. \tag{10.12}$$

Using this potential in Equation 9.17, we obtain the harmonic oscillator equation in wave mechanical form:

$$\frac{d^2\psi}{dx^2} + \left[\frac{2mE}{\hbar^2} - \left(\frac{m\omega}{\hbar} \right)^2 x^2 \right] \psi = 0. \tag{10.13}$$

It is convenient to simplify this equation with the substitutions

$$\alpha = \frac{\omega m}{\hbar} \quad \text{and} \quad \beta = \frac{2mE}{\hbar^2}. \tag{10.14}$$

Then Equation 10.13 becomes

$$\frac{d^2\psi}{dx^2} + (\beta - \alpha^2 x^2)\psi = 0. \tag{10.15}$$

The procedure[1] for solving this equation is beyond the scope of this text, but it is possible to guess at a solution in this case; if the solution satisfies the Schrödinger equation, it will be acceptable. As a possible solution let us consider the function

$$\psi = A \exp(-Cx^2). \tag{10.16}$$

Example 10-3

(a) Show that Equation 10.16 is a solution of Equation 10.15 and evaluate the constant C. (b) Find the energy corresponding to this state.

---

[1] See E. E. Anderson, op. cit., p. 176.

Solution

(a)
$$\frac{d\psi}{dx} = -2Cx\psi$$

$$\frac{d^2\psi}{dx^2} = 4C^2x^2\psi - 2C\psi$$

Then,
$$4C^2x^2\psi - 2C\psi + \beta\psi - \alpha^2x^2\psi = 0.$$

$$\therefore C = \frac{\alpha}{2} = \frac{\beta}{2} = \frac{\omega m}{2\hbar}.$$

(b) Since $\beta = \alpha$,

$$\frac{2mE}{\hbar^2} = \frac{\omega m}{\hbar}$$

$$\therefore E = \frac{\omega m \hbar^2}{2m\hbar} = \frac{1}{2}\omega\hbar.$$

When the results of Example 10-3 are substituted, the wave function suggested in Equation 10.16 becomes

$$\psi = A \exp\left(-\frac{\omega mx^2}{2\hbar}\right). \tag{10.17}$$

The energy corresponding to this state is $E = \frac{1}{2}\omega\hbar$, and we identify the state as the ground state of the one-dimensional oscillator. Since the energy levels of the harmonic oscillator are equally spaced with a separation energy of $h\nu = \omega\hbar$ (see Section 4 of Chapter 5), the energy of the nth state is given by

$$E_n = \omega\hbar\left(n + \frac{1}{2}\right), \tag{10.18}$$

where $n = 0$ corresponds to the ground state. The energy level scheme for the harmonic oscillator is shown in Figure 10-8. The wave functions for the higher

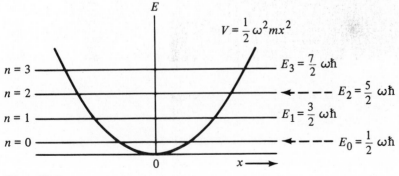

Figure 10-8

Energy levels of the harmonic oscillator.

states are more complicated. In general, they consist of a factor similar to Equation 10.17 multiplied by a polynomial in x. The probability densities for some of the higher states are shown in Figure 10-9, where they are compared with the

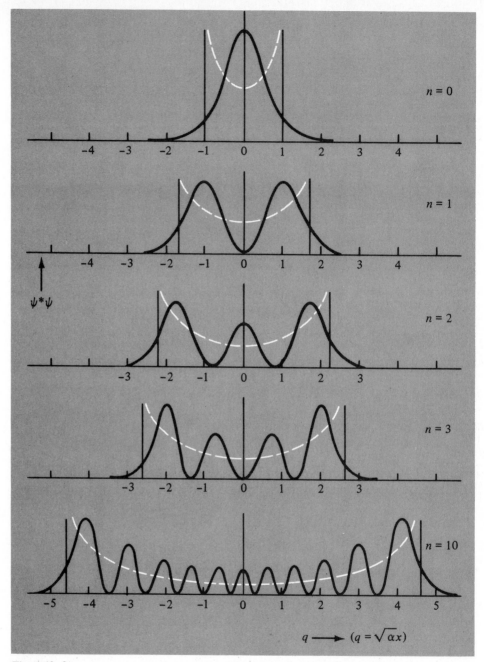

Figure 10–9

Probability densities for a few harmonic oscillator states. The dotted curves show the classical probabilities corresponding to the same energies. (From Chalmers W. Sherwin, *Introduction of Quantum Mechanics,* Holt, Rinehart and Winston, Inc., 1959. Used with permission.)

classical probabilities. In a classical oscillator, such as a simple pendulum or a mass on a spring, the mass has its maximum speed when it passes through the zero-displacement point, and it comes to rest momentarily at the end points of the motion. Therefore, it spends the least time at the center-point of its motion and the greatest time at the end points. This is consistent with the dotted curves in Figure 10–9, which show the classical probabilities. The wave picture of quantum mechanics differs somewhat from the classical view for the $n = 0$ state, but it is evident that the agreement improves considerably as n increases.

Example 10-4
   Normalize the ground state wave function for the one-dimensional harmonic oscillator so that the total probability of finding the particle is unity.

Solution
   Normalization of the wave function is discussed in Section 3 of Chapter 8. In order to normalize ψ_0 we proceed as follows:

$$1 = \int_{-\infty}^{\infty} |\psi_0|^2 \, dx = \int_{-\infty}^{\infty} |A|^2 \exp\left(-\frac{\omega m x^2}{\hbar}\right) dx$$

$$= 2|A|^2 \int_{0}^{\infty} \exp\left(-\frac{\omega m x^2}{\hbar}\right) dx$$

$$= |A|^2 \sqrt{\frac{\pi\hbar}{\omega m}}.$$

Then,

$$A = \left(\frac{\omega m}{\pi\hbar}\right)^{1/4},$$

and the normalized wave function is

$$\psi_0 = \left(\frac{\omega m}{\pi\hbar}\right)^{1/4} \exp\left(-\frac{\omega m x^2}{2\hbar}\right). \tag{10.19}$$

Example 10-5
   Find the probability density at $x = 0$ for the ground state wave function of the harmonic oscillator, Equation 10.19.

Solution

$$\rho(x) \, dx = |\psi_0|^2 \, dx = \left(\frac{\omega m}{\pi\hbar}\right)^{1/2} \exp\left(-\frac{\omega m x^2}{\hbar}\right) dx$$

$$\rho(0) \, dx = \left(\frac{\omega m}{\pi\hbar}\right)^{1/2} dx.$$

Example 10-6
   Find the expectation values of x and x^2 for the ground state of the harmonic oscillator, Equation 10.19. (Use the integrals of Appendix E.)

Solution
   Using Equation 9.29, the expectation value of x is:

$$\langle x \rangle = \frac{\int_{-\infty}^{\infty} x |\psi_0(x)|^2 \, dx}{\int_{-\infty}^{\infty} |\psi_0(x)|^2 \, dx}.$$

Since ψ_0 is normalized, the denominator is unity and we may write

$$\langle x \rangle = \left(\frac{\omega m}{\pi \hbar}\right)^{1/2} \int_{-\infty}^{\infty} x \exp\left(-\frac{\omega m x^2}{\hbar}\right) dx$$

$$= 0 \text{ (odd integrand evaluated over symmetric limits).}$$

In like manner, the expectation value of x^2 is

$$\langle x^2 \rangle = 2\left(\frac{\omega m}{\pi \hbar}\right)^{1/2} \int_{0}^{\infty} x^2 \exp\left(-\frac{\omega m x^2}{\hbar}\right) dx$$

$$= 2\left(\frac{\omega m}{\pi \hbar}\right)^{1/2} \cdot \frac{1}{4} \sqrt{\frac{\pi \hbar^3}{\omega^3 m^3}} = \frac{\hbar}{2\omega m}.$$

Example 10-7

Find the expectation values of the momentum p and its square p^2 for the ground state of the one-dimensional harmonic oscillator.

Solution

From Table 9-1 we note that the operator for p_x is $-i\hbar\, \partial/\partial x$. Then the expectation value for the momentum for the normalized function ψ_0 is:

$$\langle p_x \rangle = \int_{-\infty}^{\infty} \psi_0 \left(-i\hbar\frac{\partial}{\partial x}\right) \psi_0\, dx = -i\hbar \int_{-\infty}^{\infty} \psi_0 \frac{d}{dx} \psi_0\, dx$$

$$= -i\hbar \int_{-\infty}^{\infty} \psi_0 \left(-\frac{\omega m x}{\hbar}\right) \psi_0\, dx = i\omega m \int_{-\infty}^{\infty} x |\psi_0|^2\, dx$$

$$= 0 \text{ (odd integrand evaluated over symmetric limits).}$$

Also,

$$\langle p_x^2 \rangle = \int_{-\infty}^{\infty} \psi_0 \left(-\hbar^2 \frac{\partial^2}{\partial x^2}\right) \psi_0\, dx = -\omega m \hbar \int_{-\infty}^{\infty} \left(\frac{\omega m x^2}{\hbar} - 1\right) |\psi_0|^2\, dx$$

$$= -2\omega^2 m^2 \left(\frac{\omega m}{\pi \hbar}\right)^{1/2} \int_{-\infty}^{\infty} x^2 \exp\left(-\frac{\omega m x^2}{\hbar}\right) dx + \omega m \hbar \int_{-\infty}^{\infty} |\psi_0|^2\, dx$$

$$= -\tfrac{1}{2}\omega m \hbar + \omega m \hbar = \tfrac{1}{2}\omega m \hbar.$$

SUMMARY

In this chapter Schrödinger's wave equation was applied to several one-dimensional quantum systems, beginning with the infinite potential well in which all of the allowed energy states are bound states. Next, the finite well was treated, and the cases of both bound and free states were discussed. Tunneling through narrow barriers was introduced, and it was shown that the wave function for a particle that interacts with more than one well must be written as a linear combination of the single-well wave functions. The chapter closed with a brief introduction to the one-dimensional harmonic oscillator potential.

Additional Reading

C. H. Blanchard, C. R. Burnett, R. G. Stoner, and R. L. Weber, *Introduction to Modern Physics,* 2nd ed., Prentice-Hall, Inc., Englewood Cliffs, N.J., 1969, Chapter 7.

K. W. Ford, *Classical and Modern Physics,* Vol. 3, Xerox College Publishing, Lexington, Mass., 1974, Chapter 23.

P. A. Tipler, *Modern Physics,* Worth Publishers, Inc., New York, 1978, Chapter 6.

R. T. Weidner and R. L. Sells, *Elementary Modern Physics,* 3rd ed., Allyn & Bacon, Inc., Boston, 1980, Chapter 5.

E. H. Wichmann, *Quantum Physics,* Berkeley Physics Course, Vol. 4, McGraw-Hill Book Co., New York, 1971, Chapter 8.

10-1. Write the normalized wave functions for the three lowest states of the infinite potential well of Figure 10–1. Verify the normalization of each by integrating over the well.

10-2. Write the probability densities for the three lowest states of the infinite potential well of Figure 10–1. Find the x-value corresponding to the maximum value of the probability density in each case. Compare with Figure 10–2.

10-3. Find the expectation value of the operator x for the three states of the previous problem. How can you reconcile this result with Figure 10–2?

10-4. Find the expectation value of the momentum, $\langle p_x \rangle$, for the lowest state of the infinite well of Figure 10–1.

10-5. Find the expectation value of the momentum squared, $\langle p_x{}^2 \rangle$, for the ground state of the infinite well of Figure 10–1. Compare the kinetic energy, $\frac{1}{2m} \langle p_x{}^2 \rangle$, with Equation 10.2.

10-6. (a) Find the expectation value $\langle p_x{}^2 \rangle$ for the first two excited states of the infinite well of Figure 10–1. (b) Obtain the kinetic energies of these two states and compare with Equation 10.2.

10-7. Consider a particle of mass m performing simple harmonic motion with amplitude A along the x-axis. Show that the classical probability of finding the particle in the element dx is

$$\rho(x) \, dx = \frac{dx}{\pi \sqrt{A^2 - x^2}}.$$

10-8. Show that the quantized oscillator energies given in Equation 10.18 correspond to the classical amplitudes

$$A_n = \sqrt{\frac{\hbar}{\omega m} (2n + 1)}.$$

10-9. Find the average potential energy of a one-dimensional harmonic oscillator in its ground state. (Hint: use the result from Example 10-6.)

10-10. Find the average kinetic energy of a one-dimensional harmonic oscillator in its ground state. (Hint: use the result from Example 10-7.)

10-11. The function

$$\psi = Bx \exp\left(-\frac{\omega m x^2}{2\hbar}\right)$$

is a solution of Equation 10.15. With which quantum state is it associated, and what is the energy of the state?

10-12. Let ψ_1 and ψ_2 be two normalized wave functions in a certain space. Another possible wave function in that space is the superposition (linear combination),

$$\psi = a\psi_1 + b\psi_2,$$

where a and b are real constants. Given that the integrals $\int \psi_1^* \psi_2 \, dx$ and $\int \psi_2^* \psi_1 \, dx$ over the space are zero, show that this wave function is normalized by the factor $(a^2 + b^2)^{-1/2}$.

10-13. A wave function that is a superposition of the two lowest states in an infinite square well of width L is given by

$$\psi = A \left(\frac{2}{L}\right)^{1/2} \left(\sin \frac{\pi x}{L} + 2 \sin \frac{2\pi x}{L}\right).$$

Find the value of A that will normalize the wave function in the space defined by $0 \le x \le L$.

10-14. Using the normalized wave functions of the previous problem, find the expectation value $\langle p_x^2 \rangle$ and the expectation value of the kinetic energy for a particle of mass m.

10-15. Let ψ_1, ψ_2, and ψ_3 be the normalized wave functions for the three lowest states of the infinite well of width L as in Figures 10–1 and 10–2. A possible wave function describing a particle of mass m in this well is the linear combination

$$\psi = A(2\psi_1 + 2\psi_2 + \psi_3).$$

(a) Find the value of A that normalizes this wave function. (b) Find the expectation value of the kinetic energy.

10-16. Consider the following wave function, which describes a particle of mass m in the region between two impenetrable walls separated by a distance $2a$:

$$\psi(x) = A \left(\cos \frac{\pi x}{2a} + \sqrt{2} \sin \frac{\pi x}{a} \right).$$

(a) Normalize this wave function in the space defined by $-a \le x \le a$. (b) Find the expectation value of the kinetic energy.

10-17. Consider the following superposition of the ground state and the first excited state of the one-dimensional harmonic oscillator:

$$\psi(x) = A(\psi_0 + \psi_1) = A(1 + \sqrt{2\alpha}x) \cdot \exp\left(-\frac{\alpha x^2}{2}\right).$$

(a) Find the normalization constant A. (b) Include the appropriate time factor, $\exp\left(-\frac{iEt}{\hbar}\right)$, in each of the above terms so as to obtain the normalized wave function, $\Psi(x, t)$.

10-18. Using $\Psi(x, t)$ from problem 10-17, find the expectation value of the energy,

$$\langle E \rangle = \left\langle i\hbar \frac{\partial}{\partial t} \right\rangle.$$

10-19. Include the appropriate time factors in each of the terms of the wave function in problem 10-13 and then use the operator $\left(i\hbar \frac{\partial}{\partial t} \right)$ to find the expectation value of the energy.

10-20. If the wave function in problem 10-17 describes the motion of a particle of mass m, find (a) the expectation value of the potential energy; (b) the expectation value of the kinetic energy.

11 Application of Quantum Mechanics to Atomic Theory

Although the one-dimensional problems discussed in the previous chapter are instructive and often useful examples, one would expect that quantum mechanics would find its most extensive applications in the three-dimensional world. Here it is important to identify the symmetry of the interactions, since these are what determine the "space" for a particular problem. By the choice of an appropriate system of coordinates, the mathematical complexity of the problem can be minimized. For example, for a linear chain molecule one would expect to use the one-dimensional approach of the previous chapter; a cylindrical wave guide would normally be treated by cylindrical coordinates; an atom having spherical symmetry would best be described by spherical coordinates.

*1. THE HYDROGEN ATOM

The hydrogen atom is a two-body problem that has spherical symmetry in a relative coordinate system whose origin is fixed at one of the bodies. The only force is the Coulomb force, which is a function of the distance between the bodies. It was pointed out in Chapter 7 (see problem 7-10) that the motion of the heavier body, the proton in this case, can be neglected by using the reduced mass given in Equation 7.31 for the mass of the lighter body, the electron. Since we are concerned here only with the bound states (the energy levels) of the hydrogen atom and not the kinetic energy of the atom as a whole, we will use the reduced mass throughout this discussion.

The energy eigenequation to be solved is

$$\left(-\frac{\hbar^2}{2\mu} \nabla^2 + V \right) \psi = E\psi, \tag{11.1}$$

which is simply Equation 9.17 generalized to three dimensions. Here V is the Coulomb potential energy,

$$V = -\frac{ke^2}{r} \tag{11.2}$$

as in Equation 7.23. The Laplacian operator ∇^2, however, must now be expressed   187 in spherical coordinates as follows:

$$\nabla^2 = \left(\frac{\partial^2}{\partial r^2} + \frac{2}{r}\frac{\partial}{\partial r}\right) + \frac{1}{r^2}\left[\frac{1}{\sin\theta}\frac{\partial}{\partial\theta}\left(\sin\theta\frac{\partial}{\partial\theta}\right) + \frac{1}{\sin^2\theta}\frac{\partial^2}{\partial\phi^2}\right]. \qquad (11.3)$$

This appears, at first sight, to be an unnecessary complication, but the reader should bear in mind that this will facilitate the separation of variables required in order to obtain a solution. If ∇^2 were left in the simple form of Equation 9.3, then r in Equation 11.2 would have to be written as $r = \sqrt{x^2 + y^2 + z^2}$ and the variables would not be separable.

In order to separate the radial and the angular parts of the problem, let us write a solution as the product of an angular and a radial function of the form

$$\psi(r, \theta, \phi) = R(r) \cdot Y(\theta, \phi). \qquad (11.4)$$

Furthermore, let us represent the radial part of the Laplacian operator by $\mathscr{R}$ and the angular part by $\mathscr{L}^2$, so that Equation 11.3 becomes

$$\nabla^2 = \mathscr{R} + \frac{1}{r^2}\mathscr{L}^2. \qquad (11.5)$$

Using Equations 11.4 and 11.5, Equation 11.1 is separable as follows:

$$\frac{r^2\mathscr{R}R(r)}{R(r)} + \frac{2\mu r^2}{\hbar^2}\left(E + \frac{ke^2}{r}\right) = -\frac{\mathscr{L}^2 Y(\theta, \phi)}{Y(\theta, \phi)}. \qquad (11.6)$$

If we denote the separation constant by Λ we obtain the following two equations:

$$\mathscr{R}R(r) + \frac{2\mu}{\hbar^2}\left(E + \frac{ke^2}{r}\right)R(r) = \frac{\Lambda}{r^2}R(r) \qquad (11.7)$$

and

$$\mathscr{L}^2 Y(\theta, \phi) = -\Lambda\ Y(\theta, \phi). \qquad (11.8)$$

Since we know the differential operators $\mathscr{L}^2$ and $\mathscr{R}$, these two equations can be solved by the standard techniques for ordinary differential equations. The solution of the radial equation has the form

$$R_{n\ell} \sim \left(\frac{r}{na_0}\right)^\ell e^{-r/na_0} L_{n+\ell}^{2\ell+1}, \qquad (11.9)$$

where the $L_{n+\ell}^{2\ell+1}$ are the associated Laguerre polynomials.[1] Here n and ℓ are integers called *quantum numbers*. The *principal quantum number* n denotes a shell, corresponding to an orbit in the elementary Bohr theory. The quantum number n takes on the integral values 1,2,3 . . . . . The separation constant Λ turns out to be expressed by $\ell(\ell + 1)$, where ℓ is known as the *angular momentum quantum number*. It gets this name because $\ell(\ell + 1)\hbar^2$ is the square of the total orbital angular momentum of the electron. The allowed values of ℓ are $\ell = 0,1,2, . . . ,$ $n - 1$. The constant a_0 in Equation 11.9 is the Bohr radius given by Equation 7.21.

The solutions of Equation 11.8 are functions known as the spherical harmonics. They have the form

---

[1] See, for example, M. R. Spiegel, *Mathematical Handbook of Formulas and Tables*, New York: McGraw-Hill Book Co., 1968, p. 155.

$$Y_\ell^m \sim P_\ell^m(\cos\theta) \cdot e^{im\phi}, \tag{11.10}$$

where the $P_\ell^m(\cos\theta)$ are the associated Legendre functions.[1] Note that the quantum number ℓ appears again, as well as a new quantum number m. The latter is called the *magnetic quantum number,* and it takes on the values $0, \pm 1, \pm 2, \ldots \pm\ell$.

Now it is possible to write a general solution of the hydrogen problem in a form in which the three quantum numbers (n, ℓ, m) appear explicitly:

$$\psi_{n,\ell,m} \sim \left(\frac{r}{na_0}\right)^\ell e^{-r/na_0} L_{n+\ell}^{2\ell+1} P_\ell^m (\cos\theta) \cdot e^{im\phi}. \tag{11.11}$$

A few of the normalized wave functions for hydrogen are tabulated below:

$$\psi_{1s} = \psi_{100} = \frac{1}{\sqrt\pi}\left(\frac{1}{a_0}\right)^{3/2} \cdot e^{-r/a_0}$$

$$\psi_{2s} = \psi_{200} = \frac{1}{4\sqrt{2\pi}}\left(\frac{1}{a_0}\right)^{3/2} \cdot \left(2 - \frac{r}{a_0}\right) \cdot e^{-r/2a_0}$$

$$\psi_{2p} = \psi_{210} = \frac{1}{4\sqrt{2\pi}}\left(\frac{1}{a_0}\right)^{3/2} \cdot \frac{r}{a_0} \cdot e^{-r/2a_0} \cos\theta \tag{11.12}$$

$$\psi_{2p} = \psi_{21\pm1} = \frac{1}{8\sqrt\pi}\left(\frac{1}{a_0}\right)^{3/2} \cdot \frac{r}{a_0} \cdot e^{-r/2a_0} \sin\theta \cdot e^{\pm i\phi}$$

$$\psi_{3s} = \psi_{300} = \frac{1}{81\sqrt{3\pi}}\left(\frac{1}{a_0}\right)^{3/2} \cdot \left(27 - 18\frac{r}{a_0} + \frac{2\,r^2}{a_0^2}\right) e^{-r/3a_0}$$

*2. ENERGY LEVELS AND ELECTRON PROBABILITY DENSITIES IN THE HYDROGEN ATOM

It should be apparent from Equations 11.7 and 11.8 that the energy levels in the hydrogen atom depend only upon the radial separation and not upon angular position when all other interactions are neglected. This is the case because the potential itself is spherically symmetric and depends only upon the distance between the electron and the proton. The allowed energies turn out to be

$$E_n = -\frac{w_0}{n^2}, \tag{11.13}$$

which agrees with the Bohr theory result in Equation 7.25, except that we are now using the reduced mass μ instead of the electron mass m_0 in our calculations. The following examples should convince the reader of the validity of Equation 11.13.

Example 11-1
    Write ψ_{100} in the simplified form $\psi_{100} = Ce^{-r/a_0}$ and substitute it into Equation 11.7 in order to determine the energy E.

---

[1] E. E. Anderson, *Modern Physics and Quantum* Mechanics, Philadelphia: W. B. Saunders Co., 1971, p. 264.

Solution

Recall that the operator $\mathscr{R}$ is defined by

$$\mathscr{R} = \frac{\partial^2}{\partial r^2} + \frac{2}{r} \cdot \frac{\partial}{\partial r},$$

and that $\Lambda = \ell(\ell + 1)$. Since $\ell = 0$ for this case, after the differentiation is performed Equation 11.7 becomes:

$$\frac{1}{a_0^2} - \frac{2}{ra_0} + \frac{2\mu}{\hbar^2}\left(E_1 + \frac{ke^2}{r}\right) = 0.$$

Grouping the terms in powers of r,

$$\left(\frac{1}{a_0^2} + \frac{2\mu E_1}{\hbar^2}\right) + \frac{1}{r}\left(\frac{2\mu ke^2}{\hbar^2} - \frac{2}{a_0}\right) = 0.$$

From the coefficient of $1/r$ we obtain

$$a_0 = \frac{\hbar^2}{\mu ke^2},$$

which confirms Equation 7.21. Then, from the first set of parentheses we find that

$$E_1 = -\frac{\hbar^2}{2\mu a_0^2} = -\frac{\mu k^2 e^4}{2\hbar^2} = -w_0.$$

Example 11-2

Write $\psi_{210} = Cre^{-r/2a_0}$ and show that $E_2 = -w_0/4$.

Solution

Here $\ell = 1$, so Equation 11.7 becomes:

$$\left(\frac{\partial^2}{\partial r^2} + \frac{2}{r}\frac{\partial}{\partial r}\right)re^{-r/2a_0} + \frac{2\mu}{\hbar^2}\left(E_2 + \frac{ke^2}{r}\right)re^{-r/2a_0} = \left(\frac{2}{r}\right)e^{-r/2a_0}.$$

After differentiating we obtain:

$$\left(-\frac{2}{a_0} + \frac{2}{r} + \frac{r}{4a_0^2}\right) + \frac{2\mu}{\hbar^2}(E_2 r + ke^2) = \frac{2}{r}.$$

Grouping like powers of r:

$$\left(-\frac{2}{a_0} + \frac{2\mu ke^2}{\hbar^2}\right) + r\left(\frac{1}{4a_0^2} + \frac{2\mu E_2}{\hbar^2}\right) = 0,$$

and

$$E_2 = \frac{\hbar^2}{8\mu a_0^2} = -\frac{w_0}{4}.$$

The fact that quantum mechanics yields the same energies for hydrogen as does the simple Bohr theory derives from the assumption that the spherical Coulomb potential is the only interaction. It does not follow that the orbital radii should be the same in both theories. We now know, of course, that the electron is not constrained to an orbit in the classical sense, and that it has a small probability of being most anywhere within the atom. It will be recalled that in a one-dimensional problem the probability of finding a particle in the element dx is given by $|\psi|^2 dx$, where $\psi(x)$ is the wave function representing the particle. In three dimensions the probability is $|\psi|^2 dx\, dy\, dz$ in rectangular coordinates and $|\psi|^2 r^2 \sin\theta\, dr\, d\theta\, d\phi = |\psi|^2 r^2 dr\, d\Omega$ in spherical coordinates. Here $d\Omega$ is the element of solid angle, which is defined as $d\Omega = \sin\theta\, d\theta\, d\phi$. If we integrate this probability over

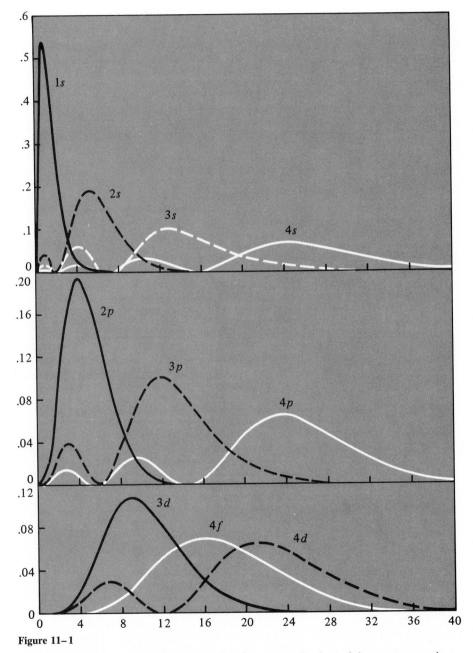

Figure 11–1

The radial probability distribution function $|rR_{n\ell}|^2$ for several values of the quantum numbers n, ℓ. (From E. U. Condon and G. H. Shortley, *The Theory of Atomic Spectra*, Cambridge University Press, Cambridge, 1953. Used with permission.)

solid angle, we obtain the following:

$$\int_\Omega |\psi_{n\ell m}|^2 r^2 dr\, d\Omega = |R_{n\ell}|^2 r^2 dr \int_\Omega |Y_\ell^m|^2\, d\Omega = |R_{n\ell}|^2 r^2 dr. \qquad (11.14)$$

It is convenient to define the product $|R_{n\ell}|^2 r^2$ as the *radial probability density*, since it expresses the probability that the electron will be found in the spherical shell bounded by r and $r + dr$. One would expect the radial probability density to increase as r^2 in isotropic space, since the volume of a spherical shell is simply $4\pi r^2 dr$. Sketches of the radial probability density for several hydrogen wave functions are shown in Figure 11-1.

The reader might well be troubled by the fact that the Bohr theory gives the correct energies for hydrogen, yet it appears to be so far off in describing the orbital radii. Is there any way to reconcile the probability model with the discrete radii, $r_n = n^2 a_0$, given by Equation 7.20? Suppose we ask the question, "At what value of r is the electron most likely to be found?" The answer is simple: the value

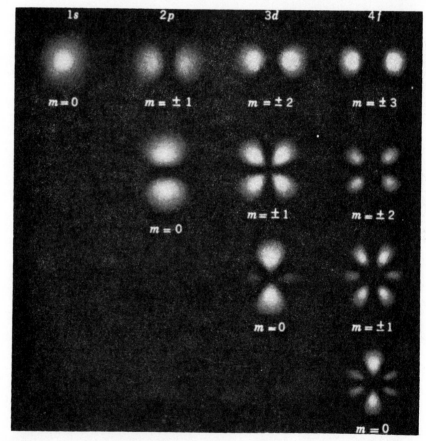

Figure 11-2

Photographic representation of the electron probability density distribution $\psi^* \psi$ for several energy eigenstates. These may be regarded as sectional views of the distributions in a plane containing the polar axis, which is vertical and in the plane of the paper. The scale varies from figure to figure. (From *Principles of Modern Physics* by R. B. Leighton. Copyright 1959 by McGraw-Hill Book Company. Used with permission.)

of r that corresponds to the peak of the plot of $|R_{n\ell}|^2 r^2$ versus r. That is to say, the most probable radius for the nth orbit is obtained by setting the derivative of $|R_{n\ell}|^2 r^2$ equal to zero. The following example will illustrate this procedure for the first shell of hydrogen.

Example 11-3

Find the most probable radial position for an electron in the ground state of hydrogen.

Solution

Ignoring the normalizing constant of ψ_{100}, we write

$$\frac{d}{dr}(r^2 R_{10}^2) = \frac{d}{dr}(r^2 e^{-2r/a_0}) = 0$$

$$2r\left(1 - \frac{r}{a_0}\right)e^{-2r/a_0} = 0$$

$$r_{max} = a_0$$

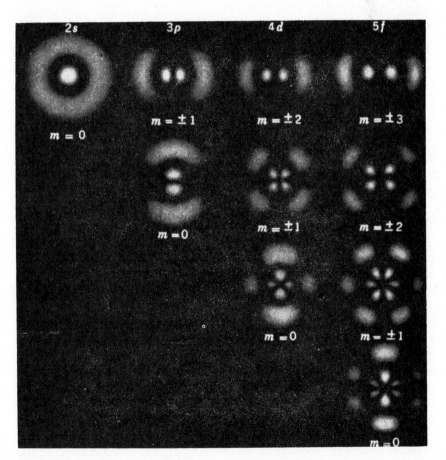

Continued from previous page

Example 11-3 shows that the *most probable* radial position for the electron in the ground state of hydrogen is a_0, the Bohr radius. It also turns out that $4a_0$, $9a_0$, and $16a_0$ are the most probable radial values for the $2p$, $3d$, and $4f$ shells, respectively. However, such nominal "agreement" with the Bohr theory results does not occur for the other subshells. A photographic representation of the probability densities for a number of hydrogen wave functions is shown in Figure 11–2.

In Chapter 9 we learned how to calculate the expectation value of an operator when the probability density is known. Thus, we can use Equation 9.29 to calculate the expectation value of the radial position of the electron in hydrogen and can compare our result with that obtained above.

Example 11-4
What is the expectation value $\langle r \rangle$ for the ground state of hydrogen?

Solution
Since ψ_{100} is normalized, we need not include the denominator of Equation 9.29. Then,

$$\langle r \rangle_{100} = \int_0^\infty \psi_{100}^* \, r \, \psi_{100} \, r^2 dr \int_\Omega d\Omega = 4\pi \int_0^\infty |\psi_{100}|^2 r^3 dr$$

$$= \frac{4}{a_0^3} \int_0^\infty r^3 e^{-2r/a_0} dr = \frac{4}{a_0^3} 3! \left(\frac{a_0}{2}\right)^4 = \frac{3}{2}(a_0).$$

$$\left[\text{Note: The integral used here is: } \int_0^\infty r^n e^{-br} dr = \frac{n!}{b^{n+1}}.\right]$$

This example illustrates the fact that the average value of a large number of measurements of the radial position of the electron in unexcited hydrogen is larger than the value predicted by Example 11-3. The reason for this is readily apparent from the asymmetry of the $1s$ function shown in Figure 11–1. Since the area to the right of the peak is greater than the area to the left, the average value of r will be greater than the value corresponding to the peak of the curve.

3.   ELECTRON SPIN AND THE PAULI PRINCIPLE

In Section 3 of Chapter 2 it was stated that the elementary particles possess an intrinsic angular momentum, which is called spin. Since the Schrödinger formulation of quantum mechanics does not include spin, it must be treated in a somewhat *ad hoc* fashion.

The necessity for an additional angular momentum in the theory arose from difficulties in explaining results in atomic spectroscopy and in magnetic beam experiments. In 1925, Uhlenbeck and Goudsmit postulated a spin angular momentum of $\frac{1}{2}\hbar$ for the electron in order to explain the fine structure splitting in the spectrum of hydrogen. It was soon noted that this assumption also explained the "anomalous" Zeeman effect (see Section 7), the doublet structure of the spectra of the alkali metals, and the anomalous magnetic moment reported by Compton, provided that one also assumed a gyromagnetic ratio (Equation 11.20, below) for spin of roughly twice the classical value. Although it must be assumed in the Schrödinger treatment, the half-integral spin of the electron arises quite naturally out of the relativistic theory of Dirac, so its validity as an additional coordinate has been established.

Figure 11–3 illustrates the right-handed convention for relating the direction of an angular momentum vector to the sense of the rotation. In the case of orbital

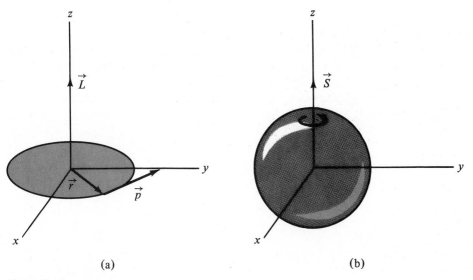

(a)   (b)

Figure 11–3

(a) Orbital angular momentum, $\vec{L} = \vec{r} \times \vec{p}$. and (b) spin angular momentum.

angular momentum, experiments confirm the hypothesis that the allowed values are always $m_\ell \hbar$, where m_ℓ is 0 or an integer. However, for the spin of the electron there are only two allowed values, $\pm\frac{1}{2}\hbar$, which are often referred to simply as "spin up" and "spin down." These two possible orientations are described by a fourth quantum number, m_s, which can be either $m_s = +\frac{1}{2}$ or $m_s = -\frac{1}{2}$. This new quantum number is incorporated in the wave function for hydrogen by writing it as

$$\psi_{n\ell m_\ell m_s} = \psi_{n\ell m_\ell} \cdot \chi_{m_s},$$

where χ_{m_s} is the spin-dependent part of the wave function. Thus, an electronic state in hydrogen is precisely defined by a given set of values for the four quantum numbers (n, ℓ, m_ℓ, m_s). Around 1925, Wolfgang Pauli enunciated a principle, now known as the *Pauli exclusion principle*, which may be stated as follows:

> No two electrons in the same atom can have the same set of four quantum numbers.

It turns out that this principle applies not only to electrons but to all spin-$\frac{1}{2}$ particles, including protons and neutrons, which comprise a group of particles known as *fermions*. A more general statement of the Pauli principle is based upon the symmetry of the wave function under particle exchange.[1]

Applying the Pauli principle to the hydrogen atom, we note that the first shell

[1] See E. E. Anderson, *Modern Physics and Quantum Mechanics*, W. B. Saunders, Philadelphia, 1971, p. 302 ff.

has just two electron states given by the quantum numbers $(1,0,0,\pm\frac{1}{2})$. The second shell has eight states given by $(2,0,0,\pm\frac{1}{2})$, $(2,1,0,\pm\frac{1}{2})$, $(2,1,1,\pm\frac{1}{2})$, and $(2,1,-1,\pm\frac{1}{2})$. The third shell has eighteen states denoted by $(3,0,0,\pm\frac{1}{2})$, $(3,1,0,\pm\frac{1}{2})$, $(3,1,1,\pm\frac{1}{2})$, $(3,1,-1,\pm\frac{1}{2})$, $(3,2,0,\pm\frac{1}{2})$, $(3,2,1,\pm\frac{1}{2})$, $(3,2,-1,\pm\frac{1}{2})$, $(3,2,2,\pm\frac{1}{2})$, and $(3,2,-2,\pm\frac{1}{2})$. This enumeration can be done for any shell. Each ℓ value corresponds to an orbital subshell of the principal shell. The $\ell = 0$ subshell is called the s subshell or s orbital, the $\ell = 1$ subshell is denoted by p, the $\ell = 2$ subshell by d, and the $\ell = 3$ subshell by f. It is sometimes convenient to use this notation for the wave functions in Equation 11.12, since $\psi_{1s} = \psi_{100}$, $\psi_{2s} = \psi_{200}$, and $\psi_{3s} = \psi_{300}$. The notation ψ_{2p} includes the three functions ψ_{210} and $\psi_{21,\pm1}$.

From the above discussion we see that there are only two allowed electron states when $n = 1$. These are the $1s$ states given by the quantum numbers $(1,0,0,\pm\frac{1}{2})$. For $n = 2$, ℓ can be either 0 or 1. As before, the zero value of ℓ corresponds to s-states, but here they are the $2s$ states given by $(2,0,0,\pm\frac{1}{2})$. The value of $\ell = 1$ can have three values of m_ℓ, namely, 1, 0, and -1, which results in the six following $2p$ states: $(2,1,1,\pm\frac{1}{2})$, $(2,1,0,\pm\frac{1}{2})$, and $(2,1,-1,\pm\frac{1}{2})$.

In a similar fashion we find for $n = 3$ that there are two $3s$ states and six $3p$ states. In addition, however, there are now ten $3d$ states.

Although hydrogen is the only neutral atom that can be treated exactly by the above model, all of the atoms in the periodic table are found to follow the same scheme of orbital electronic states as in hydrogen. Of course, the radii and the energies vary considerably with atomic number. However, in a neutral unexcited atom the electrons generally fill the lowest possible energy states in an orderly fashion consistent with the Pauli principle. Thus, in helium the second electron also occupies the ψ_{1s} orbital state, but its spin must be opposite to that of the first electron. Since these two electrons exhaust the possible electronic states for $n = 1$, we say that the first shell is full or "closed." In the case of lithium, with $Z = 3$, the third electron must go into the $2s$ subshell. For beryllium, with $Z = 4$, the fourth electron closes the $2s$ subshell. The fifth electron of boron goes into a $2p$ state, since the $1s$ and $2s$ subshells are filled. The six $2p$ states are successively filled in the elements from $Z = 5$ to $Z = 10$. Neon, for which $Z = 10$, has a completely closed shell arrangement, since the $1s$, $2s$, and $2p$ subshells are all full. The specification of the quantum states for all of the electrons of a given atom (or ion) is known as the *electronic configuration* for that atom. Thus, neon has the electronic configuration $1s^2 2s^2 2p^6$. A characteristic of a closed shell or a closed subshell is that it has a net angular momentum of zero and a net spin of zero, since all of the orbital momenta and spins cancel in pairs. Another property of the elements with closed shells is that they are so chemically inert that they have been known historically as the noble gases.

Electrons occupying a subshell beyond the outermost closed shell of an atom are called *valence* electrons. The valence electrons are the electrons that take part in the formation of chemical bonds. In a solid, these are the electrons that form the conduction bands of metals and the valence and conduction bands of semiconductors. Many aspects of the chemical and physical behavior of substances depend upon how firmly these valence electrons are bound to their atom. Figure 11–4 shows a plot of the energy required to remove the last electron of the configuration versus atomic number. Note the great stability of the electronic configurations for the noble gases, which have no valence electrons. However, the element

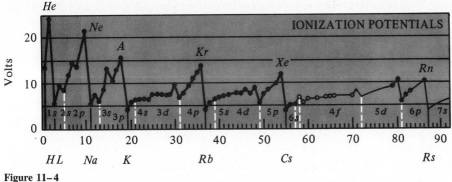

Figure 11–4

Ionization potential for the outermost electron of each element. (From *Introduction to Atomic Spectra* by Harvey E. White. Copyright 1934 by McGraw-Hill Book Company. Used with permission.)

just after each noble gas is quite active chemically because its last electron is very loosely bound. Likewise, the element just before each noble gas is active because the hole in its closed-shell configuration is easily filled. The energy given up by the atom when an electron fills the hole is called the *electron affinity*. For further details on the relationship of the properties of the elements to their electronic configurations, the reader is referred to a text on general chemistry. The ground state configuration of each element is given in Appendix D.

4. ELEMENTARY MAGNETIC MOMENTS

In the study of electromagnetism one learns that a current loop has associated with it a magnetic moment, whose magnitude is equal to the product of the current and the area enclosed by the loop, and whose direction is perpendicular to the plane of the loop (Figure 11–5). Then, for a circular Bohr orbit of radius r, we have

$$\vec{\mu}_\ell = \hat{r} \times \vec{i}\pi r^2, \tag{11.15}$$

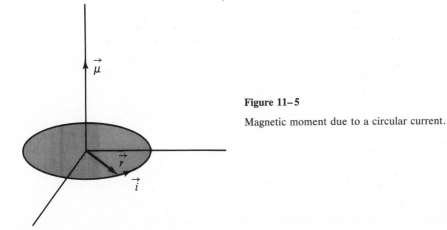

Figure 11–5

Magnetic moment due to a circular current.

where $\vec{i}$ is in amperes and the unit vector $\hat{r}$ aids in determining the correct direction for the magnetic moment.

For a circular orbit,

$$\vec{i} = \frac{-e\vec{v}}{2\pi r}, \tag{11.16}$$

where $-e$ is the charge of the electron. Then, substituting into Equation 11.15,

$$\vec{\mu}_\ell = -\frac{e}{2}\,\vec{r} \times \vec{v} = -\frac{e}{2m}\,\vec{r} \times \vec{p}, \tag{11.17}$$

where m is the mass of the electron. It should be recalled that the quantity $\vec{r} \times \vec{p}$ is the orbital angular momentum, which, according to Bohr's postulate, must be equal to an integer times $\hbar$. Replacing $\vec{r} \times \vec{p}$ with $m_\ell\hbar\hat{\ell}$, we obtain

$$\vec{\mu}_\ell = -\frac{e\hbar}{2m}\,m_\ell\hat{\ell} = -\mu_B m_\ell\hat{\ell}. \tag{11.18}$$

The minus sign tells us that the magnetic moment is directed *antiparallel* to the orbital angular momentum because of the negative charge of the electron. The quantity μ_B is called the *Bohr magneton,* and its value in mks units is approximately

$$\mu_B = \frac{e\hbar}{2m} = 9.27 \times 10^{-24} \text{ (amp-m}^2 \text{ or joule/tesla).} \tag{11.19}$$

The ratio of the magnetic moment to the orbital angular momentum is called the classical *gyromagnetic ratio.*[1] It is commonly expressed as

$$\gamma_\ell = \frac{|\vec{\mu}_\ell|}{|\vec{\ell}|} = \frac{e}{2m}. \tag{11.20}$$

Since the orbital angular momentum of the electron has a definite magnetic moment associated with it, one might expect the spin angular momentum also to produce a magnetic moment. This is indeed the case, and it turns out that spin is twice as effective as orbital momentum in terms of the magnetic moment that arises. Thus, the gyromagnetic ratio for spin is

$$\gamma_s = \frac{|\vec{\mu}_s|}{|\vec{s}|} = \frac{e}{m}, \tag{11.21}$$

and

$$\vec{\mu}_s = -\frac{e}{m}\,\vec{s}.$$

However, the magnetic moment due to electron spin is also just one Bohr magneton, since the spin is $\frac{1}{2}\hbar$ instead of $\hbar$. That is,

$$\vec{\mu}_s = -\frac{e\hbar}{2m}\,\hat{s} = -\mu_B\hat{s}. \tag{11.22}$$

Let us now consider the effect of an external magnetic field upon an elementary magnetic moment. Since such moments are not static moments like bar magnets, they must be treated dynamically by taking into account the gyroscopic

---

[1] Other names for this quantity are the *magnetogyric ratio* and *magnetomechanical ratio.*

action that occurs when a torque acts on an angular momentum. By way of illustration, assume that a magnetic field $\vec{\mathscr{B}}$ acts in the z-direction and that an orbital angular momentum is oriented at an angle θ with respect to it, as shown in Figure 11–6. The torque on $\vec{\ell}$ is given by $\vec{\mu}_\ell \times \vec{\mathscr{B}}$, which is directed into the plane of the

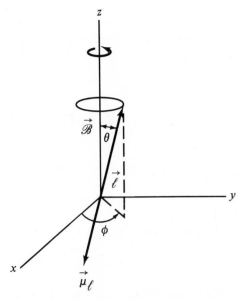

Figure 11–6

The precession of the angular momentum vector as a result of the torque produced by the action of a magnetic field on its associated magnetic moment.

page in the ϕ-direction. From elementary dynamics we know that the torque also equals the rate of change of the angular momentum, so it follows that

$$\frac{d\vec{\ell}}{dt} = \vec{\mu}_\ell \times \vec{\mathscr{B}} = \gamma_\ell \vec{\ell} \times \vec{\mathscr{B}}. \tag{11.23}$$

But $|d\vec{\ell}| = \ell \sin \theta \, d\phi$, so we may write the scalar equation

$$\ell \sin \theta \cdot \frac{d\phi}{dt} = \gamma_\ell \ell \mathscr{B} \sin \theta.$$

Defining the precessional velocity $\omega_L = d\phi/dt$, we obtain

$$\omega_L = \gamma_\ell \mathscr{B} = \frac{e\mathscr{B}}{2m}. \tag{11.24}$$

The angular velocity ω_L is often called the *Larmor frequency*. Using the Planck relation, the quantum of energy associated with this precession at the Larmor frequency is

$$\Delta E = \pm \omega_L \hbar = \pm \frac{e\hbar \mathscr{B}}{2m} = \pm \mu_B \mathscr{B}, \tag{11.25}$$

where the signs refer to the sense of the rotation. This energy difference should be recognized also as the potential energy of a magnetic dipole whose moment is one Bohr magneton. The dipolar energy in a magnetic field $\vec{\mathscr{B}}$ is given by

$$\Delta E = -\vec{\mu}_B \cdot \vec{\mathscr{B}}, \tag{11.26}$$

which becomes $+\mu_B \mathscr{B}$ for antiparallel alignment and $-\mu_B \mathscr{B}$ for parallel alignment.

Let us apply this to what we already know about the energy states of an electron in an atom. If an electron having a magnetic moment μ_B is in an energy state E_0 in the absence of a magnetic field, then it can take on one of the additional energies $E_0 \pm \mu_B \mathscr{B}$ in the presence of a magnetic field $\mathscr{B}$. The existence of the triplet of energy levels $\{E_0, E_0 \pm \mu_B \mathscr{B}\}$ for this special case has been known since 1897, and the phenomenon is referred to as the *normal Zeeman effect*. The Zeeman effect is really more complicated than the above classical model would lead one to believe. We know that electrons have spin and that the spin also has a magnetic moment associated with it. Thus, when a magnetic field is applied, both the spin and orbital angular momenta will experience precessional torques. The resulting energy levels are made more complicated by the fact that the spin is half-integral and not integral in units of $\hbar$, while the gyromagnetic ratio for spin is twice that of the orbital case. This will be treated in more detail in Section 7.

5. SPATIAL QUANTIZATION

The classical model of the precession of the magnetic moment in a magnetic field is a very useful one for describing certain aspects of modern experiments in magnetic resonance. However, it can be misleading if taken too seriously. For example, one would presume from the analysis of the previous section that the precessing vectors would relax through a continuum of precession angles, as energy is dissipated from the system, until the magnetic moment vector became aligned with the magnetic field. That is, the magnetic moment would follow a spiral path as it precessed and changed its polar angle.

However, the possibility of such a continuum of polar angles conflicts with a fundamental concept of quantum mechanics that is referred to as *spatial quantization*. Stated briefly, spatial quantization means that the projection of an angular momentum vector along any single axis in space must be one of a discrete set of allowed values. At any given time there can be only one axis of quantization; if it is altered, say by changing the direction of an applied uniaxial magnetic field, quantized states exist only along the new field direction.

The first conclusive evidence of spatial quantization was provided by the experiment first proposed by Stern and later performed with Gerlach in 1922. Silver atoms were evaporated in an oven and collimated into a narrow beam, which was passed through an inhomogeneous magnetic field as shown in Figures 11–7 and 11–8. Although a uniform magnetic field produces no net force on a magnetic dipole (only a torque), a non-uniform field can deflect a dipole with a net translational force. This can be readily seen by considering a magnetic field $\mathscr{B}$ that acts in the positive z-direction, and that is also designed to have a gradient in the positive z-direction (Figure 11–7). That is, the field intensity increases as z increases. Assuming that all other derivatives of the magnetic field are zero, we find that the translational force on a dipole oriented at an angle θ with the z-axis is

$$\vec{F} = -\vec{\nabla}E = \vec{\nabla}(\vec{\mu} \cdot \vec{\mathscr{B}}) = \vec{\nabla}(\mu \mathscr{B} \cos \theta)$$

$$= \mu \cos \theta \frac{\partial \mathscr{B}}{\partial z} \hat{k},$$

or
$$F_z = \mu \cos \theta \frac{\partial \mathscr{B}}{\partial z}. \tag{11.27}$$

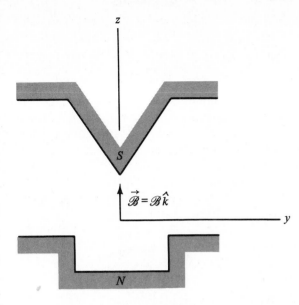

Figure 11–7

An inhomogeneous magnetic field having a large gradient in the field direction.

Thus, a dipole aligned with the field is acted upon by an upward force of magnitude $\mu(\partial \mathcal{B}/\partial z)$, while a dipole aligned antiparallel to the field is acted upon by a downward force of the same magnitude. For intermediate orientations, if allowed, the force would be reduced by the factor $\cos \theta$.

In the absence of spatial quantization, the beam of particles would contain a continuous distribution of angular orientations of the precessing dipoles, and one would expect that the effect of the non-uniform magnetic field would be to spread the narrow beam into a uniform band at the detector screen. However, instead of a continuous band, Gerlach and Stern obtained two distinct lines whose breadth was due to the spread in particle velocities rather than to a continuum of dipole orientations. This can be explained only if the dipoles are permitted just two orientational states, one state having a component in the $+z$-direction and the other having a component in the $-z$-direction. The former are deflected toward the most intense region of the field, whereas the latter are deflected toward the weakest region.

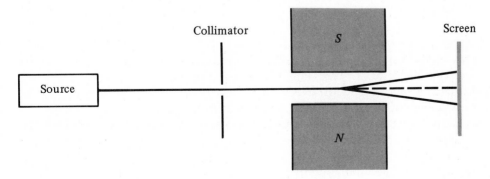

Figure 11–8

Particle trajectories in the experiment of Gerlach and Stern. The deflections of the beams are greatly exaggerated.

The spatial quantization condition underlying the Stern-Gerlach experiment may be stated as follows:

The allowed orientations of the angular momentum are such that their projections upon the axis of measurement differ successively by one unit of $\hbar$.

Thus, if the maximum projection is $1\,\hbar$ along the z-axis, there are three allowed orientations corresponding to z-components of 1, 0, and $-1\,\hbar$. In general, if the maximum component is $\ell\hbar$, there are $2\ell + 1$ allowed orientations of the angular momentum and its associated magnetic moment. Note that if the maximum component of the angular momentum is an integer, there will always be an *odd* number of allowed orientations (as illustrated in Figure 11–9) and there should also be an odd number of lines on the screen. However, for silver just two

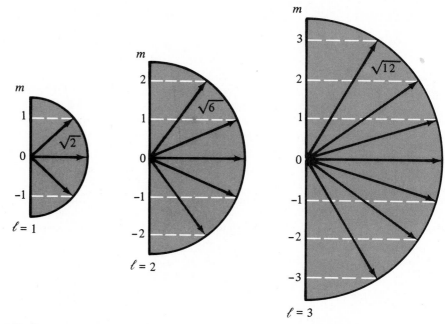

Figure 11–9

The allowed projections of the angular momentum for the cases of $\ell = 1$, 2, and 3.

lines were observed in spite of the fact that the magnetic moment was found to be one Bohr magneton. In 1927, Phipps and Taylor repeated the experiment with hydrogen, and again found just two lines. The explanation, of course, is that the moments observed in these two cases were due to spin and not orbital angular momentum. Since the maximum spin projection is $\tfrac{1}{2}\hbar$, there can be just the two allowed states $\pm\tfrac{1}{2}\hbar$ in order to satisfy the quantization condition that successive states differ by one unit of $\hbar$. Thus, atoms with half-integral spin cannot have the state corresponding to zero projections. It is interesting to note that these experiments that first demonstrated spatial quantization also showed the existence of spin angular momentum, although it was years before spin and its magnetic moment were recognized.

If the Stern-Gerlach experiment is performed with particles having one unit of angular momentum, the beam will be split into three distinct beams as required by spatial quantization. It is natural to inquire what would happen if each of these three beams were, in turn, subjected to another apparatus identical with the first. In Figure 11–10 three such beams are shown entering a second magnet with the

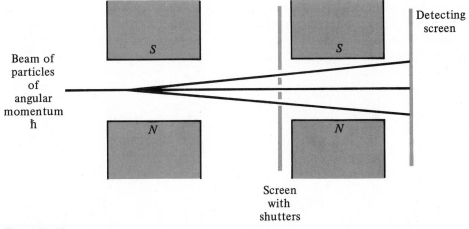

Figure 11–10

Apparatus for verifying the quantum states of particles having one unit of angular momentum.

same field orientation as the first. If only one beam at a time is permitted to enter the second magnet, just one beam will appear at the detecting screen. Hence, no further splitting will occur, regardless of which of the three beams is allowed to enter the second apparatus. The second magnet produces an additional deflection of the upper and lower beams, from which one can verify that the magnetic moments of the particles which constitute them correspond to $+\mu_B$ and $-\mu_B$, respectively. The center beam is undeflected and thus corresponds to zero projection of the magnetic moment along the vertical axis. From this result we draw the following conclusions. The first magnet establishes an axis of quantization and sorts the initial beam into three quantum states. The second magnet does not alter these quantum states but serves to confirm that the sorting of the beam into pure states has been preserved during the transit of the particles from one magnet to the other. On the other hand, if one were to alter Figure 11–10 by rotating the second magnet through an angle about the beam axis, the situation would be quite different. Each beam from the first magnet, which was a pure state in the first magnetic field, now appears as a mixture of the three quantum states in the new field orientation. Accordingly, if one beam at a time from the first magnet is allowed to enter the second, each will split into three components in the second apparatus.

*6. VECTOR COUPLING OF ANGULAR MOMENTA

Since the electron has magnetic moments associated with both its orbital motion and its spin, it is natural to expect that these magnetic moments would interact with each other. This interaction is called the *spin-orbit interaction*. A convenient way to treat the spin-orbit interaction is to use the rest frame of the electron and to calculate the interaction of its spin magnetic moment with the magnetic field produced by the orbiting proton.

We are not concerned here with the details of the interaction, but only with the fact that the orbital angular momentum $\vec{L}$ and the spin angular momentum $\vec{S}$ do exert torques on each other via their magnetic moments. However, if there is no external torque acting on the system, the *total* angular momentum must be a constant of the motion. Using the so-called *vector model*, we may express this total angular momentum as the vector sum $\vec{J} = \vec{L} + \vec{S}$. Two possible orientations of these vectors are shown in Figure 11–11(a). In the case of constant $\vec{J}$, $\vec{L}$ and $\vec{S}$ precess around $\vec{J}$ at the same angular frequency, as shown in Figure 11–11(b).

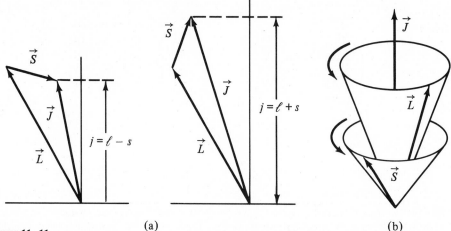

(a)   (b)

Figure 11–11

(a) The two possible orientations of the spin and orbital moments for a single electron. (b) Spin-orbit coupling causes the vectors $\vec{L}$ and $\vec{S}$ to precess about $\vec{J}$ at the same angular velocity.

When the total angular momentum $\vec{J}$ is used, it is appropriate to define a new quantum number j.

In a many electron atom the vector model provides two alternative ways of coupling the momenta of two or more electrons. When the spin-orbit interaction is large, the spin and orbital momenta of each electron combine to form $\vec{J}$ for each electron, and then the $\vec{J}$'s combine with each other vectorially. Hence this model is known as *j-j coupling*. For two electrons, with $\vec{J}_1 = \vec{L}_1 + \vec{S}_1$ and $\vec{J}_2 = \vec{L}_2 + \vec{S}_2$, the total angular momentum would be $\vec{J} = \vec{J}_1 + \vec{J}_2$.

The other form of coupling is known as Russell-Saunders coupling. Here it is assumed that the spin-orbit interaction is weaker than the interelectron interactions, so that the $\vec{L}_i$ combine to form $\vec{L}$, and the $\vec{S}_i$ combine to form $\vec{S}$. That is, we combine all of the orbital angular momenta to obtain the total orbital value $\vec{L}$, and we combine all of the spins to obtain the total spin $\vec{S}$. Then the total angular momentum vector $\vec{J}$ is obtained by combining $\vec{L}$ and $\vec{S}$. In forming these sums we also use some rules, known as Hund's rules, which are summarized below:

(1) Electron spins align so as to produce the largest value of S consistent with the Pauli principle. This means that the spins of a subshell are parallel if the shell is half full or less.

(2) In a given subshell, the states of largest m_ℓ values are filled first.

(3) The quantum number j is given by $j = \ell - s$ for less than half-filled shells, and $j = \ell + s$ for more than half-filled shells.

Historically, a special notation was developed by optical spectroscopists for labeling the quantum states derived from the Russell-Saunders coupling scheme. Following the same pattern used for the orbital angular momentum of a single electron (namely, that $\ell = 0, 1, 2, 3, \ldots$ corresponds to $s, p, d, f, \ldots$ states, respectively), the states for which the total orbital angular momentum L is $0, 1, 2, 3, \ldots$ are characterized by $S, P, D, F, \ldots$ spectroscopic *term symbols*, respectively. Furthermore, the term symbol is given a superscript, which is called the *multiplicity* of the term. It gives the number of possible values of J and is equal to $2S + 1$, where S is the total spin of the state. When $S = 0$, the multiplicity is 1 and the state is called a singlet state; when $S = \frac{1}{2}$ the state is a doublet, and when $S = 1$ the state is a triplet. The term symbol is also given a subscript corresponding to the total J value. By way of illustration, if a state has $L = 2$, $S = 1$, and $J = 3$, its term symbol is $^3D_3$. Likewise, if $L = 0$, $S = 1$, and $J = 1$, the term symbol is $^3S_1$. The following examples will serve to illustrate further the use of term symbols and the application of Hund's rules discussed above.

Example 11-5

Apply the vector coupling model to the ground state of samarium.

Solution

The electronic configuration for samarium ($Z = 62$) given in Appendix D shows that it has just one partially filled electron subshell, the 4f shell. Since an f-shell corresponds to $\ell = 3$, it can hold $2(2\ell + 1) = 14$ electrons. Since the 4f shell of samarium has 6 electrons it is less than half full, so the spins will all be parallel and the total spin quantum number will be $S = 6 \times \frac{1}{2} = 3$. According to the second rule above, the m_ℓ values must be 3, 2, 1, 0, −1, −2, which combine to give a total orbital momentum quantum number of $L = 3$. Then, from the third rule, $J = L - S = 3 - 3 = 0$. Therefore, the net angular momentum of samarium is zero in spite of the fact that it has both a large spin and a large orbital angular momentum. However, samarium does have a net magnetic moment because of the fact that spin has twice the gyromagnetic ratio of orbital momentum. Thus, the magnetic moment due to spin is 6 Bohr magnetons, while the orbital moment is 3 Bohr magnetons; since they are antiparallel, the net moment is 3 Bohr magnetons.

The term symbol for the ground state of samarium is obtained as follows: $L = 3$ corresponds to an F state; $J = 0$; and the multiplicity is $2S + 1 = 2(3) + 1 = 7$. Therefore, the ground state is denoted by $^7F_0$.

Example 11-6

Examine the ground state of iron by using the vector coupling model.

Solution

According to the table in Appendix D, iron has a closed-shell structure except for the 3d shell, which contains 6 electrons instead of the possible 10. Since the shell is more than half full by one electron, the spin of the sixth electron must be antiparallel to those of the other five. Then, $S = 5(\frac{1}{2}) - \frac{1}{2} = 2$. In order to find L we note that the first five electrons are required to have the m_ℓ values, 2, 1, 0, −1, −2, which combine to zero. The total orbital momentum, then, is due just to that of the sixth electron, which must have $m_\ell = 2$. Therefore, $L = 2$. Since the shell is more than half full, $J = L + S = 2 + 2 = 4$. The magnetic moments in this case are parallel and have the theoretical sum of 6 Bohr magnetons.

In order to find the term symbol for the ground state of iron, we note that the quantum numbers are $L = 2$, $S = 2$, and $J = 4$. The L value corresponds to a D state and the multiplicity is $2S + 1 = 5$, so that the term symbol is $^5D_4$.

*7. THE LANDÉ g FACTOR AND THE ZEEMAN EFFECT

In the previous section we defined the total angular momentum as

$$\vec{J} = \vec{L} + \vec{S}. \tag{11.28}$$

Let us now define a total magnetic moment in a similar fashion, where we use Equations 11.18 and 11.20 to write

$$\vec{\mu} = \vec{\mu}_\ell + \vec{\mu}_s = -\frac{\mu_B}{\hbar}\vec{L} - \frac{2\mu_B}{\hbar}\vec{S} = -\frac{\mu_B}{\hbar}(\vec{L} + 2\vec{S}). \tag{11.29}$$

If it were not for the factor of 2 in the parentheses, the quantities $\vec{J}$ and $\vec{\mu}$ would be collinear but antiparallel. However, because of the 2, $\vec{J}$ and $\vec{\mu}$ will not be on the same line unless either $\vec{L}$ is zero or $\vec{S}$ is zero. That is, $\vec{J}$ and $\vec{\mu}$ are antiparallel for the spin-only case and for the orbital-only case, but are never antiparallel when there are both orbital and spin angular momenta. This is illustrated schematically in Figure 11–12. Since $\vec{L}$ and $\vec{S}$ precess about $\vec{J}$ in the absence of an external magnetic field, it is apparent from the figure that $\vec{\mu}$ must also precess about $\vec{J}$. The component of $\vec{\mu}$ along $\vec{J}$ is defined as the effective magnetic moment μ_j. It may be expressed in a more convenient form by first finding the projection of the vector $\vec{\mu}$ along $\vec{J}$, as the following derivation shows:

$$\mu_j = \frac{\vec{\mu} \cdot \vec{J}}{|\vec{J}|} = -\frac{\mu_B}{\hbar}\frac{\vec{L} \cdot \vec{J} + 2\vec{S} \cdot \vec{J}}{|\vec{J}|}$$

$$= -\frac{\mu_B}{\hbar}\frac{\vec{L} \cdot (\vec{L} + \vec{S}) + 2\vec{S} \cdot (\vec{L} + \vec{S})}{|\vec{J}|}$$

$$= -\frac{\mu_B}{\hbar}\frac{\vec{L}^2 + 2\vec{S}^2 + 3\vec{L} \cdot \vec{S}}{|\vec{J}|}$$

But, in problem 11-6 it will be shown that

$$\vec{L} \cdot \vec{S} = \tfrac{1}{2}(\vec{J}^2 - \vec{L}^2 - \vec{S}^2),$$

so

$$\mu_j = -\frac{\mu_B}{\hbar}\frac{\vec{L}^2 + 2\vec{S}^2 + \tfrac{3}{2}(\vec{J}^2 - \vec{L}^2 - \vec{S}^2)}{|\vec{J}|}$$

$$= -\frac{\mu_B|\vec{J}|}{\hbar}\left(\frac{3\vec{J}^2 + \vec{S}^2 - \vec{L}^2}{2\vec{J}^2}\right)$$

$$= -\frac{\mu_B}{\hbar}|\vec{J}| \cdot \left(1 + \frac{\vec{J}^2 + \vec{S}^2 - \vec{L}^2}{2\vec{J}^2}\right)$$

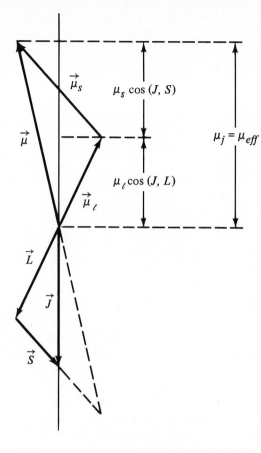

Figure 11–12

The *effective* magnetic moment of an atom is the component of $\vec{\mu}$ that lies along the direction of the total angular momentum.

$$\mu_j = -\mu_B \sqrt{j(j + 1)} \left[1 + \frac{j(j + 1) + s(s + 1) - \ell(\ell + 1)}{2j(j + 1)} \right].$$

Defining the Landé g factor as

$$g = 1 + \frac{j(j + 1) + s(s + 1) - \ell(\ell + 1)}{2j(j + 1)}, \qquad (11.30)$$

the *effective* magnetic moment becomes

$$|\mu_j| = g\mu_B \sqrt{j(j + 1)}. \qquad (11.31)$$

It should be noted that for zero spin the g factor reduces to the value $g = 1$, and for $\ell = 0$ it becomes $g = 2$. In more general cases where both spin and orbital quantum numbers are non-zero, the g factor may also take on fractional values, as is shown in the following examples.

Example 11-7

Calculate the g value for an s electron.

Solution

For an s electron, $\ell = 0$ and $s = \frac{1}{2}$, so $j = \frac{1}{2}$. Substituting these values into Equation 11.30, we see that

$$g = 1 + \frac{(\frac{1}{2})(\frac{3}{2}) + (\frac{1}{2})(\frac{3}{2})}{2 \cdot (\frac{1}{2})(\frac{3}{2})} = 1 + 1 = 2.$$

Example 11-8

Calculate the g value for a p electron.

Solution

For a p electron, $\ell = 1$ and $s = \frac{1}{2}$. But j has two possible values, $j = \ell - \frac{1}{2} = \frac{1}{2}$ and $j = \ell + \frac{1}{2} = \frac{3}{2}$.

For $j = \frac{1}{2}$,

$$g = 1 + \frac{(\frac{1}{2})(\frac{3}{2}) + (\frac{1}{2})(\frac{3}{2}) - 1(2)}{2 \cdot (\frac{1}{2})(\frac{3}{2})} = 1 - \frac{1}{3} = \frac{2}{3}$$

For $j = \frac{3}{2}$,

$$g = 1 + \frac{(\frac{3}{2})(\frac{5}{2}) + (\frac{1}{2})(\frac{3}{2}) - 1(2)}{2 \cdot (\frac{3}{2})(\frac{5}{2})} = 1 + \frac{1}{3} = \frac{4}{3}$$

Example 11-9

What are the allowed values of m_j in each of the previous examples?

Solution

For the s electron of Example 11-7 we saw that $j = \frac{1}{2}$. Then there are just two values of m_j, $m_j = \frac{1}{2}$ and $m_j = -\frac{1}{2}$.

For the p electron, $j = \frac{1}{2}$ again has just two projections, $m_j = \pm\frac{1}{2}$. However, $j = \frac{3}{2}$ has the four possible m_j values of $\frac{3}{2}, \frac{1}{2}, -\frac{1}{2}, -\frac{3}{2}$.

We are now ready to explain the so-called anomalous Zeeman effect mentioned earlier. In a weak magnetic field, the angular momentum $\vec{J}$ will precess about $\vec{\mathcal{B}}$ such that the projection of $\vec{J}$ along the magnetic field direction will be one of the allowed values, $m_j\hbar$. Taking the magnetic field direction to be the z-direction, the corresponding magnetic moment along the field direction will then be

$$\mu_z = -g\mu_B m_j. \tag{11.32}$$

It follows from Equation 11.26 that the magnetic dipolar energy is given by

$$E = -\mu_z\mathcal{B} = g\mu_B m_j\mathcal{B}. \tag{11.33}$$

Table 11–1 summarizes the quantum numbers, g values, and dipolar energies for the p and s electrons discussed above. Now let us see how this would look on an energy level diagram. On the left-hand side of Figure 11–13, the heavy line at the bottom represents the energy of an s state. Above that state, the p state corresponding to $j = \frac{1}{2}$ is at a higher energy by the amount E_2, while the $j = \frac{3}{2}$ state is higher by the amount E_1. When a magnetic field $\mathcal{B}$ is turned on, these three states split into new states corresponding to their allowed m_j values, and the magnitude of the splitting is proportional to the g value.

The double-headed arrows on the right side of Figure 11–13 correspond to the allowed electronic transitions that are accompanied by the emission of photons of light. These transitions are numbered from 1 to 10, and their relative energies are shown schematically by the spectrum given in Figure 11–14. By way of illustration, the energy of the photon emitted during the transition labeled by number 1 is $E_2 - \frac{4}{3}\mu_B\mathcal{B}$. This is arrived at by noting that the s state with $m_j = \frac{1}{2}$ is

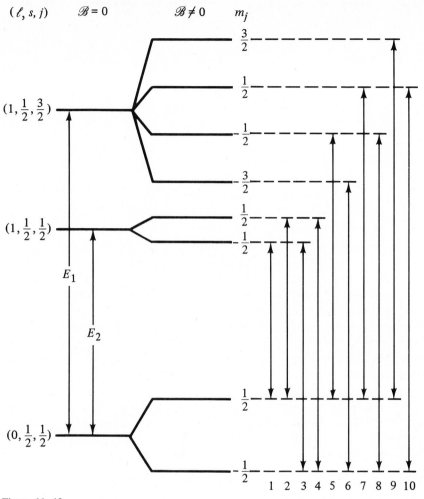

(ℓ, s, j) $\mathscr{B} = 0$ $\mathscr{B} \neq 0$ m_j

$(1, \tfrac{1}{2}, \tfrac{3}{2})$

$(1, \tfrac{1}{2}, \tfrac{1}{2})$

$(0, \tfrac{1}{2}, \tfrac{1}{2})$

E_1

E_2

$\tfrac{3}{2}$

$\tfrac{1}{2}$

$-\tfrac{1}{2}$

$-\tfrac{3}{2}$

$\tfrac{1}{2}$

$-\tfrac{1}{2}$

$\tfrac{1}{2}$

$-\tfrac{1}{2}$

1 2 3 4 5 6 7 8 9 10

Figure 11–13

Zeeman splittings for p and s states.

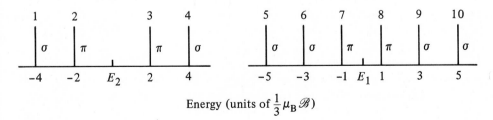

1   2   3   4   5   6   7   8   9   10

σ π π σ σ σ π π σ σ

-4 -2 E_2 2 4 -5 -3 -1 E_1 1 3 5

Energy (units of $\tfrac{1}{3}\mu_B\mathscr{B}$)

Figure 11–14

The spectral lines resulting from the electronic transitions shown in Figure 11–13. If the emitted light is observed perpendicular to the magnetic field, all of the above lines are seen (transverse Zeeman effect). However, if the emitted light is observed parallel to the field only the σ lines are seen (longitudinal Zeeman effect).

Table 11–1
Energy Splitting in a Magnetic Field

Orbital State	j	m_j	g	$\Delta E = gm_j$ (in units of $\mu_B\mathscr{B}$)
p	$\frac{3}{2}$	$\frac{3}{2}$	$\frac{4}{3}$	2
p	$\frac{3}{2}$	$\frac{1}{2}$	$\frac{4}{3}$	$\frac{2}{3}$
p	$\frac{3}{2}$	$-\frac{1}{2}$	$\frac{4}{3}$	$-\frac{2}{3}$
p	$\frac{3}{2}$	$-\frac{3}{2}$	$\frac{4}{3}$	-2
p	$\frac{1}{2}$	$\frac{1}{2}$	$\frac{2}{3}$	$\frac{1}{3}$
p	$\frac{1}{2}$	$-\frac{1}{2}$	$\frac{2}{3}$	$-\frac{1}{3}$
s	$\frac{1}{2}$	$\frac{1}{2}$	2	1
s	$\frac{1}{2}$	$-\frac{1}{2}$	2	-1

shifted upward by the external field by the amount $\mu_B\mathscr{B}$ (Table 11–1), while the p state with $j = \frac{1}{2}$ and $m_j = \frac{1}{2}$ is shifted downward by $\frac{1}{3}\mu_B\mathscr{B}$; thus, the zero-field transition energy, E_2, is reduced by $\frac{4}{3}\mu_B\mathscr{B}$. The spectral lines corresponding to transitions in which m_j changes by ± 1 are labeled σ and those corresponding to $\Delta m_j = 0$ are labeled π. The π lines are plane polarized, with the directions of polarization parallel to the field. The σ lines are circularly polarized when observed parallel to the field, and linearly polarized (perpendicular to the field) when observed at right angles to the field.

It is evident from the above that what is termed the "anomalous" Zeeman effect is, in reality, what would normally be expected for an electron having half-integral spin in a weak magnetic field. In fact, the "normal" Zeeman effect mentioned in Section 4 of this chapter cannot occur for a single electron because of its spin. However, in atoms in which the spins are paired so that the total spin is zero, the g value for all spectroscopic states is the classical value of 1. In such a case the spacings between the split states would be the same for all quantum numbers, and each allowed transition in the zero-field spectrum would produce three spectral lines in the presence of a magnetic field. The light associated with the pair of lines whose energies are shifted by $\pm\mu_B\mathscr{B}$ is circularly polarized, while that of the unshifted line is plane polarized.

Example 11-10
Show that the spectrum of Figure 11–13 would reduce to the normal Zeeman effect if the splittings were equal in all three energy levels.

Solution
Line 1 would have energy $E_2 - \mu_B$; lines 2 and 3, E_2; line 4, $E_2 + \mu_B$; lines 5 and 6, $E_1 - \mu_B$; lines 7 and 8, E_1; lines 9 and 10, $E_2 + \mu_B$.

SUMMARY

In this chapter Schrödinger's equation was applied to the hydrogen atom. First it was shown that the two-body problem in three dimensions is separable in the coordinates of the center of mass of the system and in the relative coordinates that give the position of the electron with respect to the proton. For a spherically symmetric potential, the Schrödinger equation can be separated further into a radial equation and an angular equation. The differential operator of the angular equation is shown to be the square of the total angular momentum operator. Then it follows that the angular momentum eigenfunctions are the spherical harmonics. The solutions of the radial equation are the associated Laguerre polynomials. Hence the total wave function for an electron in a hydrogen-like atom (neglecting spin) may be written as the product of a spherical harmonic, a Laguerre polynomial, and a time factor that depends upon the energy of the state. Each wave function is characterized by three spatial quantum numbers, n, ℓ, and m.

When it is assumed that the Coulomb force between the electron and proton is the only interaction, the energy levels obtained for hydrogen agree with those predicted by the Bohr theory. However, if spin-dependent forces or other interactions are included, the inverse-square nature of the central force is altered and the Bohr energies are no longer correct. Hydrogen is the only atom for which exact solutions of the Schrödinger equation are obtained. However, more complex atoms may be treated in the Schrödinger theory by employing sophisticated approximation methods in which hydrogen-like wave functions are used as the starting point.

The discrete radii of the Bohr theory are not obtained in the Schrödinger theory, since the electron has a non-zero probability of being almost anywhere within the atom. However, the first four Bohr radii correspond to the *most probable* radial positions based on the Schrödinger solutions for the $1s$, $2p$, $3d$, and $4f$ shells, respectively.

Since the spin of the electron is not included in the Schrödinger theory, the wave function obtained must be multiplied by an additional factor—a spin function—in order to define the state completely. Thus, the spin quantum number and the three spatial quantum numbers are required in order to specify a quantum state for an electron in an atom. According to the Pauli exclusion principle, no two electrons in the same atom can have the same set of four quantum numbers. The application of this principle accounts for the number of electrons allowed in each shell or subshell, and thus explains the main features of the periodic chart as well as many of the chemical properties of the elements.

The elementary magnetic moment, the *Bohr magneton,* was derived and its behavior in a magnetic field was discussed. When the magnetic moment is associated with an angular momentum, as in the elementary particles, the torque produced by an external magnetic field causes the angular momentum to precess. The frequency of this precession is known as the *Larmor frequency.* When photons corresponding to the precession frequency are absorbed by the precessing moments, the process is called *magnetic resonance* absorption. A precessing magnetic moment can have only one of a discrete set of orientations in a magnetic field. That is, the angle between the magnetic moment and the magnetic field direction must change suddenly from one value to another with the absorption or emission of a photon corresponding to the resonant (Larmor) frequency. The energy of the photon is also equal in magnitude to the difference in magnetic dipole energy between the final and the initial orientations in the field. The term *spatial quantization* is used to express the fact that just a few precession angles are possible, in contrast with the classical concept of a continuum of angles from 0 to π radians. Spatial quantization requires that the projection of an angular momentum vector along a single axis in space be one of a discrete set of allowed values that differ from one another by $\hbar$. Since the magnetic moment of an elementary particle is collinear with its angular momentum vector, this constraint on the angular momentum also affects the magnetic moment. Spatial quantization is confirmed experimentally by beam experiments of the Stern-Gerlach type and by the Zeeman effect, both of which are discussed in some detail in the chapter.

The Russell-Saunders and the *j-j* vector coupling models for the addition of angular momenta were introduced and illustrated in connection with the anomalous Zeeman effect.

Wolfgang Pauli in 1931. (Courtesy of the American Institute of Physics Niels Bohr Library, S. A. Goudsmit Collection.)

M. Alonso and E. J. Finn, *Fundamental University Physics,* Vol. III, Addison-Wesley Publishing Co., Reading, Mass., 1968, Chapter 3.

E. E. Anderson, *Modern Physics and Quantum Mechanics,* W. B. Saunders Co., Philadelphia, 1971, Chapter 7.

A. Beiser, *Perspectives of Modern Physics,* McGraw-Hill Book Co., New York, 1968, Chapters 9, 10, and 11

H. D. Young, *Fundamentals of Optics and Modern Physics,* McGraw-Hill Book Co., New York, 1968, Chapter 8.

Additional Reading

PROBLEMS

11-1. Show that ψ_{200} as given in Equation 11.12 is a solution of Equation 11.1 corresponding to the energy eigenvalue $E = -w_0/4$.

11-2. Show that ψ_{300} as given in Equation 11.12 is a solution of Equation 11.1. What is the value of the energy eigenvalue E?

11-3. The wave function for the (3,2,0) state of hydrogen is

$$\psi_{320} = Cr^2 \cdot (3\cos^2\theta - 1)\, e^{-r/3a_0}$$

(a) What is the value of Λ, the eigenvalue for the square of the total angular momentum operator? (b) Show that the wave function is a solution of Equation 11.1 and find the energy eigenvalue E.

11-4. What is the most probable radial position for an electron in the ψ_{210} state of hydrogen?

11-5. Find the value of r for which the radial probability density is a maximum for the ψ_{320} state of hydrogen, using the wave function given in problem 11-3.

11-6. Find the expectation value $\langle r \rangle$ for the ψ_{200} state of hydrogen.

*11-7. Find the expectation value $\langle r \rangle$ for the ψ_{210} state of hydrogen.

*11-8. Find the expectation value $\langle r \rangle$ for the ψ_{320} state of hydrogen. The wave function is given in problem 11-3; the normalization constant C has the value $C = (81\, a_0^3\, \sqrt{6\pi a_0})^{-1}$.

11-9. Find the expectation value of the potential energy for an electron in the ψ_{210} state of hydrogen. Use the operator

$$\langle V \rangle = \left\langle -\frac{ke^2}{r} \right\rangle.$$

*11-10. Find the expectation value of the potential energy for an electron in the ψ_{320} state of hydrogen, using the operator given in problem 11-9. (See problems 11-3 and 11-8 for the wave function and its normalization constant.)

*11-11. (a) Find the expectation value of the kinetic energy for an electron in the ψ_{210} state of hydrogen. Use the operator

$$\langle T \rangle = \frac{1}{2\mu}\langle p^2 \rangle = \frac{-\hbar^2}{2\mu}\langle \nabla^2 \rangle.$$

(b) Use the result from problem 11-9 to show that

$$\langle E \rangle = \langle T \rangle + \langle V \rangle = -\frac{w_0}{4}.$$

11-12. (a) What is the difference in energy between an electron whose spin magnetic moment is aligned parallel to a magnetic field and one that is antiparallel to the field if the field strength is 0.9 tesla? (b) What is the frequency of the photon associated with a transition between these two states?

11-13. An electron in a circular orbit has an angular momentum $\hbar$. What is its Larmor frequency in a magnetic field of 0.5 tesla?

11-14. A spectral line having a wavelength of 5500 Å shows a normal Zeeman splitting of 1.1×10^{-2} Å. What is the magnitude of the magnetic field causing the splitting?

11-15. What will be the separation in angstroms due to the normal Zeeman splitting of the 4916 Å line in the mercury spectrum in a magnetic field of 0.5 tesla?

11-16. What are the frequencies of the three photons that can be observed in the normal Zeeman splitting of a 4500 Å line in a magnetic field of 0.4 tesla?

11-17. In hydrogen there is a difference in dipolar energy between the aligned and anti-aligned states, that is, when the spins of the proton and electron are antiparallel or parallel, respectively. A spin-flip transition between these states is accompanied by the emission or absorption of a photon of wavelength 21 cm. Such 21 cm radiation is prevalent in intergalactic space and is studied by radio-astronomers. What is the effective magnetic field experienced by the spin magnetic moment of the electron?

11-18. Calculate the *nuclear magneton,* $\mu_N = \dfrac{e\hbar}{2m_p}$, where m_p is the proton mass.

Express the answer in joules per tesla. The nuclear magneton is a convenient unit for expressing the magnetic moments of nuclei.

11-19. In electron spin resonance, the magnetic moment of the electron is flipped from a low energy state to a higher energy state by the absorption of a photon. What is the photon frequency required for electron spin resonance in a magnetic field of 7150 gauss? (10^4 gauss = 1 tesla)

11-20. What are the possible values of j and m_j for a d-electron? What are the g values?

11-21. A beam of silver atoms traveling at 450 m/s is sent through a Stern-Gerlach apparatus such as that shown in Figure 11-8. Calculate the angular deflection of the beam if the path length through the inhomogeneous magnetic field is 4 cm and the average field gradient is 500 tesla/meter.

11-22. If the silver atoms in problem 11-21 strike a screen that is 10 cm beyond the edge of the inhomogeneous magnetic field, what is the separation of the two beams observed at the screen?

11-23. Using Russell-Saunders coupling, obtain the spectral term for the ground state of Gd^{3+}, which has 7 electrons in the unfilled $4f$ shell. Calculate the g value and the effective magnetic moment.

11-24. Using Russell-Saunders coupling, obtain the spectral term for the ground state, the g value, and the effective magnetic moment for Nd^{3+}, which has 3 electrons in the unfilled $4f$ shell.

11-25. Using Russell-Saunders coupling, obtain the g value and the effective magnetic moment for Ho^{3+}, which has 10 electrons in the unfilled $4f$ shell.

*11-26. Two electrons that occupy different atoms or different orbits of the same atom automatically satisfy the Pauli principle. Such electrons are called *inequivalent* electrons. Find the 10 quantum states resulting from the coupling of two inequivalent p-electrons using (a) the Russell-Saunders coupling scheme and (b) the j-j coupling scheme. (c) Show that 36 states are produced by turning on a magnetic field in each of these models.

12 APPLICATIONS TO SOLID STATE PHYSICS

1. INTRODUCTION

In many respects a solid is like a giant molecule. The forces that hold it together are electrical in nature, and their collective effect is called the *cohesive force* of the solid. The total energy of the aggregate solid must be less than the total energy of all of the separated atoms, and this energy is called the binding energy or cohesive energy of the solid. Everyday experience tells us that the binding energy varies greatly from one substance to another, as evidenced by the great disparities in hardness and melting temperatures. If all of the binding forces are Coulomb forces, how can we account for these large differences in binding energies? Why do some solids cleave? Why is graphite a good lubricant?

The answers to such questions are found after a detailed study of the bonds of the solid. A crystal that cleaves can be split cleanly along certain planes without shattering. This implies that the bonds perpendicular to the cleavage plane can be broken without disturbing the other bonds in the solid. In graphite the bonds are extremely anisotropic; those which lie within a certain set of parallel planes are very strong, while those which hold these planes together are so weak that the planes easily slide against each other. This slipping accounts for the good lubricating properties of graphite.

Solids may be divided into four distinct classes according to the model used for the distribution of the uncompensated charges that participate in the bonding. These classes are molecular solids, ionic solids, covalent solids, and metals. Some of the general properties of representatives of these classes are summarized in Table 12–1.

The *molecular bond* or van der Waals attraction occurs between atoms or molecules which are electrically neutral but which have induced dipole moments that fluctuate rapidly in time. Any neutral atom can have a fluctuating dipole moment as a consequence of its zero-point motion, which is required by the uncertainty principle. An instantaneous dipole moment of one atom will induce a polarization in a second atom, which results in an attractive force between them. Although the time average of such a polarization is zero, there is always an instan-

			Physical Properties			
Bond Type	Binding Force	Examples	Mechanical	Thermal	Electrical	Optical
Molecular (van der Waals) 0.01–0.1 eV/atom	Attraction of induced oscillating dipoles and multipoles	Argon, paraffin, calomel	Soft, deformable	Low melting points	Insulators; dissolve in non-ionizing solvents	Transparent
Ionic 1–10 eV	Coulomb force between ions of opposite charge	NaCl, calcite	Cleave; hardness increases with ionic charge	High melting points	Insulators; conduct in solution	Transparent or colored
Covalent 3–10 eV	Covalent bond; hydrogen bond	Diamond, carborundum	Cleave; very hard	Very high melting points	Semiconductors, except diamond	Opaque, or transparent with high index of refraction
Metallic 1–5 eV	Electrostatic attraction between positive ion cores and electron gas	Copper, iron, sodium	Tough, malleable, ductile (except tungsten)	Moderately high melting points; good thermal conductors	Electrical conductors; soluble in acids	Opaque and reflecting

Table 12–1
Characteristics of the Four Types of Bonds in Solids

taneous attractive force. The molecular bond is relatively weak, typically having an energy in the range of 0.01 to 0.1 eV per atom.

An *ionic solid* is composed of positive and negative ions arranged so that the Coulomb attraction between unlike charges is greater than the Coulomb repulsion between like charges. The ions are formed by electron transfer so that each ion has a closed shell or rare gas configuration. For example, in sodium chloride the Na^+ ion has the neon configuration and the Cl^- ion has the argon configuration. In LiF the Li^+ ion has the helium configuration and the F^- ion has the neon configuration. Ionic bonds are much stronger than van der Waals bonds, and the binding energy per ion is of the order of 1 to 10 eV.

The *covalent bond* accounts for the binding of many organic compounds and most of the semiconductors. This form of bonding is often described as electron sharing, such as that which occurs in the formation of the hydrogen molecule. Since the two electrons that form the covalent bonds share the same spatial quantum state, they must have opposite spins in order to satisfy the Pauli principle. Covalent bonds are strong, typically 3 to 10 eV per atom in covalent crystals.

In simplest terms, the *metallic bond* results from the electrostatic attraction between the positive metal ions—which are assumed to be fixed at the atomic positions—and the "gas" of free electrons that pervades the metal. The binding energy of metals is typically 1 to 5 eV per atom.

2. THE FREE ELECTRON THEORY OF METALS

One of the earliest fruitful models of a metal was that of Sommerfeld, in which he represented a metal by the square well potential shown in Figure 12–1.

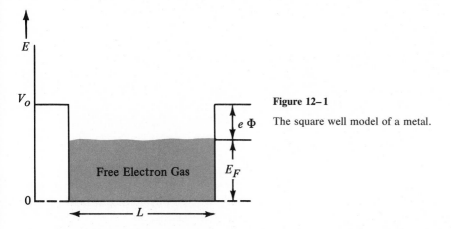

Figure 12–1

The square well model of a metal.

He assumed that each metal atom gives up at least one valence electron and that these "free" electrons may roam throughout the crystal lattice, since they are no longer identified with a particular atom site. The periodic fluctuations in the potential due to the regular array of positive ion cores is ignored. Instead it is assumed that each electron moves in a constant potential due to the average effect of the metal ions and the other free electrons. All direct interactions between electrons are neglected. The potential energy E_F in the figure is called the *Fermi energy;* it is the energy of the highest filled electronic state with respect to the bottom of the well. The potential energy $e\Phi$ is the work function of the metal as determined from the photoelectric effect or from thermionic emission. It is evident that $e\Phi$ is simply the energy required to carry an electron from the highest filled state to the surface of the metal where it would be truly free of the metal. The energy $E_F + e\Phi$, which represents the depth of the well, can be determined experimentally from the diffraction of slow electrons.[1]

If we take the bottom of the well as the zero point for energy, then Schrödinger's equation for an electron has the form

$$-\frac{\hbar^2}{2m}\left(\frac{d^2}{dx^2} + \frac{d^2}{dy^2} + \frac{d^2}{dz^2}\right)\psi_n = E_n\psi_n, \qquad (12.1)$$

where ψ_n is given by

$$\psi_n = \left(\frac{2}{L}\right)^{3/2} \sin k_x x \cdot \sin k_y y \cdot \sin k_z z. \qquad (12.2)$$

---

[1] C. Davisson and L. Germer, *Phys. Rev. 30,* 705 (1927).

This should be recognized as the problem of a particle in a three-dimensional box, and the treatment that follows is identical to the calculation of Jeans' number in Section 3 of Chapter 5. The allowed k-values are $k_x = (\pi/L)n_x$, $k_y = (\pi/L)n_y$, and $k_z = (\pi/L)n_z$. Defining

$$k^2 = k_x{}^2 + k_y{}^2 + k_z{}^2$$

and

$$n^2 = n_x{}^2 + n_y{}^2 + n_z{}^2,$$

we find that $k^2 = (\pi^2/L^2)n^2$ and

$$E_n = \frac{\hbar^2 k^2}{2m} = \frac{\hbar^2 \pi^2 n^2}{2mL^2} = \frac{h^2 n^2}{8mL^2}. \tag{12.3}$$

We know that the electronic states will be so closely packed that they will appear to form a continuum of energies. However, it is not at all evident that these states will be spaced uniformly. In order to determine their spacing it is convenient to calculate the quantity $g(E)$, which is called the *density of states* per unit volume. We know the density of states in λ from Equation 5.4, namely,

$$g(\lambda) = \frac{8\pi}{\lambda^4}. \tag{12.4}$$

The factor of 2 included in Equation 5.16 for the two polarization states in a transverse wave is required here also, because of the two possible spin states for an electron. In order to obtain $g(E)\,dE$ we use the relationships

$$g(E)\,dE = g(\lambda)\,d\lambda \tag{12.5}$$

and

$$E = \frac{\hbar^2 k^2}{2m} = \frac{h^2}{2m\lambda^2}. \tag{12.6}$$

Then,

$$\left|\frac{dE}{d\lambda}\right| = \frac{h^2}{m\lambda^3} \tag{12.7}$$

and

$$g(E) = \frac{g(\lambda)}{\dfrac{dE}{d\lambda}} = \frac{8\pi m}{h^2 \lambda} = 4\pi \left(\frac{2m}{h^2}\right)^{3/2} E^{1/2}. \tag{12.8}$$

Whenever the energy is proportional to the square of the wave vector, the density of states will have the parabolic form shown in Figure 12–2.

Equation 12.8 gives us the density of available electronic states but says nothing about the actual number of electrons occupying these states. If $f(E)$ is the distribution function that describes which states are occupied, then the number of electrons per energy interval dE is given by

$$n(E)\,dE = f(E) \cdot g(E)\,dE. \tag{12.9}$$

The distribution function for electrons must take the Pauli principle into account, that is, it must allow only one electron in a quantum state. This function is known as the Fermi-Dirac distribution function, and it has the following form[1]:

$$f(E) = \frac{1}{e^{(E-E_F)/kT} + 1} \tag{12.10}$$

---

[1] This is derived in E. E. Anderson, *Modern Physics and Quantum Mechanics*, W. B. Saunders Co., Philadelphia, 1971, pp. 333–334.

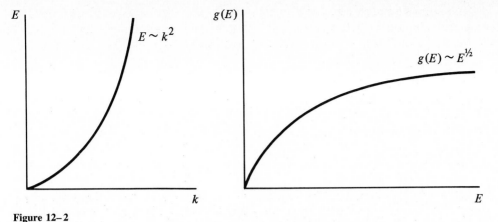

Figure 12–2

A parabolic density of states curve follows from a parabolic dispersion curve.

where k is the Boltzmann constant. The plus sign in the denominator is the distinguishing feature between the Fermi-Dirac function and the Bose-Einstein distributions for photons that was given in Equation 5.26. Let us examine the properties of Equation 12.10 at the absolute zero of temperature. For energies below the Fermi energy, that is, for $E < E_F$, the exponential term goes to $e^{-\infty} = 0$ and $f(E) = 1$. However, for $E > E_F$, the exponential term goes to e^{∞} and $f(E) = 0$. Then, at absolute zero all states below the Fermi energy are fully occupied and all states above the Fermi energy are empty. This is shown in Figure 12–3. Also

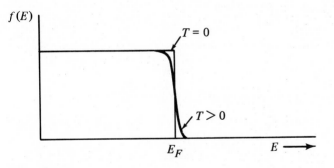

Figure 12–3

The Fermi-Dirac function for $T = 0$ and for $T > 0$.

shown in the figure is a sketch of the function at a relatively high temperature. Note that some rounding of the function occurs below E_F and a small tail appears above E_F due to the thermal excitation of electrons from states below E_F to empty states above E_F. This is not a large effect, since only electrons that are within an energy range equal to kT from the Fermi energy can be excited. At room temperature, for example, $kT \sim 1/40$ eV while E_F is normally about 5 to 10 eV; that is, the depth of thermal excitation is about one-half of one percent of the Fermi energy. Figure 12–4 shows a plot of Equation 12.9 where the Fermi distribution function has been incorporated with the density of states curve of Figure 12–2.

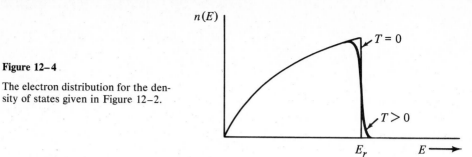

Figure 12–4

The electron distribution for the density of states given in Figure 12–2.

Example 12-1

Calculate the Fermi energy at the absolute zero of temperature for a metal having $N = 8.38 \times 10^{28}$ electrons per cubic meter.

Solution

From Equations 12.8 and 12.9,

$$n(E)\, dE = f(E) \cdot g(E)\, dE = f(E) \cdot 4\pi \left(\frac{2m}{h^2}\right)^{3/2} E^{1/2}\, dE.$$

But $f(E) = 1$ below E_F, so

$$N = \int_0^{E_F} n(E)dE = \int_0^{E_F} g(E)dE = 4\pi \left(\frac{2m}{h^2}\right)^{3/2} \int_0^{E_F} E^{1/2}\, dE = \frac{8\pi}{3}\left(\frac{2m}{h^2}\right)^{3/2} E_F^{3/2}$$

$$E_F = \left[\frac{3N}{8\pi}\left(\frac{h^2}{2m}\right)^{3/2}\right]^{2/3} = \left(\frac{3N}{8\pi}\right)^{2/3} \cdot \frac{h^2}{2m}$$

$$= 4.64 \times 10^{18} \times 2.41 \times 10^{-37}\ \text{J} = 11.18 \times 10^{-19}\ \text{J} = 7\ \text{eV}.$$

Example 12-2

Find the average energy per electron at absolute zero, using the free electron model.

Solution

At 0 K, electrons fill all of the states from the lowest up to the Fermi level. Then the average energy per electron is given by

$$\langle E \rangle = \frac{\displaystyle\int_0^{E_F} E \cdot f(E) \cdot g(E) \cdot dE}{\displaystyle\int_0^{E_F} f(E) \cdot g(E) \cdot dE} = \frac{\displaystyle\int_0^{E_F} E \cdot g(E) \cdot dE}{\displaystyle\int_0^{E_F} g(E) \cdot dE}.$$

Using Equation 12.8 and the results of Example 12-1,

$$\langle E \rangle = \frac{\displaystyle\int_0^{E_F} 4\pi \left(\frac{2m}{h^2}\right)^{3/2} E^{3/2}\, dE}{\dfrac{8\pi}{3}\left(\dfrac{2m}{h^2}\right)^{3/2} E_F^{3/2}} = \frac{3}{2}\left(\frac{1}{E_F}\right)^{3/2} \int_0^{E_F} E^{3/2}\, dE = \frac{3}{5} E_F.$$

Example 12-3

Assuming that an electron at the Fermi level has kinetic energy equal to $E_F = 7.0$ eV, find its momentum, its velocity, and its temperature equivalent in the kinetic theory of ideal gases.

Solution

The *Fermi momentum* of an electron is given by the classical expression,

$$p_F = \sqrt{2mE_F} = (2 \times 9.1 \times 10^{-31} \text{ kg} \times 1.1 \times 10^{-18} \text{ J})^{1/2}$$

$$= 1.4 \times 10^{-24} \text{ kg-m/s}.$$

The *Fermi velocity* is

$$v_F = p_F/m = \frac{1.4 \times 10^{-24} \text{ kg-m/s}}{9.1 \times 10^{-31} \text{ kg}} = 1.5 \times 10^6 \text{ m/s}.$$

The *Fermi temperature* is given by $kT_F = E_F$, or

$$T_F = E_F/k = \frac{1.1 \times 10^{-18} \text{ J}}{1.38 \times 10^{-23} \text{ J/K}} = 80{,}000 \text{ K}.$$

3. THE BAND THEORY OF METALS

Since the metal ions occupy regular positions in the crystal lattice, the potential affecting each free electron would not have a constant value as was assumed in the previous section. Instead, the potential would vary periodically in three dimensions in a manner similar to that depicted for one dimension in Figure 12–5. Each of the wells in Figure 12–5 represents the potential of an electron in the vi-

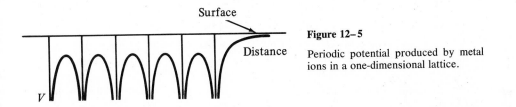

Surface

Distance

Figure 12–5

Periodic potential produced by metal ions in a one-dimensional lattice.

V

cinity of a single metal atom. An electron that is so tightly bound that it is confined to just one atom will occupy one of the energy levels of that atom. When the electron is more loosely bound so that it can interact with a neighboring atom, each atomic level is split into two closely spaced energy levels, as in the case of the interacting potential wells discussed in Section 4 of Chapter 10. In general, if an electron interacts with N wells, each atomic level is split into N levels. In a metal, N becomes a very large number, since each free electron can interact with every metal atom of the crystal, so each atomic energy level forms a band or continuum of energy states. The process of band formation is illustrated as a function of atomic separation for solid hydrogen in Figure 12–6.

It is customary to represent the periodic potential of Figure 12–5 as a first approximation by the multiple square well potential shown in Figure 12–7. Calculations of the bands based on this model are quite simple and are reasonably good. Since the potential function must have the period of the lattice, that is, $V(x + d) = V(x)$, we expect the wave function to have the same periodicity. It turns out that the wave functions are plane waves as before, but they are "modulated" by the periodicity of the lattice. That is, the solutions have the form

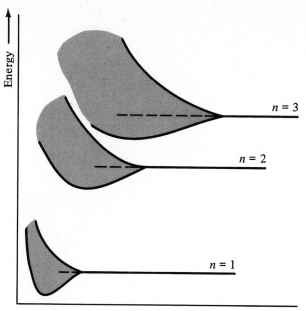

Figure 12–6

At small atomic separations the electron energy levels form continuous bands in solid hydrogen. At large atomic separations the bands shrink to the discrete levels of atomic hydrogen.

$$\psi(x) = e^{ikx}u(x), \tag{12.11}$$

where $u(x + d) = u(x)$ and $k = 2\pi nx/Nd$; N is the total number of wells, which is a very large number (the order of Avogadro's number). It follows that

$$\psi(x + d) = e^{ik(x+d)}u(x + d) = e^{ikd}\psi(x). \tag{12.12}$$

The last result enables one to obtain the solution anywhere in the lattice once it is known for a single well.

If we denote the width of each well by a, the thickness of the barrier by b, and the height of the barrier by V_0, then we may define the propagation constants

$$\hbar k_1 = \sqrt{2mE} \quad \text{and} \quad \hbar K_2 = i\hbar k_2 = \sqrt{2m(V_0 - E)},$$

where k_1 holds within the well boundaries and K_2 within the barrier boundaries. Although the procedure is straightforward, the algebra is too laborious to repro-

Figure 12–7

A square well representation of a linear lattice. Here $d = a + b$.

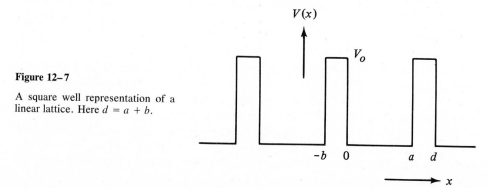

duce here. After substituting Equation 12.11 into Equation 9.17 and employing the boundary conditions, Equation 10.7, we obtain:

for $E > V_0$,

$$\left(1 + \frac{V_0^2}{4E(E - V_0)} \sin^2 k_2 b\right)^{1/2} \cos (k_1 a - \delta) = \cos kd$$

for $E < V_0$,

$$\left(1 + \frac{V_0^2}{4E(V_0 - E)} \sinh^2 k_2 b\right)^{1/2} \cos (k_1 a - \delta) = \cos kd \qquad (12.13)$$

Since the right-hand side of each of these equations is restricted to the range of values between 1 and -1, solutions exist only for those energies for which the magnitude of the left-hand side does not exceed unity. As a consequence of this restriction, the energies that correspond to physically realizable solutions form continuous bands, which are separated by *forbidden* bands of energies for which no solutions exist. This is shown in Figure 12–8, where the left-hand sides of

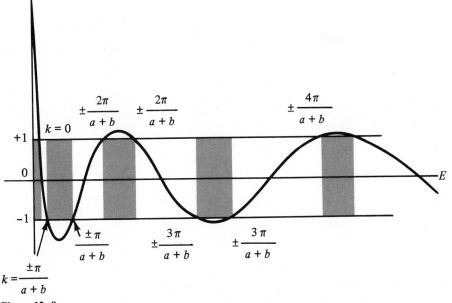

Figure 12–8

The left-hand side of Equation 12.13 plotted against energy. The shaded regions correspond to forbidden bands.

Equation 12.13 are plotted versus energy. The forbidden bands are shown by the shaded regions.

Note that the breadth of the allowed bands increases as E increases, approaching the limit of a continuum of energies in the free particle state. At the other extreme we see that as the well gets stronger (that is, for V_0 very large), the bands shrink and the allowed energies reduce to the discrete energy levels of the particle in the infinite well, Equation 9.1. The latter corresponds to the electron being tightly bound to a single positive ion.

Although the details of the energy band structure of a solid require a much more refined model than that used here, it is instructive to see that the general quantum mechanical behavior of the electrons can be predicted by just using the elementary concepts of wells and barriers developed in the previous sections.

4. METALS, INSULATORS, AND SEMICONDUCTORS

The characteristic of a solid that determines whether it is a metal or an insulator is the position of the Fermi energy relative to the energy bands of the valence electrons. For example, in a metal the Fermi energy is below the top of an energy band, as in Figure 12–9(a), so there are adjacent empty states that are readily ac-

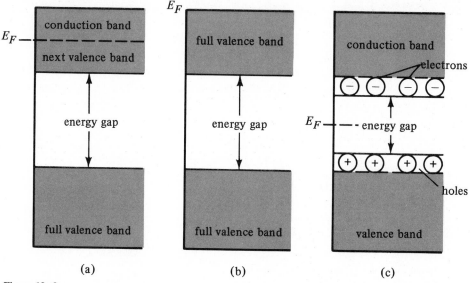

Figure 12–9

(a) The Fermi level separates the conduction band and the valence band in a metal. (b) In an insulator the Fermi level is the top of the valence band. (c) In a semiconductor the gap is narrow enough to permit thermal or optical excitation to the conduction band.

cessible to electrons near the Fermi level. At any temperature above absolute zero, thermal excitations will scatter some electrons into these states. This continuum of states above and adjacent to the Fermi energy is called the *conduction band* because electrons that occupy these states will contribute to the electrical conductivity of the substance when an electric field is applied. Thus, the conductivity of a metal should be directly proportional to the number of electrons at the Fermi level (the density of states at E_F) and the number of available empty states in the conduction band (the breadth of the conduction band).

In Figure 12–9(b) the Fermi energy coincides with the top of the valence band, so there are no empty states available. Since electrons have nowhere to go, there can be no conduction and the material is an insulator. The energy gap between E_F and the next allowed band is much too large for the excitation of electrons to the upper band optically, thermally, or by ordinary electric fields. In a sufficiently large electric field, electrons may be torn from a filled band and thrust into a higher empty band. The material then ceases to be an insulator, since the holes in the lower band permit conduction to occur there while the electrons in the upper band form a conduction band. This phenomenon is known as *dielectric breakdown*.

In the materials known as *semiconductors,* the situation depicted in Figure 12–9(c) is achieved without a catastrophic event like dielectric breakdown. Instead, the electrons are excited across the gap thermally. This is achieved either by using a material with a very narrow forbidden energy gap or by introducing impurities that form intermediate states (like stepping stones) within the gap. Since the electrical conductivity is proportional to the number of electrons in the conduction band and this number increases exponentially with temperature, the electrical properties of semiconductors are extremely temperature-dependent. The conductivity of a semiconductor *increases* rapidly with temperature, while the conductivity of a metal decreases as the temperature rises.

The foregoing discussion is highly oversimplified in view of the real world of three-dimensional crystals. In the first place, the Fermi energy in Figure 12–9 becomes a *Fermi surface* in three dimensions. This surface is a sphere in an isotropic crystal but is a very complicated surface in some real materials. Further, the band gap often varies widely with *k*-value, and sometimes two or more bands will overlap as in Figure 12–6. The detailed knowledge of such features of real materials has led to the development of numerous solid state circuit elements and devices.

5. TYPES OF SEMICONDUCTORS

Semiconductors are broadly classified in two groups, namely, *intrinsic* semiconductors and *impurity* semiconductors. Intrinsic semiconductors are pure materials in which the charge carriers are created by the thermal or optical excitation of electrons from the valence band to the conduction band. Since each conduction electron leaves a hole in the valence band, the number of holes is always equal to the number of conduction electrons in an intrinsic semiconductor. The process of electrical conduction is illustrated in Figure 12–10. An impurity (or *extrinsic*) semiconductor is one in which the majority of free charge carriers are provided by impurity atoms situated in lattice sites of the crystal. Each impurity atom replaces one of the original atoms of the crystal; it does not occupy an interstitial site. Of

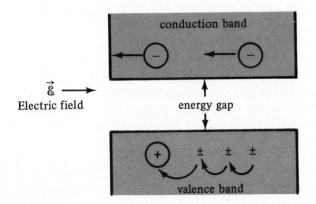

Figure 12–10

Conduction of electrons and holes in a semiconductor. If an electric field acts toward the right, electrons in the conduction band will move toward the left. A hole in the valence band moves progressively to the right as successive electrons move to the left to fill it.

course, the intrinsic process of thermal excitation of an electron across the entire band gap still occurs in an impurity semiconductor; it may, in fact, be the dominant process at very low temperatures. However, at operating temperatures the carrier concentrations of an impurity semiconductor are primarily determined by the type of impurity and its concentration. If each impurity atom provides an extra electron above the number required to complete the covalent bonds in the host crystal, the majority charge carriers are electrons and the semiconductor is denoted as *n-type* (negative majority carriers). On the other hand, if the impurity has an insufficient number of electrons to complete all of the covalent bonds of the host, there will be a free hole at the impurity site and the semiconductor is called *p-type* (positive majority carriers). See Figure 12–11.

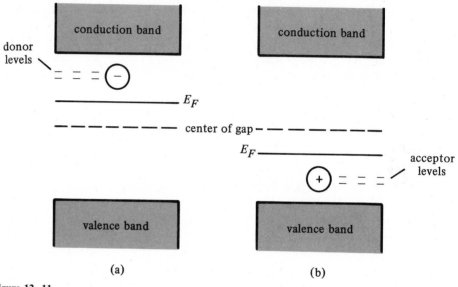

(a)   (b)

Figure 12–11

(a) In an *n*-type semiconductor the Fermi level E_F is above the midpoint of the gap near the donor levels. (b) In a *p*-type semiconductor the Fermi level is below the midpoint of the gap near the acceptor levels.

The most widely studied intrinsic semiconductors are crystals of the pure elements, the prototypes of which occupy Group IV of the table of the elements. Carbon, silicon, germanium, and grey tin have the diamond structure and are characterized by tetrahedral bonds. If an impurity from Group V, such as phosphorus or arsenic (see Table 12–2 and Figure 12–12), replaces an atom such as silicon, there will be an extra electron and the semiconductor will be *n*-type. However, if a silicon atom is replaced by a Group III atom such as boron or gallium, the lattice site is deficient one electron and the semiconductor will be *p*-type. An impurity atom that produces an *n*-type semiconductor is called a *donor* atom, while an impurity that produces a *p*-type material is called an *acceptor*. Although this terminology occurs frequently in the literature, it is somewhat ambiguous. For example, Ge is a donor atom when it is an impurity in Ga, it is an acceptor when it is an impurity in As, it is a host atom (neither donor nor acceptor) when impurities are added to the germanium crystal itself, and it is an intrinsic semiconductor in a pure germanium crystal.

	II	III	IV	V	VI
		B	C		
		Al	Si	P	S
	Zn	Ga	Ge	As	Se
	Cd	In	Sn	Sb	Te
			Pb	Bi	

**Table 12–2
The Elements That Comprise the Vast Majority
of the Semiconductors now in Use**

Many useful semiconductors have been developed from compounds in which the average number of electrons per atom is four as in the Group IV elements. Examples are the III-V compounds such as GaP, GaAs, and InSb, and the II-VI compounds such as ZnS, CdTe, CdS, and CdSe.

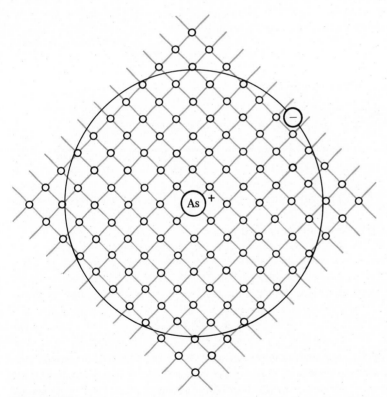

Figure 12–12

A silicon or germanium host lattice, with an arsenic impurity atom in a lattice site. Four of the arsenic valence electrons form covalent bonds with the nearest neighbors in the host lattice. The fifth electron is loosely bound in a "Bohr" orbit at low temperatures, but enters the conduction band at higher temperatures. Holes are bound in a similar manner when the lattice is doped with acceptor atoms such as Ga or In instead of As.

6. BINDING ENERGIES OF IMPURITY STATES

At all but extremely high temperatures the impurity atom itself is an immobile ion, fixed at a lattice site in the impurity semiconductor. A donor impurity is a positive ion to which the extra electron is weakly bound, while an acceptor impurity is a negative ion to which a hole is weakly bound. We will now examine the nature of this binding. It is known to be quite weak, since even at very low temperatures the thermal activation energy is sufficient to produce large increases in the carrier concentration.

The model we will use is the simple Bohr model. That is, we visualize the carrier to be bound in a large Bohr orbit about the impurity center, (see Figure 12–12), the orbit having a radius of 10 to 100 atomic spacings. This problem could be treated quantum mechanically by hydrogenic wave functions and a screened Coulomb potential. However, we get a reasonably good answer by using the simple Bohr expression, Equation 7.18,

$$\frac{m^* v^2}{r} = \frac{ke^2}{\kappa r^2},$$

(12.14)

where the dielectric constant κ takes into account the screening of the Coulomb force, and the effective mass m^* provides an average correction due to the periodic potential. Using the Bohr quantum condition,

$$m^* v r = n\hbar,$$

(12.15)

the quantized orbits for the bound carrier are

$$r_n = \frac{\kappa n^2 \hbar^2}{k m^* e^2} = \kappa n^2 a_0 \left(\frac{m_0}{m^*} \right),$$

(12.16)

where a_0 is the Bohr radius for hydrogen. Thus, the ground state ($n = 1$) has an orbital radius that is $\kappa(m_0/m^*)$ times as large as the Bohr radius. In germanium, for example, $\kappa \sim 16$ and $m^* \sim 0.25 m_0$, so the ground state orbit has a radius of about 34 Å. This is roughly 10 to 20 atomic spacings, and it indicates that the use of the dielectric constant to account for the screening of the Coulomb interaction is a reasonable one. In order to find the binding energy we write

$$E = T + V = \frac{1}{2} m^* v^2 - \frac{ke^2}{\kappa r} = \frac{ke^2}{2\kappa r} - \frac{ke^2}{\kappa r} = -\frac{ke^2}{2\kappa r}.$$

(12.17)

From Equation 12.16, with $n = 1$,

$$E = \frac{-ke^2}{2\kappa^2 a_0} \cdot \left(\frac{m^*}{m_0} \right) = -\frac{w_0}{\kappa^2} \left(\frac{m^*}{m_0} \right),$$

(12.18)

where w_0 is the ionization energy of hydrogen. With the values for germanium given above, the binding energy is $\sim 10^{-3} w_0$ or about 0.01 eV. Since kT is greater than this binding energy, even considerably below room temperature, it is easy to see why such states are ionized at ordinary temperatures. The meaning of the term "ionization" as used here includes either of the following processes: (1) electrons are elevated from donor states to the conduction band, or (2) electrons are elevated from the valence band to acceptor states, leaving holes in the valence band.

Figure 12–13 shows a schematic diagram of the impurity levels in an extrinsic semiconductor.

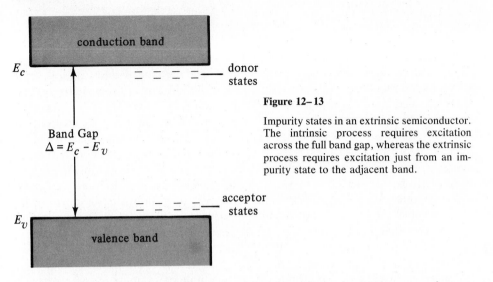

Figure 12–13

Impurity states in an extrinsic semiconductor. The intrinsic process requires excitation across the full band gap, whereas the extrinsic process requires excitation just from an impurity state to the adjacent band.

Some experimental values for donor and acceptor ionization energies are given in Table 12–3. Note that there is some dependence upon the nature of the impurity, which is notable in the case of indium in silicon, but for most impurities the values are in quite good agreement for a given host.

		Si	Ge
Table 12–3			
Experimental Values for Impurity Ionization Energies (eV)			
Donors	P	0.045	0.0120
	As	0.049	0.0127
	Sb	0.039	0.0096
Acceptors	B	0.045	0.0104
	Al	0.057	0.0102
	Ga	0.065	0.0108
	In	0.16	0.0112

7. THE *p-n* JUNCTION

In its simplest form, a *p-n* junction is a single crystal of silicon or germanium that contains n and p regions separated by an interface, as indicated in Figure 12–14. Ideally, the transition region should be quite narrow, but this is difficult to achieve in practice. The junction cannot be formed by simply pressing the two materials together because surface recombination of electrons and holes would prevent the diffusion of majority carriers across the interface.

The Fermi level in a semiconductor lies somewhere within the gap between the valence and conduction bands. For example, in an intrinsic semiconductor the

Figure 12–14

A *p-n* junction and its circuit symbol.

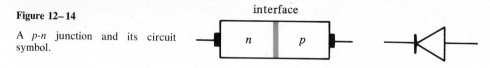

Fermi level is at the middle of the gap, because the number of electrons above the gap in the conduction band is exactly equal to the number of holes below the gap in the valence band. In an impurity semiconductor, however, this is not the case, as Figure 12–11 showed. In an *n*-type semiconductor the majority carriers are electrons, and there are more electrons in the conduction band and in donor levels than there are holes in the valence band. Hence the Fermi level is above the middle of the gap. In a *p*-type semiconductor, the Fermi level is below the middle of the gap. The actual position of the Fermi level depends, of course, upon the impurity concentration in each case. In an energy diagram for a *p-n* junction, the Fermi level at equilibrium must be constant across the junction, so the bands must bend in the region of the junction; this is shown in Figure 12–15.

In the absence of any external potential, there are two competing processes whose rates must be equal in order to achieve dynamic equilibrium. One of these processes is the *thermal generation* of electron-hole pairs, in which an electron is elevated to the conduction band and a hole is created in either the valence band or a donor level. The other is *carrier recombination,* which occurs when an electron encounters a hole and they recombine. Charge neutrality in each region is preserved by the fact that the number of free carriers is equal to the number of oppositely charged immobile ions or impurity centers. In the junction region the abrupt change in carrier densities results in large concentration gradients, which set up diffusion currents of the majority carriers. Thus, electrons will diffuse from the *n*-material to the *p*-side of the junction, and holes will diffuse in the opposite direction. Recombination will take place in the transition region, resulting in a depletion of electrons on the *n*-side and a depletion of holes on the *p*-side. This depletion region produces a dipole layer consisting of the immobile charges of the ionized sites, which in turn sets up an internal electric field as shown in Figure 12–15. The electric field builds up to the point at which it prevents further increase in the recombination current. It is evident that both electrons from the *n*-region and holes from the *p*-region must surmount an energy barrier in order to maintain the recombination current. Let us denote the potential of this barrier by ϕ.

The competing process is the motion of thermally generated *minority* carriers across the junction. Thus, electrons that are raised to the conduction band in the *p*-material by thermal excitation readily "slide down" the potential hill into the *n*-region, while thermally generated holes in the valence band of the *n*-region readily move in the direction of the electric field into the *p*-region. The effect of these two thermally generated currents is to reduce the potential ϕ, which in turn would increase the recombination currents sufficiently to restore the equilibrium value of ϕ. If we use the subscripts *g* for the thermally generated currents, *r* for the recombination currents, *e* for electrons, and *h* for holes, the conditions for equilibrium are as follows:

$$j_{er} = j_{eg} \\ j_{hr} = j_{hg}. \Big\}$$

(12.19)

and

Since j_{eg} depends only upon the number of electrons thermally excited from the Fermi level to the conduction band in the *p*-material, this is simply determined by the Boltzmann factor, that is,

$$j_{eg} \sim \exp\left(-\frac{E_1}{kT}\right),$$

(12.20)

where $E_1 = E_C - E_F$ as shown in Figure 12-15. Now the equal and opposite cur-

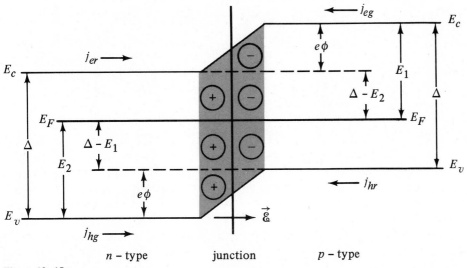

Figure 12-15

The bending of the energy bands in an unbiased *p-n* junction. E_c is the bottom of the conduction band, E_v is the top of the valence band, and $\Delta = E_c - E_v$ is the intrinsic gap energy. The shaded area is the depletion region, in which a dipole layer produces a potential barrier $e\phi$ that opposes the recombination currents (see Equation 12.19).

rent, j_{er}, is determined by the number of conduction electrons in the *n*-material multiplied by a factor that gives the probability that the electron can surmount the energy barrier $e\phi$.

$$j_{er} \sim n_e \exp\left(-\frac{e\phi}{kT}\right).$$

(12.21)

But $n_e \sim \exp\left(-\dfrac{\Delta - E_2}{kT}\right)$, so

$$j_{er} \sim \exp\left(-\frac{\Delta - E_2 + e\phi}{kT}\right) = \exp\left(-\frac{E_1}{kT}\right).$$

(12.22)

It is evident from Equations 12.20 and 12.22 that $j_{eg} = j_{er}$ as we assumed. Likewise, for the hole currents we find that

$$j_{hg} \sim \exp\left(-\frac{E_2}{kT}\right),$$

(12.23)

and
$$j_{hr} \sim n_h \exp\left(-\frac{e\phi}{kT}\right)$$

$$\sim \exp\left(-\frac{\Delta - E_1}{kT}\right) \cdot \exp\left(-\frac{e\phi}{kT}\right)$$

$$\sim \exp\left(-\frac{\Delta - E_1 + e\phi}{kT}\right)$$

$$\sim \exp\left(-\frac{E_2}{kT}\right) = j_{hg}. \tag{12.24}$$

8. THE *p-n* JUNCTION AS A RECTIFIER

It is interesting to note that the thermally generated currents depend only on the position of the Fermi energy relative to the width of the gap Δ. Hence, j_{eg} and j_{hg} are not affected by external emf's. On the other hand, the recombination currents are strongly affected by external potentials, since they must surmount the potential hills. If the positive terminal of a bias voltage is connected to the *p*-material, as shown in Figure 12–16(a), the effective barrier potential is reduced to

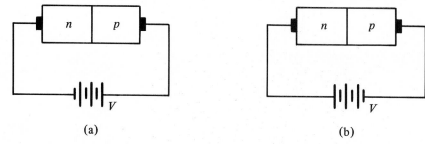

(a)   (b)

Figure 12–16

(a) Forward bias. (b) Reverse bias.

$(\phi - V)$ and the impedance of the junction is reduced. This is known as *forward bias*. However, if the polarity is as shown in Figure 12–16(b), the impedance is greatly increased and the junction is said to have *reverse bias*.

Whenever an external bias is applied, Equations 12.19 no longer hold at equilibrium. Instead, we find that a net current flows across the junction due to the altered recombination currents. Equations 12.19 now become:

$$j_{er} \sim \exp\left(\frac{-E_1}{kT}\right) \cdot \exp\left(\pm\frac{eV}{kT}\right) = j_{eg} \cdot \exp\left(\pm\frac{eV}{kT}\right) \tag{12.25}$$

and
$$j_{hr} \sim \exp\left(\frac{-E_2}{kT}\right) \cdot \exp\left(\pm\frac{eV}{kT}\right) = j_{hg} \cdot \exp\left(\pm\frac{eV}{kT}\right), \tag{12.26}$$

where the plus sign is used for forward bias and the minus sign for reverse bias. Then the currents for these cases may be obtained from the following simple analysis. The net current to the left in Figure 12.15 is:

$$I = Ce(j_{er} - j_{eg} + j_{hr} - j_{hg})$$

$$= Ce\left[j_{eg} \cdot \exp\left(\pm\frac{eV}{kT}\right) - j_{eg} + j_{hg} \cdot \exp\left(\pm\frac{eV}{kT}\right) - j_{hg}\right]$$

$$= Ce(j_{eg} + j_{hg}) \cdot \left[\exp\left(\pm\frac{eV}{kT}\right) - 1\right]$$

$$I = I_0\left[\exp\left(\pm\frac{eV}{kT}\right) - 1\right], \tag{12.27}$$

where C is a constant of proportionality and the plus sign is used for forward bias. Figure 12–17 shows a typical plot of I versus V for a *p-n* junction. Note that I_0 is a "saturation current" under reverse bias. (It is also known as the "dark current"

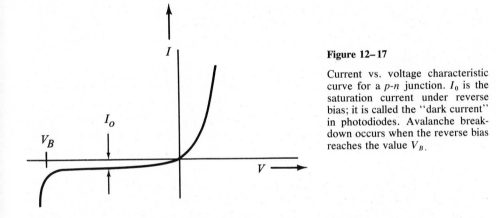

Figure 12–17

Current vs. voltage characteristic curve for a *p-n* junction. I_0 is the saturation current under reverse bias; it is called the "dark current" in photodiodes. Avalanche breakdown occurs when the reverse bias reaches the value V_B.

in photodiodes.) At a large value of reverse bias, breakdown occurs. (See Section 9 below.) This voltage is called the "avalanche breakdown" voltage, V_B. Rectification of an alternating emf occurs because of the large difference in the forward and reverse currents. Typical values are shown in Figure 12–18.

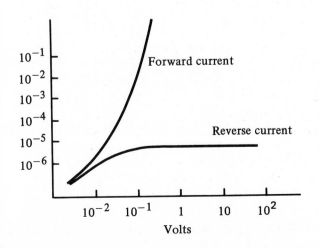

Figure 12–18

Typical values for forward and reverse currents across a *p-n* junction. The efficiency of a diode as a rectifier depends upon a large difference in forward and reverse currents.

9. APPLICATIONS OF *p-n* JUNCTIONS

The use of the *p-n* junction as a rectifier or diode has been discussed in the previous section. Recall that in such a device the generation current, which is thermally excited, is fixed, and it is the recombination current which varies widely with bias voltage.

In another class of devices it is the generation current that determines the operation of the device. In such devices the minority carriers may be excited by photons—visible, x-rays, or gammas—or by energetic particles. Thus they can be used as photodiodes, photovoltaic cells, or radiation detectors. The silicon or gallium arsenide solar cell is one of the important photovoltaic devices now in use.

The avalanche breakdown under a large reverse bias (see Figure 12–17) is so abrupt that it has an application in a voltage-regulating diode. The avalanche current is caused by impact ionization. That is, minority carriers acquire so much energy between collisions from the large electric field in the depletion layer that they create electron-hole pairs when they collide with lattice atoms. These newly created carriers rapidly acquire sufficient energy to produce additional pairs, and an avalanche current results.

Still another type of diode is the Zener or tunnel diode. Unlike the diodes discussed above, the doping in a tunnel diode is so extreme that the Fermi energy lies within the conduction band of the *n*-type material and within the valence band of the *p*-type material. Furthermore, the transition region is so narrow that an electron wave function can span the barrier (recall Section 4 of Chapter 10). A tunnel current will flow, and it is subject to a wide range of control by small changes in bias voltage.

The recombination process itself has device applications. When an electron and a hole recombine, it is possble for the energy of recombination to appear as a radiated photon. This process forms the basis for semiconductor lasers.[1] Unfortu-

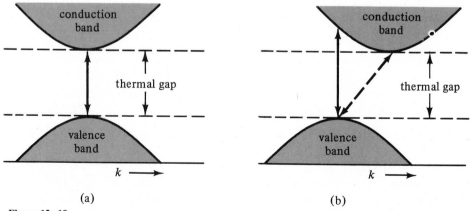

Figure 12–19

(a) In a direct optical transition the photon energy is equal to the gap energy. (b) Photon energy is greater than the gap energy for the vertical transition indicated by the solid arrow. The dashed arrow shows a phonon-assisted transition in which the *k*-value changes but the photon energy again equals the gap energy.

[1] J. I. Pankove, *Optical Processes in Semiconductors,* Prentice-Hall, Englewood Cliffs, N. J., 1971, Chapter 9.

nately, the recombination energy can be carried off non-radiatively as a competing process to optical emission. Another difficulty in the development of efficient semiconducting lasers is associated with the shapes of the energy bands. In Figure 12–19(a), note that the minimum energy of the conduction band occurs at the same k-value as the maximum energy of the valence band. Such a semiconductor is said to have a *direct intrinsic gap*. The energy of a recombination photon exactly equals the thermal energy kT of the gap. However, it is possible to have an *indirect gap*, shown in Figure 12–19(b). In the latter, a vertical transition involving a constant k-value does not correspond to the thermal energy of the gap. The other transition, shown by the arrow to the conduction band minimum, must correspond to the gap energy plus the energy of a *phonon*[1] created to conserve momentum. Transitions of this kind are known as *phonon-assisted transitions*.

10. JUNCTION TRANSISTORS

The most important application of the *p-n* junction is the junction transistor. These devices actually contain two junctions, and they may be designated either as *p-n-p* or *n-p-n* transistors according to the type of doping in the three regions. Basically, a transistor is a semiconductor device having an *emitter* electrode to inject excess majority carriers into a semiconductor crystal (*base*) and a *collector* electrode to recover them. A transistor is often used as an amplifier, in which the emitter is generally biased in the forward direction and the collector in the reverse direction, as shown in Figure 12–20. The *n-p* emitter-base junction behaves just

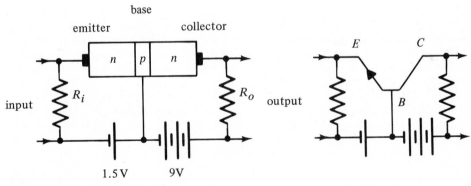

Figure 12–20

Schematic circuit diagram of an *n-p-n* transistor used as an amplifier.

like the forward-biased junction discussed previously. That is, the recombination current flowing in the input circuit will depend upon the algebraic sum of the input voltage and the 1.5 V battery. When the input signal is in the forward direction the current will be large, and when it is in the reverse direction the current will be decreased. Since the base-collector junction is reverse biased, there is an intense electric field across it, which is directed from right to left in the figure. If the section of *p*-material were thick, the collector circuit would not be affected by the

---

[1] A *phonon* is the quantum unit associated with thermal vibrations of the atoms in a crystal lattice.

behavior of the emitter circuit. On the other hand, if the p section is thin, many of the electrons entering it from the emitter will be caught in the strong electric field of the collector junction and will produce a current flow through the output resistor R_o. In practice, the base is only a few tenths of a millimeter thick, and the emitter and collector are a few millimeters wide. If the transistor is properly designed, the collector current will be nearly equal to the emitter current. The amplification will then be approximately R_o/R_i, which is typically of the order of 100.

Transistors are also used as oscillators and as switches. They are particularly useful in the latter capacity since they are nearly instantaneous in their action. They require very little power and no filament heating, and they operate at low voltages. They are mechanically rugged and can be made very small. Some of their disadvantages are that their operating characteristics can be seriously altered by temperature changes or by radiation damage.

SUMMARY

Crystalline solids may be divided into four classes according to the nature of the bonds that hold the atoms or molecules together. These classes are molecular crystals, ionic crystals, covalent crystals, and metals. Many of their physical properties may be readily understood by simply applying the Coulomb electrostatic interaction to the appropriate crystalline model.

The free electron theory of metals is used to obtain the density of electron states and the allowed energy states in an ideal metal. The energy of the highest filled state at very low temperatures is called the Fermi energy. It is typically of the order of 5 eV. A serious shortcoming of the free electron theory is that it ignores the fact that an electron moves in a potential that varies periodically due to the positive metal ions, which remain fixed at the lattice sites. The qualitative effect of a periodic potential may be readily learned by assuming a periodic potential consisting of rectangular steps as a first approximation. With this model we obtain bands of allowed energies, which are separated by gaps or bands of forbidden energies. Hence the atomic s-, p-, d-, and f-states of isolated atoms are broadened into energy bands in a solid. In real materials these bands may be separated or they may even overlap, depending upon temperature, pressure, and other parameters.

The band model of solids is very useful in explaining the differences in electrical properties of metals, insulators, and semiconductors. Semiconductors may be intrinsic, or they may derive their properties from impurities. In either case the electrons and holes that contribute to the electrical current may be thermally generated. In intrinsic semiconductors, electrons are thermally excited from the valence band across the gap to the conduction band. The holes left in the valence band, as well as the electrons in the conduction band, contribute to the current flow. In impurity semiconductors the intrinsic process also occurs, but the dominant process is the excitation of either electrons or holes from impurity states that lie within the energy gap. In n-type semiconductors, electrons are bound to impurity states that lie just below the conduction band; a small excitation energy will put an electron into the conduction band. In p-type semiconductors, holes are bound to impurity states that lie just above the valence band; a small excitation energy will create a hole in the valence band by elevating an electron to the acceptor state. In either case an estimate of the binding energy of an impurity state may be calculated from the Bohr model for hydrogen-like binding.

In discussing the p-n junction, the competing processes of thermal generation of electron-hole pairs and carrier recombination are described in order to understand how equilibrium is achieved. Applications of p-n junctions to rectification, detection, and transistor amplification are briefly treated.

J. S. Blakemore, *Solid State Physics*, 2nd ed., W. B. Saunders Co., Philadelphia, 1974, Chapter 4.

C. Kittel, *Introduction to Solid State Physics*, 5th ed., John Wiley & Sons, New York, 1976.

M. A. Omar, *Elementary Solid State Physics: Principles and Applications*, Addison-Wesley Publ. Co., Reading, Mass., 1975, Chapters 4, 6, 7.

C. A. Wert and R. M. Thomson, *Physics of Solids*, 2nd ed., McGraw-Hill Book Co., New York, 1970, Chapters 9, 10, 12, 13.

ADDITIONAL
READING

PROBLEMS

12-1. (a) Calculate the atomic mass of each of the following metals: Na, K, Cs, Cu, Ag, Au. (b) For each of these metals, find the number of free electrons per cubic meter, assuming that one free electron is provided by each atom.

12-2. Using the information obtained in the previous problem and the free electron model, calculate the Fermi energies for Na, K, and Cs.

12-3. Using the information obtained in problem 12-1 and the free electron model, calculate the Fermi energies of Cu, Ag, and Au.

12-4. The Fermi energy of aluminum is 11.63 eV. (a) Assuming the free electron model, calculate the number of free electrons per unit volume of aluminum at very low temperatures. (b) Determine the valence of aluminum by dividing the answer found in (a) by the number of aluminum atoms per unit volume as calculated from the density and the atomic volume.

12-5. Lead has a Fermi energy of about 9.4 eV. Determine its valence by comparing the electron density from the free electron model with the number of atoms per unit volume.

12-6. (a) Find the average energy per electron for metallic copper, using the free electron model. (b) At what temperature would the average energy per molecule of an ideal gas equal the energy obtained in (a)? (c) Using the result in (a) and the number of electrons per unit volume, how much energy is stored in electronic energy levels in one cubic centimeter of copper?

12-7. (a) What is the Fermi velocity for free electrons in silver at 0 K? (b) Show that the Fermi velocity is identical with the rms velocity of electrons in an ideal gas in which $\langle E \rangle$ equals the Fermi energy.

12-8. How does the Fermi temperature compare with the temperature of an ideal gas whose average energy per molecule equals the Fermi energy?

12-9. (a) From what depth below the Fermi level can electrons be raised to excited states thermally at a temperature of 200 K? (b) If the density of states were uniform and independent of temperature, approximately what percentage of the electrons in silver would be in excited states at 200 K?

12-10. The energy gaps at room temperature for several semiconductors are as follows: (a) diamond 6 eV, (b) silicon 1.1 eV, (c) indium arsenide 0.33 eV, (d) germanium 0.66 eV, (e) gallium arsenide 1.4 eV. Are they transparent to visible light? Explain.

12-11. A germanium crystal is doped with small quantities of (a) indium, (b) phosphorus, (c) arsenic, (d) gallium, (e) aluminum, and (f) antimony. Explain why each of these impurity semiconductors is *n*-type or *p*-type, as the case may be.

12-12. Assume that the extra electron of a phosphorus impurity atom in germanium is bound to the phosphorus nucleus in a Bohr orbit. Assuming an effective mass of 0.17 m_0 for the electron and a dielectric constant of 16, find the binding energy and the orbital radius.

12-13. The extra electron of an arsenic impurity in silicon is bound to the arsenic atom in a Bohr orbit. If the effective mass of an electron in silicon is 0.31 m_0 and the dielectric constant is 12, find the binding energy and the orbital radius.

12-14. From Table 12–3, the binding energy of a hole due to an aluminum impurity in germanium is about 0.01 eV. Assuming the Bohr model, find an approximate value for the effective mass of the hole. (Assume that $\kappa = 16$ for germanium.)

12-15. The binding energy of an electron to an antimony donor atom in silicon is given in Table 12–3 as about 0.04 eV. Find the effective mass of the electron, assuming the Bohr model and a dielectric constant of 12.

12-16. The energy gap in germanium at room temperature is 0.66 eV. What is the threshold wavelength for producing an electron-hole pair by photon absorption?

12-17. The distribution function in Equation 12.10 gives the probability that an electronic state of energy E will be occupied at temperature T. (a) What is the probability of thermally exciting an electron to the conduction band of germaniun, whose energy gap is 0.66 eV, at a temperature of 300 K? (b) What temperature would be required in order to produce an appreciable number of electron-hole pairs by thermal excitation? (Assume that the gap energy does not change with temperature.)

12-18. An impurity state lies 0.01 eV above the Fermi level in a semiconductor. What is the probability that this state will be ionized (a) at 10 K? (b) at 100 K?

12-19. A solar cell using a silicon p-n junction converts solar radiation to an electric current. If the energy band gap is 1.1 eV, what is the longest wavelength that will produce a photocurrent?

12-20. Germanium is frequently used as a gamma ray detector. Assuming a gap energy of 0.66 eV, how many electron-hole pairs could be produced by a photon having an energy of 1.02 MeV?

13 The Atomic Nucleus

In our discussion of atomic structure in Chapter 7, we treated the atomic nucleus as a point mass and a point charge. This is a reasonable approximation for hydrogen, in which the nucleus consists of a single elementary particle, the proton; the distributions of mass and charge are spherically symmetric in the proton, as nearly as we can tell. However, for all other elements the nucleus must be a composite of several elementary particles. Its disparity from a structureless point increases with increasing atomic number. One of the goals of nuclear physics is to understand the internal structure of the nucleus. Another goal that is even more challenging is to understand the ''glue'' that holds the nucleus together. In the pursuit of these goals nuclear physicists study collisions and scattering interactions between the elementary particles themselves, between particles and nuclei, and between nuclei. (For a review of scattering, see Section 2 of Chapter 7.) In addition, much useful information is obtained from the study of the products of radioactive decay, both in nature and in the laboratory.

Particles of extremely high energy are required in order to pursue much of the research mentioned above. A variety of accelerating devices have been built, including linear accelerators, cyclotrons, betatrons, and synchrotrons, in order to meet this need. The earliest devices produced particles having kinetic energies of the order of a few MeV, whereas the latest operate in the GeV (10^9 eV) region. For example, proton energies of 400 GeV are expected soon at Brookhaven National Laboratory on Long Island, and 1000 GeV protons are anticipated at Fermilab in Illinois. Accelerators have grown from devices that would fit in an ordinary laboratory to monstrous machines that require many acres of ground. For example, the Stanford Linear Accelerator, which produces 20 GeV electrons, is 2 miles long! The feasibility of increasing this energy to 50 GeV is currently under study there. The Alternating Gradient Synchrotron at Brookhaven National Laboratory is 800 feet in diameter. At Fermilab in Illinois, the main accelerator is 4 miles in circumference!

There is evidently a practical upper limit to the energies that can be achieved in man-made devices of this kind, because of the enormous costs and the engineering problems involved in construction. Perhaps the next high-energy physics laboratory will be a space platform for the observation of nuclear events that cannot be duplicated in the laboratory.

239

1. NUCLEAR STRUCTURE

The nucleus of the atom was once thought to be composed of a number of protons equal to its atomic mass number A, and a number of electrons equal to $(A - Z)$, that is, the difference between the mass number A and the atomic number Z. Such a composition would not only explain the mass and the charge of the nucleus but would also account for the existence of the electrons that are emitted by those nuclei which are known to be β-emitters. However, Rutherford rejected this hypothesis and predicted the existence of the neutron. Its subsequent discovery by Chadwick established the neutron as a nuclear constituent and put to rest the notion that electrons exist within the nucleus.

We now refer to both neutrons and protons as *nucleons,* and the mass number A is often called the *nucleon number.* Since Z is the atomic number or proton number, we may define the neutron number as $N = A - Z$. The atomic or proton number Z determines the chemical properties of the element. It is well known that many chemical elements have nuclei with more than one value of mass, A. Thus, for $Z = 1$ we have hydrogen, deuterium, and tritium, where hydrogen has $A = 1$, $Z = 1$, $N = 0$; deuterium has $A = 2$, $Z = 1$, $N = 1$; and tritium has $A = 3$, $Z = 1$, $N = 2$. Such nuclei having the same Z but different values of N are called *isotopes*. The symbol used for a nuclear species (often called a *nuclide*) consists of the chemical symbol corresponding to the value of Z, with the A-value written as a superscript and the Z-value written as a subscript. Thus, we write hydrogen, deuterium, and tritium as ^1_1H, ^2_1H, ^3_1H. (Deuterium is sometimes written ^2_1D).

In a similar manner, nuclei of *different* elements that have the same mass number A are called *isobars*. That is, isobars have different Z and different N, but the sum, $A = Z + N$, is the same. Examples are carbon-14 and nitrogen-14, where $A = 14$ for both elements. Some isobars of special interest are the so-called *mirror nuclei*. Mirror nuclei are pairs of nuclei having odd A such that Z and N differ by one. This means that the two elements must be adjacent in the periodic table. For example, $^{15}_8\text{O}$ ($Z = 8$, $N = 7$) and $^{15}_7\text{N}$ ($Z = 7$, $N = 8$) are mirror nuclei containing 15 nucleons.

Approximately 1400 different nuclides are now known to exist, but only about one-fifth of them are stable. Figure 13–1 shows a plot of N versus Z for a number of stable nuclides. Note that N must be greater than Z in order to achieve stability except for some of the very light nuclei. Thus we conclude that all nucleons contribute to the attractive forces that hold the nucleus together, whereas the instability is due to the repulsive Coulomb forces between the protons. As Z increases, an excess of N over Z is required in order to achieve stability.

The majority of the stable nuclides have even A, and all but eight of them are even-even nuclides, that is, both Z and N are even numbers. There are 110 odd A nuclides, of which about half have even Z–odd N and half have odd Z–even N. Certain values of Z and N seem to produce such unusual stability that they have been named *magic numbers*. The magic numbers are

$$Z \text{ or } N = 2, 8, 20, 28, 50, 82, 126.$$

For example, there are six stable isotopes for $Z = 20$ and there are 10 stable isotopes for $Z = 50$. The alpha particle, with both Z and $N = 2$, is remarkably stable.

Rutherford determined from α-particle scattering experiments that the nu-

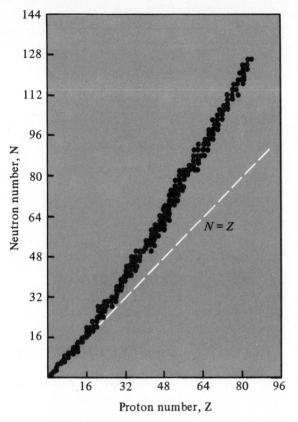

Figure 13–1

Plot of N versus Z, showing the majority of the stable nuclides.

clear radius is about 10^{-14} m (see Chapter 7). Subsequent experiments have led to the empirical relationship

$$r = r_0 A^{1/3}, \tag{13.1}$$

which expresses the approximate radius of any nucleus in terms of its nucleon number and the constant r_0. Equation 13.1 expresses the important experimental fact that the volume of a nucleus is directly proportional to the number of nucleons contained in it, just as if the nucleons were tightly packed spheres as shown in Figure 13–2.

There are a number of ways of determining the constant r_0, among which are the scattering of high-energy electrons (~ 10 GeV) and the scattering of energetic

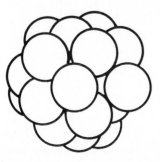

Figure 13–2

A nucleus visualized as a cluster of tightly packed spherical nucleons.

neutrons (~100 MeV). Since neutrons are not affected by the Coulomb force, they are deflected only by the short-range force between the neutron and the nucleons of the nucleus. The values of r_0 obtained by these two methods differ somewhat, but a reasonable compromise is to take

$$r_0 \sim 1.2 \times 10^{-15} \text{ m.}$$

Since distances $\sim 10^{-15}$ m play such an important role in nuclear physics, a new unit of length, called the *fermi*, is defined as

$$1 \text{ fermi} = 1 \times 10^{-15} \text{ m.}$$

From Equation 13.1 and the mass numbers of known nuclides, it is evident that there is an upper limit of about 10 fermis for the radius of the largest nucleus.

An important consequence of Equation 13.1 is that all nuclear matter has roughly the same density, given by

$$\rho_{\text{nuc}} = \frac{A}{V_{\text{nuc}}} = \frac{A}{\frac{4}{3}\pi r_0^3 A} \sim \frac{1}{4r_0^3} \sim 2 \times 10^{17} \text{ kg/m}^3. \tag{13.2}$$

In order to compare this with the density of atomic matter, the reader should recall that atomic dimensions are given approximately by a_0, the Bohr radius. Then atomic densities are roughly

$$\rho_{\text{at}} \sim \frac{A}{\frac{4}{3}\pi a_0^3},$$

and

$$\frac{\rho_{\text{nuc}}}{\rho_{\text{at}}} \sim \left(\frac{a_0}{r_0}\right)^3 \sim \left(\frac{5 \times 10^{-11} \text{ m}}{1.2 \times 10^{-15} \text{ m}}\right)^3 \sim 7 \times 10^{13}.$$

Thus, if atomic matter could be squeezed sufficiently to remove all of the empty space in atoms, the density would increase by a factor of about 10^{13}. This is thought to account for the enormous densities of stellar matter, such as in white dwarfs and neutron stars, where the gravitational forces can be sufficiently great to approach nuclear densities.

Example 13-1

Use the Heisenberg uncertainty principle to make a convincing argument against the notion that electrons are contained in the nucleus.

Solution

The uncertainty principle states that the product of the uncertainty in a coordinate and the uncertainty in momentum must be at least as large as h. If an electron were to exist in the nucleus, its uncertainty in x-position would be about 10 fermis or 10^{-14} m. Then, from Equation 6.20, the minimum momentum it can have in the x-direction is

$$\Delta p_x = \frac{h}{\Delta x} = \frac{6.6 \times 10^{-34} \text{ J-s}}{2 \times 10^{-14} \text{ m}} = 3.3 \times 10^{-20} \text{ kg-m/s},$$

and similarly for Δp_y and Δp_z. Then

$$\Delta p = \sqrt{3}\Delta p_x = 5.7 \times 10^{-20} \text{ kg-m/s}$$

$$\Delta p \cdot c = (5.7 \times 10^{-20} \text{ kg-m/s}) \cdot (3 \times 10^8 \text{ m/s}) = 1.7 \times 10^{-11} \text{ J} \sim 110 \text{ MeV.}$$

The total energy from Equation 4.7 is

$$E = \sqrt{(pc)^2 + (m_0c^2)^2} = \sqrt{(110 \text{ MeV})^2 + (0.5 \text{ MeV})^2} \sim 110 \text{ MeV},$$

and the kinetic energy from Equation 4.5 is

$$T = E - m_0c^2 = 110 \text{ MeV} - 0.5 \text{ MeV} \sim 110 \text{ MeV}.$$

In order for a 110-MeV electron to be bound to a nucleus, the attractive potential energy would have to be greater than 110 MeV. The only known force between an electron and a proton is the Coulomb force. Even as close as 0.1 fermi from a proton (1/100 the nuclear diameter), the Coulomb energy binding the electron to the proton would be no more than

$$|V| = \frac{ke^2}{r} = \frac{9 \times 10^9 \times (1.6)^2 \times 10^{-38}}{1 \times 10^{-16}} \text{ J} \sim 10 \text{ MeV}.$$

Such an electron would escape from the nucleus with a net kinetic energy of 100 MeV! This is in sharp contrast to the energies actually observed in β decay, which rarely are greater than a few MeV.

2. NUCLEAR BINDING ENERGY

In any bound system of particles the potential energy of the system is negative. It follows that the total energy of the bound system is *less* than the total energy of the separated particles, and this energy difference is called the *binding energy* of the system. In the case of the nuclear constituents, the binding energy, E_B, is so great that the energy difference also produces an observable mass difference, ΔM, between the sum of the masses of the nucleons and the actual mass of the nucleus. From Equation 4.5 we may write

$$\left. \begin{aligned} E_B \text{ (joules)} &= (\text{Mass difference } \Delta M \text{ in kg}) \times c^2 \\ E_B \text{ (joules)} &= (\text{Mass difference } \Delta M \text{ in atomic units}) \times 1.49 \times 10^{-10} \text{ J/u} \\ E_B \text{ (MeV)} &= (\text{Mass difference } \Delta M \text{ in atomic units}) \times 931.5 \text{ MeV/u} \end{aligned} \right\} \quad (13.3)$$

In order to find the binding energy of the deuteron, for instance, we first find the mass difference as follows:

$$
\begin{aligned}
\text{Mass of proton} &= 1.007\,276\,470 \text{ u} \\
+ \text{ Mass of neutron} &= \underline{1.008\,665\,012 \text{ u}} \\
& 2.015\,941\,482 \text{ u} \\
- \text{ Mass of deuteron} &= \underline{2.013\,553\,215 \text{ u}} \\
\Delta M &= 0.002\,388\,267 \text{ u}
\end{aligned}
$$

$$\text{Binding energy, } E_B = 0.002\,388\,267 \text{ u} \times 931.5 \text{ MeV/u} = 2.225 \text{ MeV}$$

It should be noted that the binding energy of the only two-particle nucleus, the deuteron, is about one million times as great as the binding energy of the only two-particle atom, hydrogen. As a general rule, the outermost electrons of atoms are bound by energies of 1 to 10 eV, whereas each nucleon is bound to the nucleus by an energy of 1 to 10 MeV.

Example 13-2

Calculate the binding energy of the alpha particle, $^4_2$He. What is the binding energy per nucleon?

Solution

The alpha particle consists of two neutrons and two protons. So the total mass of the separate particles is:

$$2(m_p + m_n) = 2 \times 2.015\ 941\ 482\ u$$
$$= \quad 4.031\ 882\ 964\ u$$
$$-\text{Mass of the } \alpha \text{ particle} = \quad 4.001\ 506\ 106\ u$$
$$\Delta M = \quad 0.030\ 376\ 858\ u$$
$$\text{Total binding energy, } E_B = \quad 0.030\ 376\ 858\ u \times 931.5\ \text{MeV/u} = 28.30\ \text{MeV.}$$

Since the nucleus consists of four nucleons, the binding energy per particle is $E_B/4 = 7.075$ MeV/nucleon.

Figure 13–3 shows a plot of the binding energy per nucleon, E_B/A, as a function of A for the stable nuclides. Note that the average binding energy per nucleon

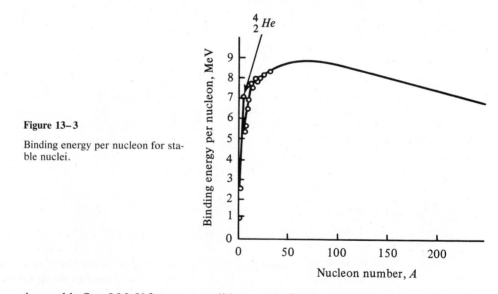

Figure 13–3

Binding energy per nucleon for stable nuclei.

is roughly 7 or 8 MeV for most nuclides except for the lighter elements. The nearly constant nuclear binding energy per particle contrasts sharply with the dependence of electronic binding energy per electron on Z in atomic potentials. This behavior is evidence that the nuclear force *saturates*. That is, each nucleon can interact only with a limited number of other nucleons, namely, its adjacent neighbors (see Figure 13–2). Saturation is further proof of the extremely short range of the nuclear force. By way of contrast, the Coulomb and gravitational forces are infinite in range, so that a given charge or mass interacts with every other charge or mass in an aggregate, regardless of the distance of separation.

Another force that saturates is the short-range chemical force (the van der Waals attraction) that holds atoms together in some substances and that holds molecules together in liquid drops.

The sharp increase in binding energy per particle for $A = 4$ and the concept of saturation account for the unusual stability of the alpha particle. There appears to be little or no attraction between an alpha particle and another nucleon; hence there is no stable nuclide with $A = 5$. The gradual decrease in binding energy per nucleon for the heavy nuclei is due to the fact that the Coulomb repulsion increases with Z. This is partly compensated for by increasing the N/Z ratio as Z increases (see Figure 13–1).

Finally, the binding energy of any nuclide, $_Z^A X$, may be expressed as follows:

$$E_B(\text{MeV}) = [Zm_p + (A - Z)m_n - \text{mass of } _Z^A X] \times 931.5 \text{ MeV/u}, \quad (13.4)$$

where all of the masses are in atomic units.

Example 13-3

Assume that the nucleus is a sphere of radius R with charge $+Ze$ distributed uniformly throughout its volume. Find the total electrostatic potential energy of the ball of charge.

Solution

The charge density of the sphere is simply the total charge divided by the volume of the sphere. Then,

$$\rho = \frac{Ze}{\frac{4}{3}\pi R^3}.$$

Now consider a portion of the ball, having radius $r < R$. The charge within this smaller sphere is

$$q = \tfrac{4}{3}\pi r^3 \rho.$$

The potential energy at its surface is

$$V = \frac{kq}{r},$$

where $k = 1/4\pi\epsilon_0 = 9 \times 10^9$ N-m²/coul². Suppose we now add a tiny layer of charge to this sphere, that is, a spherical shell of thickness dr. Then the added charge is

$$dq = \rho \, dv = \rho(4\pi r^2 \, dr).$$

The electrostatic potential energy of this shell is then

$$dE = Vdq = (k/r)q \, dq = (k/r)(\tfrac{4}{3}\pi r^3 \rho)(4\pi r^2 \rho \, dr) = (k/3)(4\pi\rho)^2 r^4 \, dr$$

The total electrostatic energy is found by integrating from $r = 0$ to $r = R$. That is,

$$E_{\text{Coul}} = \int_{r=0}^{R} dE = 3k \left(\frac{4\pi\rho}{3}\right)^2 \int_0^R r^4 \, dr = \left(\frac{3kR^5}{5}\right)\left(\frac{4\pi\rho}{3}\right)^2.$$

Substituting the value of ρ into this, we obtain

$$E_{\text{Coul}} = \left(\frac{3kR^5}{5}\right)\left(\frac{Ze}{R^3}\right)^2 = \frac{3k(Ze)^2}{5R}.$$

For large Z the assumption that the charge within the nucleus is continuous does not introduce a large error. On the other hand, when Z is small a correction must be made for the discreteness of charge. For example, when $Z = 1$, the Coulomb energy should be zero. This result can be obtained by replacing Z^2 with $Z(Z-1)$. Then the corrected energy becomes

$$E_{\text{Coul}} = \frac{3ke^2Z(Z-1)}{5r_0A^{1/3}} = 0.72\frac{Z(Z-1)}{A^{1/3}} \text{ MeV}. \tag{13.5}$$

Example 13-4

Calculate the Coulomb energy of the nuclide $^{18}_{8}O$.

Solution

$$E_{\text{Coul}} = 0.72 \text{ MeV} \times \frac{8\cdot7}{18^{1/3}} = 0.72 \text{ MeV} \times \frac{56}{2.62} = 15.4 \text{ MeV}.$$

Example 13-5

Calculate the minimum energy required to remove a neutron from $^{43}_{20}Ca$.

Solution

Removing a neutron from $^{43}_{20}Ca$ would form $^{42}_{20}Ca$. Then the separation energy for the last neutron is:

$$S_n = (\text{Mass of } ^{42}_{20}Ca + M_n - \text{Mass of } ^{43}_{20}Ca) \times 931.5 \text{ MeV/u}$$

$$= (41.958\ 625 + 1.008\ 665 - 42.958\ 780) \times 931.5 \text{ MeV}$$

$$= (0.008\ 510) \times 931.5 \text{ MeV}$$

$$= 7.93 \text{ MeV}.$$

Example 13-6

Calculate the minimum energy required to remove a proton from $^{42}_{20}Ca$.

Solution

Removal of a proton from $^{42}_{20}Ca$ would form $^{41}_{19}K$. Then, the separation energy for the last proton is:

$$S_p = (\text{Mass of } ^{41}_{19}K + M_p - \text{Mass of } ^{42}_{20}Ca) \times 931.5 \text{ MeV/u}$$

$$= (40.961\ 832 + 1.007\ 825 - 41.958\ 625) \times 931.5 \text{ MeV}$$

$$= (0.011\ 032) \times 931.5 \text{ MeV}$$

$$= 10.28 \text{ MeV}.$$

3. THE NUCLEAR FORCE AND THE PION

As was stated in Chapter 7, Rutherford observed that the Coulomb law failed to describe his results for the scattering of alpha particles for impact parameters less than 1 fermi. He predicted that a hitherto unknown short-ranged, attractive force dominated the Coulomb repulsion at distances of about 1 fermi. Subsequent studies of the scattering of protons from protons and neutrons from protons show that the force has a range of about 2 fermis and that it is *charge independent*. This means that the *p-p*, *n-n*, and *n-p* interactions are essentially the same except for the extra contribution from the Coulomb repulsion in the case of the *p-p* interaction.

In 1935 Yukawa proposed that the strong nuclear force could be represented

An aerial view of the principal components of the accelerator system at the Fermi National Accelerator Laboratory, Batavia, Illinois. The largest circle is the Main Accelerator, 4 miles in circumference. The smaller circle, with cooling water pond in the center, is the Booster Accelerator. Experimental areas lie at a tangent to the left. (Courtesy of the Fermi National Accelerator Laboratory.)

by a potential of the form

$$V \sim \frac{e^{-r/b}}{r}, \tag{13.6}$$

where b is the range of the nuclear force, about 1 fermi. This expression is not valid for $r < 0.4$ fermi, where a strong repulsive term dominates. For $r = b = 1$ fermi, V is of the order of 10 MeV. The potential falls off rapidly for $r > b$; for example, when $r = 10b$, V is about 100 eV. A schematic diagram of the nuclear potential seen by a nucleon is shown in Figure 13–4. The depth of the well is typically

Figure 13–4

Schematic diagram of the nuclear potential for an incoming nucleon. Since the exact shape of the well is not known, it is approximated by a square well about 40 MeV deep. The potential for a proton would be the sum of the Yukawa and Coulomb potentials. A neutron would be affected only by the Yukawa potential. For very small values of r (less than a few tenths of 1 fermi) the potential is strongly repulsive.

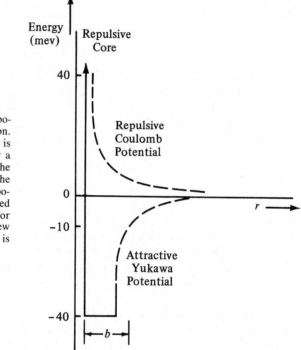

about 40 MeV. Since its exact shape is not known, it is approximated by a square well for $r < b$. The potential for a proton would be the sum of the Yukawa potential and the repulsive Coulomb potential. A neutron would experience only the Yukawa potential.

In order to explain the strong nuclear force, Yukawa proposed that it results from the virtual exchange of particles in a manner similar to that discussed for the Coulomb force in Section 6 of Chapter 6. However, instead of postulating massless photons as the exchange particles, he assumed the exchange particle to have a rest mass M intermediate between that of the electron and the nucleon. Hence it was given the name *meson* (intermediate). It is interesting to see how Yukawa arrived at the mass of the meson. The energy required to create a particle of mass M is

$$\Delta E = Mc^2.$$

If this particle is to be created without adding energy to the nucleon—that is, if it is to be a virtual particle—then it must be created and reabsorbed within a time Δt given by the uncertainty principle as

$$\Delta E \cdot \Delta t \sim \hbar. \qquad (13.7)$$

If we take the range of the nuclear force as the distance that the exchanged meson must move, then the transit time is at least $\Delta t \sim b/c$. Using this value of Δt in Equation 13.7,

$$\Delta E = Mc^2 = \frac{\hbar c}{b}. \qquad (13.8)$$

Using $b = 1.4$ fermi,

$$Mc^2 = 140 \text{ MeV}$$

and

$$M = 270\, m_0.$$

Thus the mass of the Yukawa meson is 270 times the mass of the electron. Under the right conditions a detectable meson can be emitted if sufficient energy is added to create it and to provide its kinetic energy. Particles of the right size were soon discovered in cosmic radiation, but it was not until quite recently that they could be produced and studied in the laboratory. The Yukawa particle is called the π-*meson* or *pion*.

Before leaving this topic, it should be noted from Equation 13.8 that the range of the force and the mass of the exchange particle are inversely related. The very short range nuclear force requires an exchange particle that is massive. The Coulomb force, on the other hand, has an infinite range ($b \to \infty$), so the exchange particle—the photon—must have zero rest mass. The gravitational force also has an infinite range, so it is predicted that its exchange particle must have zero rest mass and must travel at the speed of light. This exchange particle has been named the *graviton,* but it has thus far eluded discovery.[1]

4. NUCLEAR SPIN, ENERGY LEVELS, AND MAGNETIC MOMENTS

In Section 3 of Chapter 2, spin angular momentum was introduced as an intrinsic property of the elementary particles. A further discussion of the spin of the electron occurred in Section 3 of Chapter 11, where it was stated that its observed value is always $\frac{1}{2}\hbar$.

It happens that positive and negative electrons, the neutron, the proton, the antiproton, muons, and perhaps a number of still unknown particles, all display spins of $\pm\frac{1}{2}\hbar$ along the direction of an external magnetic field. They are members of a class called *fermi particles* or *fermions*. Although they have little else in common, fermions share the distinctive property that each of these particle species obeys the Pauli exclusion principle discussed earlier for electrons. A neutron does not exclude a proton, nor does an electron exclude a muon, since the exclusion principle holds only for identical fermions. But only one fermi particle of a given species can occupy a quantum state. This means that there can be two neu-

---

[1] J. Weber, *Physics Today* 21, 34 (1968).

trons, for example, in a given nuclear energy level provided that one neutron has spin $+\frac{1}{2}\hbar$ and the other has spin $-\frac{1}{2}\hbar$. This same level can also hold two protons having opposite spins, making four particles in all. Thus, $^1_1$H, $^2_1$H, $^3_2$He, and $^4_2$He could be accommodated by a single energy level, as shown schematically in Figure 13–5.

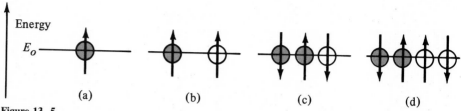

Figure 13–5

Hypothetical arrangement of protons (solid circles) and neutrons (open circles) in hydrogen and helium nuclei. (a) Proton, (b) deuteron, (c) $^3_2$He, (d) the alpha particle.

The total spin of a nuclide is the algebraic sum of the spins of the constituent particles. For example, in Figure 13–5, the spin is $\frac{1}{2}$ in (a), 1 in (b), $\frac{1}{2}$ in (c), and 0 in (d). All nuclides having half-integral total spin are also fermions and must obey the Pauli principle. Thus, *all nuclides having odd mass number A are fermions.* Nuclides having zero or integral spin—that is, nuclides with even mass number A—are called *bosons.* There is no limit on the number of bosons that can occupy the same quantum state at the same time.

It is evident from Figure 13–5 that nuclides having more than four nucleons will require additional energy levels in order to accommodate more nucleons. Therefore, we must conclude that some sort of energy level scheme exists within the nucleus itself. We may use the infinite square well in order to get a crude estimate of the magnitude of these energies relative to the bottom of the well. The one-dimensional square well was discussed in Section 1 of Chapter 10. We may generalize this to three dimensions by assuming that the width of the well is L in each of the directions x, y, z. Then the conditions for standing waves in these direction are independent and may be stated as follows:

$$k_x L = n_x \pi$$

$$k_y L = n_y \pi$$

$$k_z L = n_z \pi$$

where n_x, n_y, and n_z are integers. Then Equation 10.1 becomes

$$E(n_x, n_y, n_z) = \frac{\hbar^2}{2m}(k_x^2 + k_y^2 + k_z^2) = \frac{h^2}{8mL^2}(n_x^2 + n_y^2 + n_z^2). \quad (13.9)$$

It is important to note that n_x, n_y, and n_z cannot take on the value of zero. For example, if n_x were 0, then k_x and p_x would be 0 and the uncertainty principle would be violated in the x-direction. Consequently, the ground state must be given by $n_x = n_y = n_z = 1$, or

$$E_0 = E_{1,1,1} = \frac{3h^2}{8mL^2}. \quad (13.10)$$

Example 13-7
If the nucleus is regarded as a three-dimensional infinite square well of width 20 fermis, what is the approximate energy of the ground state of a nucleon relative to the bottom of the well?

Solution

$$E_0 = E_{111} = \frac{3h^2}{8mL^2} = \frac{3 \times (6.6 \times 10^{-34})^2 \text{ J}^2\text{-sec}^2}{8 \times 1.67 \times 10^{-27} \text{ kg} \times 2^2 \times 10^{-28} \text{ m}^2}$$

$$= 2.45 \times 10^{-13} \text{ J} \sim 1.5 \text{ MeV}.$$

Example 13-8
What are the energies of the first two excited states in the previous example?

Solution
The first excited state may be characterized by the values (n_x, n_y, n_z) of $(2, 1, 1)$, $(1, 2, 1)$, or $(1, 1, 2)$. That is, the state may be obtained in three different ways, but each has the same energy given by

$$E_1 = E_{211} = \frac{6h^2}{8mL^2} = 2E_0 \sim 3 \text{ MeV}.$$

The second excited state corresponds to $(2, 2, 1)$, $(2, 1, 2)$ or $(1, 2, 2)$. Its energy is

$$E_2 = E_{221} = \frac{9h^2}{8mL^2} = 3E_0 \sim 4.5 \text{ MeV}.$$

Although the energy levels calculated in the above examples are for the infinite well, the general features of the energy level scheme will be retained in a well having a finite depth. A hypothetical well of depth V_0 is shown schematically in Figure 13–6.

In Chapter 11, Section 4, it was stated that the spin of the electron is accompanied by a magnetic moment. It should not be surprising, then, to learn that other elementary particles having spin also possess magnetic moments. The unit used for nuclear moments is called the *nuclear magneton,* and its symbol is μ_N. It is defined in a manner similar to the Bohr magneton, Equation 11.19, except that the mass of the proton replaces the mass of the electron in the denominator. Thus the nuclear magneton is

$$\mu_N = \frac{e\hbar}{2m_p} = 5.05 \times 10^{-27} \text{ J/tesla}, \qquad (13.11)$$

which is smaller than the Bohr magneton by a factor of roughly a thousand.
The magnetic moment of the proton is not simply μ_N, but is

$$\mu_p = 2.7928 \, \mu_N \approx 1.41 \times 10^{-26} \text{ J/tesla}. \qquad (13.12)$$

Surprisingly enough, the neutron also has a magnetic moment. Since the moment is directed opposite to the neutron's spin direction, it is written with a minus sign, that is,

$$\mu_n = -1.9135 \, \mu_N. \qquad (13.13)$$

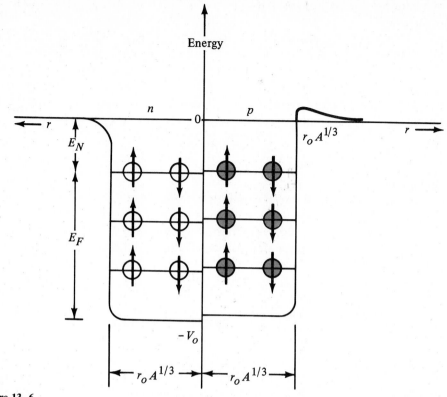

Figure 13-6

A square well of depth V_0 containing 12 nucleons. The energy of the highest filled state is called the Fermi energy, E_F, and $E_N = V_0 - E_F$ is the binding energy of the "last" nucleon. The levels for protons (solid circles) are actually slightly higher than those for neutrons (open circles) because of the Coulomb potential. This effect becomes more pronounced as Z increases. At about $Z = 22$, the nth proton level becomes higher than the $(n + 1)$th neutron level.

We know that the spins of the neutron and proton are parallel in the deuteron, since its observed spin is always unity. It does not follow, however, that the magnetic moment of the deuteron is exactly equal to the algebraic sum of Equations 13.12 and 13.13. It turns out to be $\mu_d = 0.8576 \, \mu_N$, which is a discrepancy of about 2.5%.

Nuclear moments can be made to precess in external magnetic fields in a manner similar to that described for electrons. This phenomenon, known as nuclear magnetic resonance, is now one of the important ways of studying nuclides and their interactions with their neighbors in molecules, liquids, and solids.

It is worth noting that the intrinsic magnetic moments of the elementary particles have provided the strongest argument against the existence of electrons in the nucleus. The electron, having a magnetic moment roughly 600 times greater than any known nuclear moment, could hardly remain undetected within the nucleus.

5. MODELS OF THE NUCLEUS

In the previous sections we have discussed the nuclear constituents, the force that holds them together, and the enormous binding energy of the nucleus, but we have said nothing about how the nucleons are arranged within the nucleus. We do not know, of course; and the evidence is so conflicting that no single model can currently account for all of it. The knowledge that the nuclear volume is directly proportional to the number of nucleons and that the densities of many nuclei are approximately the same (Equations 13.1 and 13.2) makes the so-called *liquid drop model* appealing.

Another similarity between liquids and nuclei derives from the fact that the binding energy per nucleon is nearly constant (see Figure 13–3). Thus the total binding energies of both nuclei and liquids (the heat of vaporization) are roughly proportional to their masses. In a detailed treatment of the model, correction terms to the binding energy are included that account for the surface energy, the Coulomb energy, and the pairing energy. The nuclear drop has surface tension, just like a liquid, and the surface energy reduces the total binding energy. The pairing energy term attempts to account for the extra binding energy associated with even Z and even N.

Although the liquid drop model is instructive, it cannot account for some of the very important properties of nuclides, such as the magic numbers. Another difficulty is that a cluster of nucleons like that shown in Figure 13–2 seems quite unrealistic when one considers the kinetic energy requirement of the uncertainty principle. This is an appreciable amount of energy, even for a nucleon, so it is quite unlikely that it could be achieved by the vibration of a nucleon about an equilibrium position in a cluster. It seems more reasonable to assume that some sort of independent motions exist within the nucleus, in spite of the difficulty of envisioning such motion in a tightly packed nucleus.

This approach leads first to the *Fermi gas model,* which is essentially the same as the free electron gas model of a metal discussed in Section 2 of Chapter 12. It is assumed that each nucleon moves in an average potential due to the net effect of all of the other nucleons. This potential is represented by a three-dimensional square well, and the energy states are calculated from Equation 13.9 as illustrated by Examples 13-7 and 13-8. As in the case of a metal, the energy of the highest filled state may be called the *Fermi energy, E_F*. The energy required to remove a nucleon from the Fermi level is evidently $E_N = V_0 - E_F$, where V_0 is the depth of the well. The quantity E_N is known as the binding energy of the last nucleon, and it is shown schematically in Figure 13–6.

A natural development from the Fermi gas model is to seek solutions of the orbital states of nucleons by solving Schrödinger's equation in a manner analogous to the atomic theory for electronic states. This is known as the *shell model* of the nucleus. Each nucleon has a principal quantum number and a total angular momentum quantum number, j. The quantum number j results from the vector coupling of the orbital and spin angular momenta of the nucleon, just as in the case of electrons in atoms (see Section 6 of Chapter 11). The *j-j* coupling scheme is used to obtain the allowed quantum states for the nuclide as a whole.

One of the striking successes of the shell model is that it predicts the magic numbers. It also gives the correct ground state angular momentum for almost all nuclides. However, the shell model ignores such factors as the saturation of the nuclear force, the nearly constant nuclear density, and the behavior of the binding energy as illustrated in Figure 13–3. Furthermore, it predicts the wrong results for the scattering of neutrons from nuclei. It turns out that neutrons interact so strongly with nuclei for certain discrete energies (sometimes spaced only a few hundred eV apart) that they can be trapped for a while before being ejected by the target nucleus. Such trapping events are called *resonances*.

The model that has been proposed to account for these resonances and still preserve the best features of the shell model is called the *collective model of the nucleus*. Here the nucleus is viewed as having a core of filled shells surrounded by "loose" nucleons, which do not form closed shells. The core is not rigid like the inner electronic shells of atoms because the net force on a nucleon, being saturated, is not always directed toward the center of nucleus as is the case with core electrons in atoms. Consequently, significant deformations can occur, and these can often be detected through experiments that measure the electric quadrupole moment or the moment of inertia of the nucleus.

In the collective model the nucleons can be viewed as constituting a many-body system analogous to the lattice vibrations in a solid. The energy states of the system are the excitation energies of the collective system—like normal modes—which form a set of discrete but closely-spaced levels. Hence, a neutron having one of these energies will, in a collision with a nucleus, immediately share its energy with the whole system. A considerable time interval will elapse (compared with the transit time through a distance equal to the nuclear diameter) before a neutron accumulates enough energy to emerge as the "scattered" particle.

SUMMARY

The concept of the atomic nucleus has evolved rapidly from that of a structureless hard sphere to a submicroscopic dynamic system that rivals the electronic structure of the atom itself in complexity. It is composed of neutrons and protons, which are collectively called nucleons. The total number of nucleons is given by the mass number A. In particular, $A = Z + N$, where Z is the number of protons and N is the number of neutrons. A nuclear species corresponding to given values of A and Z is called a nuclide; it is denoted by $_Z^A X$, where X represents the chemical symbol for the appropriate element. The nuclear volume is directly proportional to A, which leads to the important result that the density of nuclear matter is essentially constant for all nuclides.

Nucleons interact with each other through the strong nuclear force, which is extremely short-ranged and which is much greater than the Coulomb interaction. As a result of the very short range of this force, nucleons within the nucleus interact primarily with their nearest neighbors and we say that the nuclear force saturates. An important consequence of this is that the net force on each nucleon is not directed toward the center of the nucleus as would be the case for a long-range force. This explains why the nucleus, though very tightly held together, can assume a shape that departs considerably from spherical symmetry. Another consequence of saturation is that the binding energy per nucleon has a nearly constant value of 7 to 8 MeV for all but the lightest of nuclides. The total binding energy of a nuclide, then, is approximately $7A$ MeV. This energy is sufficiently great to produce a detectable change in the mass of a submicroscopic particle. Therefore, we may define the binding energy, E_B, by the expression

$$E_B = \Delta Mc^2,$$

where ΔM is the difference between the total mass of the nucleons as separate particles and the mass of the nuclide after it is formed. Improvements on this value may be made by correcting for the Coulomb energy, the surface energy, and the pairing of nucleon spins.

Nuclei are studied by observing their interactions with other nuclei, elementary particles, photons, electric fields, and magnetic fields. Experimenters were at first limited to those reactions that occurred naturally through radioactive decay and to those that could be induced using natural decay products. With the development of particle accelerators, however, the opportunities for innovative experiments have become enormous.

We now have a vast amount of experimental information about the sizes and shapes of nuclei, their total spins, their magnetic moments, their binding energies, and their energy levels. A number of different models of the nucleus have been proposed to account for these facts. The liquid drop model can explain the saturation of the nuclear force, the constancy of nuclear densities, and the dependence of the total binding energy on the mass number A. However, this model cannot explain the total spin of the nuclide, the magic numbers, and the known energy level schemes, nor can it accommodate the enormous kinetic energies required of the nucleons by the uncertainty principle. The Fermi gas model provides a crude estimate of the energy levels of the nucleus by incorporating the Pauli exclusion principle, which holds for all fermi particles, with the energy levels of the three-dimensional square well. The shell model accounts for the greatest number of experimental facts, since it yields the correct values for the total spins of all nuclides, it explains the magic numbers, and it provides for large nucleon kinetic energies as they circulate within the nucleus. The shell model has shortcomings, though, since it cannot explain saturation, the constancy of nuclear density, and resonances that occur in neutron scattering. The collective model treats the nucleus as a many-body system where the normal modes are the allowed states of the system. This model does account for the neutron resonances but it leaves many other properties unexplained. Additional successes and failures of some of these models will be mentioned in the following chapter in connection with specific nuclear reactions.

Finally, there is no single theory of the nucleus that consistently accounts for all of the known facts. The quest for such a theory continues to be the greatest single challenge in all of physics.

M. Baranger and R. A. Sorensen, "The Size and Shape of Atomic Nuclei," *Scientific American, 221,* 58 (August 1969).

J. P. Blewett, "Resource Letter PA-1 on Particle Accelerators," *American Journal of Physics 34,* 742 (Sept. 1966).

C. H. Blanchard, C. R. Burnett, R. G. Stoner, and R. L. Weber, *Introduction to Modern Physics,* Prentice-Hall, Inc., Englewood Cliffs, N.J., 1969, Chapter 12.

H. W. Kendall and W. Panofsky, "The Structure of the Proton and the Neutron," *Scientific American, 224,* 60 (June 1971).

W. K. H. Panofsky, "Needs Versus Means in High-Energy Physics," *Physics Today 33,* 24 (June 1980).

H. Semat and J. R. Albright, *Introduction to Atomic and Nuclear Physics,* 5th ed., Holt, Rinehart and Winston, Inc., New York, 1972, Chapter 12.

P. A. Tipler, *Modern Physics,* Worth Publishers, Inc., New York, 1978, Chapter 11.

R. T. Weidner and R. L. Sells, *Elementary Modern Physics,* 3rd ed., Allyn and Bacon, Inc., Boston, 1980, Chapter 10.

Additional Reading

PROBLEMS

13-1. At what kinetic energy will an electron have a de Broglie wavelength of 1 fermi?

13-2. What energy in eV is required in order for a neutron to have a de Broglie wavelength comparable to a nuclear diameter?

13-3. Calculate the total binding energy of the nuclide $^{12}_{6}$C. What is the binding energy per nucleon?

13-4. Calculate the binding energy per nucleon in $^{34}_{16}$S.

13-5. Which nuclide has the greater binding energy, tritium ($^{3}_{1}$H) or helium-3 ($^{3}_{2}$He)? Show that the Coulomb energy can nearly account for the difference.

13-6. What energy is required to remove a neutron from $^{17}_{8}$O?

13-7. Find the minimum energy required to remove a neutron from $^{44}_{20}$Ca.

13-8. Calculate the binding energy per nucleon for $^{202}_{80}$Hg.

13-9. What is the difference in Coulomb energy for the two mirror nuclei $^{15}_{8}$O and $^{15}_{7}$N? How does this compare with the difference in their binding energies?

13-10. What is the energy of the third excited state in the square well of Example 13-7?

13-11. A spherical square well of radius 5 fermis contains four nucleons. Compare the ground state energies if (a) the four nucleons are all protons, or (b) two are protons and two are neutrons.

13-12. Use the square well model to estimate the Fermi energy of the nucleus shown in Figure 13-6.

13-13. If all of the nucleons in Figure 13-6 were suddenly released from the constraint of the Pauli principle, they could drop to the ground state. How much energy would be released if this were to occur?

13-14. In the Fermi gas model of the nucleus, the Fermi energy for nucleons is expressed by

$$E_F = \frac{h^2}{8m}\left(\frac{3\rho}{\pi}\right)^{2/3},$$

where m is the nucleon mass and ρ is the nucleon density. (a) Calculate the Fermi energy for protons in ^{44}Ca. (b) Calculate the Fermi energy for neutrons in ^{44}Ca.

13-15. (a) Using the expression given in problem 13-14, calculate the Fermi level for neutrons in ^{17}O. (b) Using the neutron separation energy of 4.1 MeV obtained in problem 13-6, estimate the depth of the nuclear well in ^{17}O.

13-16. State the spin of each of the following nuclides: $^{1}_{1}$H, $^{2}_{1}$H, $^{3}_{1}$H, $^{3}_{2}$He, $^{4}_{2}$He, $^{6}_{3}$Li, $^{7}_{3}$Li, $^{9}_{4}$Be.

13-17. Calculate the minimum energy required to remove a proton from $^{21}_{10}$Ne.

13-18. Calculate the Coulomb energy of ^{175}Lu.

13-19. Calculate the binding energy per nucleon in ^{126}Te.

13-20. Find the difference in binding energies for the mirror nuclei $^{23}_{11}$Na and $^{23}_{12}$Mg.

*13-21. (a) What is the least energy required to separate a neutron from ^{40}Ca? (b) What is the least energy required to separate a proton from ^{40}Ca? Why does this differ from the answer in (a)? (c) Using the expression in problem 13-14, find the Fermi level for the ^{40}Ca nucleus. (d) From the above results, estimate the depth of the nuclear well in ^{40}Ca.

14 Nuclear Transformations

Nearly 1200 of the known nuclides are unstable. However, only a small number of these occur naturally at the present time. It is believed that the cosmological event that created our universe ten or so billion years ago produced all of the known nuclides, both stable and unstable, but that only those with long lifetimes have survived.

The unstable nuclides found in nature decay by the emission of alpha particles or beta particles, which are sometimes accompanied by gamma rays. Since an alpha particle has a mass of four and a charge of two, the *daughter* nucleus

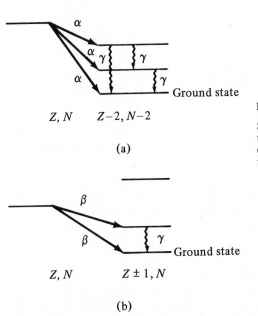

Figure 14-1

Schematic energy level structure for hypothetical nuclei. An α-emitter (a) or a β-emitter (b) may decay directly to the ground state or to an excited state of the daughter nucleus.

(decay product) formed by alpha emission will have a mass number four units less than the parent nuclide and its atomic number will be less by two units. The daughter may also be unstable and undergo further decay. Beta emission has no effect upon the mass number A, but it does change Z by ± 1, depending upon whether the emitted particle is an electron or a positron. It frequently happens that alpha or beta emission leaves the daughter nucleus in an excited state, and that one or more gamma rays are then emitted as it reverts to the ground state. Since gamma rays are massless photons, gamma emission has no effect upon either A or Z. These processes are shown schematically in Figure 14–1.

The radioactive elements occurring in the earth's crust lie within the range of atomic numbers from $Z = 81$ to $Z = 92$. They may be grouped into three decay series, each beginning with a nuclide having a very long lifetime and each ending with a stable isotope of lead. These series are: the *uranium-radium series,* the *thorium series,* and the *actinium series.* It is speculated that a fourth series, whose end product is a stable isotope of bismuth, once occurred on earth before the disappearance of neptunium ($Z = 93$), plutonium ($Z = 94$), and americium ($Z = 95$). This series is called the neptunium series. Elements having $Z > 92$ are called *transuranic elements.* To the best of our knowledge there are no stable nuclides beyond uranium.

1. THE RADIOACTIVE DECAY LAW

It is a known fact that the rate at which a particular decay process occurs is directly proportional to the number of parent nuclides present. Thus, if N is the number of unstable nuclides of a given species, then the number of disintegrations that will occur in time dt is given by

$$dN = -\lambda N \, dt, \tag{14.1}$$

where the minus sign indicates the the population N decreases as time increases. The proportionality constant λ is called the *decay constant.*

There is no way of predicting when a particular nucleus will decay, nor is there any known way to stimulate or inhibit its decay. As far as we can tell, the decay rate is not affected by extreme temperatures or pressures, by strong electromagnetic fields, or by its past history. Regardless of how long a particular parent has survived, *the probability that it will decay in the next time interval dt is given by* $\lambda \, dt$.

If we say that the number of parent nuclei at time $t = 0$ was N_0, then we may integrate Equation 14.1 in order to find the population N at any time t. Thus,

$$\int_{N_0}^{N} \frac{dN}{N} = - \int_{0}^{t} \lambda \, dt$$

$$\ln \frac{N}{N_0} = -\lambda t \tag{14.2}$$

or

$$N = N_0 e^{-\lambda t}. \tag{14.3}$$

Equation 14.3 is the *radioactive decay law.* Note that the population of parent nuclei decreases exponentially with time. If the daughter nuclide is stable, then the population of daughter atoms will increase exponentially according to the equation

$$N_{\text{daught}} = N_0 - N = N_0(1 - e^{-\lambda t}). \qquad (14.4)$$

These processes are illustrated in Figure 14–2.

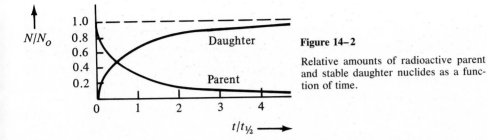

Figure 14–2

Relative amounts of radioactive parent and stable daughter nuclides as a function of time.

It was stated above that it is impossible to predict how long a given parent will live before undergoing spontaneous decay. There are, however, two statistical measures of the lifetime in common use, the *half-life* and the *average lifetime*. The half-life, $t_{1/2}$, is the time required for one-half of the initial population to disintegrate. Mathematically,

$$\frac{N}{N_0} = 1/2 = e^{-\lambda t_{1/2}},$$

or

$$\lambda t_{1/2} = \ln 2,$$

and also

$$t_{1/2} = \frac{0.693}{\lambda}. \qquad (14.5)$$

Thus the half-life can be determined as soon as the decay constant is known. After the expiration of n half-lives, the population will be reduced to $1/2^n$ of its original value. Naturally occurring radioactive nuclei have half-lives that vary from 10^{-14} seconds to 10^{11} years, a range of 32 powers of 10!

The average lifetime or mean life, t_{av}, may be obtained as follows. The number of nuclei with lifetimes between t and $t + dt$ is the number that decays in time dt, which, from Equation 14.1, is $\lambda N\, dt$. Then, the sum of the lifetimes of all N_0 nuclei is

$$\int_0^\infty t\lambda N\, dt,$$

and the average lifetime is

$$t_{\text{av}} = \frac{1}{N_0}\int_0^\infty t\lambda N\, dt = \int_0^\infty \lambda t e^{-\lambda t}\, dt = \frac{1}{\lambda}, \qquad (14.6)$$

where Equation 14.3 has been used. The average lifetime may also be expressed in terms of the half-life as follows:

$$t_{\text{av}} = \frac{1}{\lambda} = \frac{t_{1/2}}{0.693} = 1.44 t_{1/2}. \qquad (14.7)$$

The *activity* of a given radioactive specimen is defined as the number of disintegrations per second. From Equation 14.1 we see that the activity is simply

$$\text{Activity} = \lambda N = \lambda N_0 e^{-\lambda t}, \tag{14.8}$$

where λ is expressed in s^{-1}. Two common units for measuring the activity are:

$$1 \text{ curie (Ci)} = 3 \times 10^{10} \text{ disintegrations/s}$$

$$1 \text{ rutherford (Rd)} = 1 \times 10^6 \text{ disintegrations/s}$$

The activity of a radioactive sample is readily measured by means of a wide variety of modern detectors. If the activity is then plotted as a function of time on semilogarithm paper, as shown in Figure 14–3, both the half-life and the decay constant can be obtained graphically.

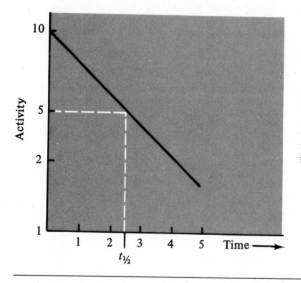

Figure 14–3

Semilogarithmic plot of activity as a function of time. The slope of the line is the decay constant, λ. The half-life is the time for which the activity drops to $\frac{1}{2}$ of its initial value.

Example 14-1
The activity of a radioactive sample decreases by a factor of 10 in 5 min. (a) Find the decay constant. (b) Find the half-life and the average lifetime.

Solution
(a) Using Equation 14.8,

$$\frac{\lambda N_1}{\lambda N_2} = \frac{e^{-\lambda t_1}}{e^{-\lambda t_2}} = e^{\lambda(t_2-t_1)}$$

$$10 = e^{\lambda \cdot 300s}$$

$$\lambda = \frac{\ln 10}{300 \text{ s}} = \frac{2.302}{300 \text{ s}} = 7.7 \times 10^{-3} \text{ s}^{-1}$$

(b) $$t_{1/2} = \frac{0.693}{7.7 \times 10^{-3} \text{ s}^{-1}} = 90 \text{ s}$$

$$t_{av} = 1.44 t_{1/2} = 129.6 \text{ s}.$$

Example 14-2
The half-life of $^{238}_{92}U$ is 4.51×10^9 years. What percentage of the $^{238}_{92}U$ nuclides that existed 10 billion years ago still survive?

Solution

$$\lambda = \frac{0.693}{t_{1/2}} = \frac{0.693}{4.51} = 0.154 \text{ per } 10^9 \text{ years,}$$

$$\frac{N}{N_0} = \exp\left(-0.154 \times 10\right) = \exp\left(-1.54\right) = 0.214.$$

That is, 21.4% of the original number still survive.

2. MULTIPLE DECAY PROCESSES

In the natural radioactive series, and frequently in man-made radioactivity, the daughter nuclides are themselves unstable and undergo decay. The same can be said for granddaughters, great-granddaughters, and so on. The decay particles may be different; for instance, ^{216}Po emits an α particle to become ^{212}Pb, which in turn emits a β^- to become ^{212}Bi. Also, the half-lives may vary considerably, as will be seen in Example 14-4.

Suppose that N_1 is the population of a parent nucleus and that N_2 is the population of its unstable daughter at time t. The parent has decay constant λ_1 and the daughter has decay constant λ_2. Then the rate of change of N_2 with respect to time is simply the difference in activities of the parent and daughter, namely,

$$\frac{dN_2}{dt} = \lambda_1 N_1 - \lambda_2 N_2. \tag{14.9}$$

Let us assume that the population of the parent is N_{10} and the population of the daughter is $N_2 = 0$ at $t = 0$. Then Equation 14.9 can be integrated to obtain[1]:

$$N_2 = \frac{\lambda_1 N_{10}}{\lambda_2 - \lambda_1} \left(e^{-\lambda_1 t} - e^{-\lambda_2 t}\right). \tag{14.10}$$

It is evident that the maximum value of N_2 will also be the equilibrium value, since the two decay processes will occur at the same rate when $dN_2/dt = 0$. The maximum value of N_2 is therefore

$$N_2 = \frac{\lambda_1}{\lambda_2} N_1 = \frac{\lambda_1}{\lambda_2} N_{10} e^{-\lambda_1 t}. \tag{14.11}$$

Example 14-3
How much time must elapse before the equilibrium value of N_2 is reached?

Solution
Substituting Equation 14.11 into Equation 14.10, we obtain

$$\frac{\lambda_1 N_{10}}{\lambda_2} e^{-\lambda_1 t} = \frac{\lambda_1 N_{10}}{\lambda_2 - \lambda_1} \left(e^{-\lambda_1 t} - e^{-\lambda_2 t}\right)$$

$$e^{-\lambda_1 t} = \frac{\lambda_2}{\lambda_2 - \lambda_1} \left(e^{-\lambda_1 t} - e^{-\lambda_2 t}\right)$$

$$\lambda_1 e^{-\lambda_1 t} = \lambda_2 e^{-\lambda_2 t}$$

---

[1] R. Gautreau and W. Savin, *Theory and Problems of Modern Physics*, Schaum's Outline Series in Science, McGraw-Hill Book Co., New York, 1978, p. 195.

$$\frac{\lambda_1}{\lambda_2} = e^{(\lambda_1 - \lambda_2)t}$$

$$(\lambda_1 - \lambda_2)t = \ln\left(\frac{\lambda_1}{\lambda_2}\right)$$

$$t = \frac{1}{\lambda_1 - \lambda_2} \ln\left(\frac{\lambda_1}{\lambda_2}\right). \qquad (14.12)$$

Example 14-4

Radium-226 decays by alpha emission to radon-222, which in turn decays by alpha emission to polonium-218. The half-life of ^{226}Ra is 1620 years, while the half-life of ^{222}Rn is 3.83 days. (a) What is the equilibrium ratio of the number of ^{226}Ra nuclei? (b) How much time was initially required to establish that equilibrium?

Solution

(a) $\lambda_1 = \dfrac{0.693}{1620 \text{ yr}} = 1.17 \times 10^{-6} \text{ days}^{-1}$

$\lambda_2 = \dfrac{0.693}{3.83 \text{ days}} = 1.81 \times 10^{-1} \text{ days}^{-1}$

$\dfrac{N_1}{N_2} = \dfrac{\lambda_2}{\lambda_1} = \dfrac{1.81 \times 10^{-1}}{1.17 \times 10^{-6}} = 1.55 \times 10^5$

(b) From Equation 14.12,

$$t = \frac{1}{\lambda_2 - \lambda_1} \ln\left(\frac{\lambda_2}{\lambda_1}\right)$$

$$= \frac{1}{0.18} \ln (1.55 \times 10^5) \text{ days}$$

$$= 66 \text{ days}$$

When a sample contains more than one species of radioactive nuclei, a semilogarithmic plot of activity versus time will not be a straight line as was the case in Figure 14–3. However, after a sufficiently long time, the nuclide with the longest half-life will dominate and the curve will approach a straight line. The slope of that line will correspond to the decay constant of the nuclide having the longest half-life. The extrapolation of the line back to $t = 0$ will give the activity of this long-lived nuclide at any time t. If this activity is subtracted from the total activ-

Figure 14–4

The total activity of a mixture of radioactive nuclides as a function of time. The decay constant for the nuclide having the longest half-life may be determined from the slope of the line that fits the decay curve for large t.

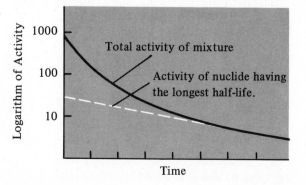

ity and the new values replotted, then the decay constant of the nuclide having the next longest half-life may be obtained in the same way. Proceeding in this fashion, it is possible to analyze an assortment of radionuclides. (See Figure 14–4.)

3. RADIO-CARBON DATING

Carbon-14 is produced in the upper atmosphere as a result of nuclear reactions caused by cosmic rays. The ratio of ^{14}C to ^{12}C is about 1.3×10^{-12} in the CO_2 of our atmosphere. The same ratio occurs in all living organisms since they continually exchange CO_2 with their environments. When an organism dies, however, the ratio of ^{14}C to ^{12}C nuclei changes due to the radioactive decay of ^{14}C through beta emission. The half-life for this process is 5730 years.

Assuming that the ratio of ^{14}C to ^{12}C in the atmosphere has remained essentially constant for thousands of years, it is then possible to estimate how long ago a sample of organic material died by measuring its activity per unit mass today.

Example 14-5

What is the activity due to ^{14}C decay in a gram of carbon in a living organism?

Solution

The decay constant of ^{14}C is:

$$\lambda = \frac{0.693}{t_{1/2}} = \frac{0.693}{5730 \text{ yr} \times 5.26 \times 10^5 \text{ min/yr}} = 2.30 \times 10^{-10} \text{ disint/min.}$$

The number of ^{14}C nuclei in one gram of carbon is:

$$N = \frac{6.02 \times 10^{23}}{12} \times 1.3 \times 10^{-12} = 6.52 \times 10^{10} \text{ g}^{-1}.$$

Then,

$$\text{Activity} = \lambda N = 2.30 \times 10^{-10} \times 6.52 \times 10^{10} = 15.0 \text{ disint/min/g.}$$

Example 14-6

A piece of charcoal weighing 50 g was found in an ancient ruin. If it shows a ^{14}C activity of 480 disint/min, how long has the tree been dead?

Solution

$$\frac{\text{Present activity}}{\text{Original activity}} = e^{-\lambda t} = \exp\left(-0.693t/t_{1/2}\right)$$

$$\frac{480 \text{ min}^{-1}}{50 \text{ g} \times 15 \text{ min}^{-1} \text{ g}^{-1}} = \exp\left(-0.693t/5730 \text{ yr}\right)$$

$$\ln 1.5625 = 0.693\, t/5730 \text{ yr}$$

$$t = \frac{5730 \text{ yr} \times \ln 1.563}{0.693} = 3690 \text{ yr}$$

In order to estimate the age of organic matter by measuring the activity of ^{14}C, a sample must contain at least a few grams of carbon. It should be mentioned here that a more sensitive method of determining the ratio of ^{14}C to ^{12}C now exists which does not require the measurement of the decay rates.[1] This method requires only 15 mg of carbon and is expected to extend the dating range back to 100,000 years.

4. GAMMA RAY EMISSION

Gamma rays are photons just as are visible light and x-rays; they differ only in their origin and their energies. Generally speaking, photons of visible light are emitted or absorbed during transitions of the valence electrons in atoms, discrete x-rays correspond to energy differences in the inner electronic shells, and gamma ray energies correspond to the differences between nuclear energy levels. In all these cases the photon energy is given by $h\nu = \Delta E$, where ΔE is the energy difference between the two states. Photons of visible light have energies of the order eV, x-rays of the order of keV, and gamma rays are in the MeV to GeV range. Photons can, of course, be created in ways other than just transitions between energy levels. For example, we have previously noted the production of gammas by the annihilation of antiparticle pairs and the creation of continuous x-rays from the kinetic energy of electrons. However, the study of the gammas emitted and absorbed by nuclei—which is called *nuclear spectroscopy* or *gamma ray spectroscopy*—is as important to our understanding of the nucleus as optical spectroscopy is to our knowledge of atomic energy states.

Any disturbance of a nucleus, whether it is the absorption of a photon or particle, radioactive decay, fission, or a high-energy collision, can leave the nucleus in one of its excited states. It will then probably return to its ground state by the emission of one or more gamma photons. A nucleus that emits a photon must suffer a recoil in order to conserve linear momentum, and the recoil kinetic energy reduces the energy carried by the photon. This is also true for a visible photon, but the recoil energy of the atom is often negligible in such a case. The recoil energy is not negligible, however, in the case of an energetic gamma photon. We express the conservation of energy as

$$\Delta E = h\nu + \frac{(Mv)^2}{2M}, \tag{14.13}$$

where ΔE is the actual energy difference between the levels, $h\nu$ is the photon energy, and the last term is the recoil kinetic energy of the nucleus. (The classical expression is suitable here since the rest mass energies of nuclei are of the order of GeV.) In order to conserve linear momentum we require that

$$\frac{h\nu}{c} = Mv. \tag{14.14}$$

Substituting for Mv in Equation 14.13,

$$\Delta E = h\nu + \frac{(h\nu)^2}{2Mc^2} = h\nu \left(1 + \frac{h\nu}{2Mc^2}\right).$$

Then,
$$h\nu = \Delta E \left(1 + \frac{h\nu}{2Mc^2}\right)^{-1}.$$

---

[1] R. A. Muller, "Radioisotope Dating with Accelerators," *Physics Today* 32, 23 (Feb. 1979).

Now we will make two approximations. When $h\nu \ll Mc^2$, then $h\nu$ will not differ greatly from ΔE, so let us replace $h\nu$ with ΔE inside the parentheses. Then we will use the binomial theorem to expand the parenthetical expression as follows:

$$h\nu = \Delta E \left(1 + \frac{\Delta E}{2Mc^2}\right)^{-1}$$

$$h\nu \approx \Delta E \left(1 - \frac{\Delta E}{2Mc^2}\right). \tag{14.15}$$

Thus, the energy of an emitted photon can be reduced by as much as $(\Delta E)^2/2Mc^2$ by the recoil of the nucleus. This effect *broadens* the spectral line; that is, the gammas emitted from a given transition will have a small spread of energies instead of one precise energy.

Example 14-7

Calculate the recoil energy for the emission of a 14.4 keV gamma photon from a free atom of excited ^{57}Fe.

Solution

From Equation 14.15, the recoil energy is:

$$\frac{(\Delta E)^2}{2Mc^2} = \frac{(14.4 \times 10^{-3} \text{ MeV})^2}{2 \times 57 \times 931.5 \text{ MeV}} = 1.95 \times 10^{-9} \text{ MeV} \approx 2 \times 10^{-3} \text{ eV}.$$

The recoil energy amounts to about 1 part in 10^7 in this particular case.

Although the recoil broadening in the previous example appears to be quite small, it is many times larger than the natural breadth of the spectral line. Rudolph Mössbauer was awarded the Nobel prize in 1961 for showing that recoilless gamma emission is possible under certain special conditions. Recoilless emission (and absorption) can occur when the emitting (or absorbing) atom is bound to its neighbors in a crystal lattice with sufficient energy to prevent its recoil. There are also certain conditions on the lattice vibration spectrum of the crystal, so the sample must usually be cooled to very low temperatures. When recoilless emission occurs, the line has essentially its natural linewidth and the photon may be absorbed by an unexcited nucleus of the same species. This phenomenon is now known as the *Mössbauer effect*.

A frequently used source for Mössbauer studies is $^{57}_{27}$Co, which captures an electron from its own *K*-shell and then decays to an excited state of $^{57}_{26}$Fe. When ^{57}Fe returns to its ground state, it emits the 14.4 keV photon discussed in Example 14-7. Under recoilless emission its linewidth is only about 1×10^{-7} eV, which is about one part in 10^{11}. Thus, a sample of ^{57}Co provides a source of radiation of exceptional spectral purity, and unexcited ^{57}Fe becomes an equally sharp detector. To get some appreciation for the sharpness of this line, it has been shown that a relative velocity of only a few millimeters per second between source and detector will produce a Doppler shift great enough to prevent the resonant absorption!

Example 14-8

Calculate the Doppler shift in the energy of the 14.4 keV photon emitted by ^{57}Fe if the relative velocity between the source and the detector is 3 cm/s.

Solution

First,

$$\beta = \frac{3 \text{ cm/s}}{3 \times 10^{10} \text{ cm/s}} = 1 \times 10^{-10}$$

From Equation 4.20,

$$\Delta E = h \, \Delta \nu = h(\nu_0 - \nu) = h\nu_0 \left[\frac{1 \pm \beta}{(1 - \beta^2)^{1/2}} - 1 \right] \simeq \pm \beta h \nu_0,$$

since $\beta^2 = 1 \times 10^{-20}.$

Then $\Delta E = \pm 1 \times 10^{-10} \times 1.4 \times 10^4 \text{ eV} = \pm 1.4 \times 10^{-6} \text{ eV}.$

Since this shift is about 7 times the natural breadth of the emitted line, resonant absorption will not occur.

5. ALPHA DECAY

Certain of the heavier unstable nuclides decay by means of the emission of alpha particles. For example,

$$^{234}_{92}\text{U} \rightarrow {}^{230}_{90}\text{Th} + {}^{4}_{2}\text{He}$$

$$^{226}_{88}\text{Ra} \rightarrow {}^{222}_{86}\text{Rn} + {}^{4}_{2}\text{He}$$

$$^{212}_{83}\text{Bi} \rightarrow {}^{208}_{81}\text{Tl} + {}^{4}_{2}\text{He}$$

In all cases we see that the total charge and the total nucleon number are conserved, in addition to the familiar conservation of linear and angular momentum and total energy. The conservation of nucleon number is often referred to as the conservation of *baryon number,* since nucleons are now known to be a part of a larger class of particles called baryons.

In alpha decay, the emitted alpha particle usually has a kinetic energy of the order of a few MeV and a speed that is no greater than one-tenth the speed of light. The daughter nuclide must also receive some kinetic energy, since it must recoil in order to conserve linear momentum. If we denote the masses of the parent, the daughter, and the alpha particle by M_P, M, and M_α, respectively, and the recoil speed of the daughter by v, then we have the conservation equations,

$$M_P c^2 = (M + M_\alpha)c^2 + \tfrac{1}{2}Mv^2 + \tfrac{1}{2}M_\alpha v_\alpha^2 \qquad (14.16)$$

and $$Mv = M_\alpha v_\alpha, \qquad (14.17)$$

where it is assumed that the parent nucleus was initially at rest.

We define the disintegration energy or the *reaction energy* by the symbol Q, where

$$Q = (M_P - M - M_\alpha)c^2 = \tfrac{1}{2}Mv^2 + \tfrac{1}{2}M_\alpha v_\alpha^2. \qquad (14.18)$$

It is evident from Equation 14.18 that Q must be positive in order for the reaction to occur spontaneously. That is, the mass of the parent must be greater than the total mass of the reaction products. Furthermore, the reaction energy is transformed to kinetic energy of the decay products. Eliminating v from Equations 14.17 and 14.18,

$$Q = \tfrac{1}{2} M_\alpha v_\alpha{}^2 \left(1 + \frac{M_\alpha}{M} \right). \tag{14.19}$$

To a close approximation, $M_\alpha/M = 4/(A - 4)$, where A is the mass number of the parent. Then we may express the reaction energy as

$$Q \approx \frac{A}{A - 4} E_\alpha, \tag{14.20}$$

where E_α is the kinetic energy of the alpha particle. This tells us that the alpha particles from a given reaction are essentially monoenergetic, since the reaction energy is a fixed quantity. Measurements of alpha particle energies (for example, by bending their trajectories in magnetic fields) have confirmed this result.

Example 14-9

Use Equation 14.20 to calculate the kinetic energy of alpha particles emitted by the decay of $^{222}_{86}$Rn.

Solution

From Equation 14.18,

$$Q = (\text{Mass of } {}^{222}\text{Rn} - \text{Mass of } {}^{218}\text{Po} - M_\alpha)c^2$$

$$= (222.017\,531 - 218.008\,930 - 4.002\,603) \times 931.5 \text{ MeV}$$

$$= 5.587 \text{ MeV}.$$

From Equation 14.20, the kinetic energy of the alpha particle, E_α, is

$$E_\alpha = \frac{A - 4}{A} Q = \frac{218}{222} \times 5.587 \text{ MeV} = 5.486 \text{ MeV}.$$

This is in excellent agreement with the experimental value.

The first successful explanation of alpha particle emission was by Gamow in 1928 and Gurney and Condon in 1929. They applied the concept of barrier penetration or *tunneling* (discussed in Section 4 of Chapter 10) to a well such as that shown in Figure 13–6. Suppose, for example, that the well is about 40 MeV deep and that the Coulomb barrier is roughly 16 MeV high. Let the breadth of the barrier be about 10 fermis. A typical value for the kinetic energy of an emitted alpha particle is 6 MeV, which is considerably below the summit of the barrier. A calculation[1] of the probability that a 6 MeV alpha particle can tunnel through the barrier when it strikes it turns out to be only one chance in 10^{25}! Although this is a very small probability, the high speed of a 6 MeV alpha particle trapped in such a well would enable it to strike the barrier about 10^{21} times per second. Thus, the

---

[1] E. E. Anderson, *Modern Physics and Quantum Mechanics*, W. B. Saunders Co., Philadelphia, 1971, p. 165.

probability per second that a tunneling event would occur is $P = 10^{-25} \times 10^{21} \text{ s}^{-1} = 10^{-4} \text{ s}^{-1}$, and the probable decay time would be $t = 1/P = 10^4 \text{ s}$, or about 3 hours. In order to see how critically the probable decay time depends upon the strength of the barrier, suppose that the width of the barrier were increased by 20 percent and that all other quantities remain the same. This slight change would decrease the tunneling probability to one chance in 10^{30}, and the decay time would increase to 100 years! Therefore, in spite of the crudeness of this model, it is able to account for the experimental fact that the mean lifetimes of alpha emitters vary from 1 microsecond to 10^{10} years.

6. BETA DECAY

In contrast to alpha particle emission, in which discrete energies are obtained, beta particles are emitted over a continuous range of energies as shown in Figure 14–5. It is significant that the maximum kinetic energy observed for a given

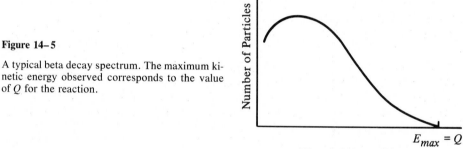

Figure 14–5

A typical beta decay spectrum. The maximum kinetic energy observed corresponds to the value of Q for the reaction.

reaction corresponds to the Q value for that reaction. Therefore, if we assume that the energy Q is released each time a decay occurs, then some other particle must share an indefinite fraction of the released energy with the beta particle. Also, studies of the recoil of the nucleus showed that linear momentum could not be conserved by the nucleus and the beta particle alone, since their momenta were frequently not even colinear. Furthermore, the emitted beta particle is known to have an angular momentum of $\frac{1}{2}\hbar$, which is not compensated for by an equivalent change in the angular momentum of the nucleus.

In 1931 Pauli suggested that a third particle must be present in order to conserve linear and angular momentum and carry away the extra reaction energy. This particle was named the *neutrino*, since it had to be neutral and have little or no rest mass. Although the neutrino's existence was hard to dispute, it eluded detection for another 25 years because it interacts so rarely with matter. It has been estimated that a neutrino could pass through a slab of iron 130 lightyears thick before interacting.

Neutrinos are produced in beta decay by one of the following reactions:

$$n \rightarrow p + \beta^- + \bar{\nu} \tag{14.21}$$

$$p \rightarrow n + \beta^+ + \nu. \tag{14.22}$$

Here ν represents the neutrino produced in positron emission, and $\bar{\nu}$ represents the antineutrino that accompanies the emission of an electron. Each of these neutrinos has a spin of $\frac{1}{2}\hbar$, but they differ in sense. The neutrino is said to be left-handed; that is, if the thumb of the left hand points in the direction of its motion, the fingers give the sense of the spin (counterclockwise when viewed along the direction of motion). The antineutrino is right-handed; its spin would be in the clockwise sense when viewed along the direction of its motion. The following inverse reactions provide the only means of detecting these particles:

$$\nu + n \rightarrow p + \beta^- \tag{14.23}$$

$$\bar{\nu} + p \rightarrow n + \beta^+ \tag{14.24}$$

Clyde Cowan and Frederick Reines made the first direct observations of antineutrinos in 1953. They used a reactor that produced a very large flux of β^- and which, from Equation 14.21, should be accompanied by an equally large supply of antineutrinos. In order to detect the antineutrinos, the beam was allowed to enter a liquid scintillator having a very high density of protons (such as liquid hydrogen). From Equation 14.24, if a proton captures an antineutrino we should then obtain a neutron and a positron. In order to confirm this reaction, it is necessary to detect both of these particles. The neutron detector consisted of ^{113}Cd, which decays as follows:

$$^{113}Cd + n \rightarrow {}^{114}Cd + \gamma. \tag{14.25}$$

That is, ^{114}Cd is in an excited state and emits a gamma photon when it returns to the ground state. The energy of the photon and the time-delay between neutron capture and gamma emission are known precisely enough to identify this event. The positron is detected by the gamma photons produced when it annihilates with an electron. In summary, the energies of all of the gammas produced are known and the time delays are known, so it is possible to identify an antineutrino capture unambiguously.

7. LOW-ENERGY NUCLEAR REACTIONS

The reaction energy Q as defined in Equation 14.18 is actually the negative of the binding energy of the parent nucleus. When Q is positive, the reaction can occur spontaneously and energy will be released by it, although a small activation energy might be required to initiate it. On the other hand, when Q is negative, sufficient energy must be provided in order to overcome the binding energy of the nucleus. Such external energy is generally provided by means of particle kinetic energy or photon absorption. In the center-of-mass coordinate system the kinetic energy required to initiate the reaction would be equal to $|Q|$. However, in the laboratory system additional energy is required in order to insure that the momentum of the incoming particle is conserved. If the energy of an incident particle is less than 50 MeV, classical mechanics may be used and the reaction is said to be a low-energy reaction. When the energies of incident particles exceed 100 MeV, they must be treated relativistically, and such reactions are a part of high-energy physics.

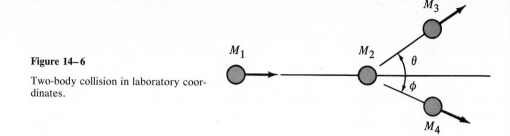

Figure 14-6

Two-body collision in laboratory coordinates.

Consider a low-energy reaction in which mass M_1 with kinetic energy E_1 strikes mass M_2, which is initially at rest. The reaction products are M_3 and M_4, having energies E_3 and E_4 and directions of motion as shown in Figure 14–6. All of the above quantities are measured in the laboratory. M_4 is the larger reaction product.

From the conservation of energy,

$$(M_1 + M_2)c^2 + E_1 = (M_3 + M_4)c^2 + E_3 + E_4. \tag{14.26}$$

However,

$$Q = (M_1 + M_2 - M_3 - M_4)C^2,$$

so Equation 14.26 becomes

$$E_1 + Q = E_3 + E_4. \tag{14.27}$$

Conservation of linear momentum yields the two following equations:

$$\sqrt{2M_1E_1} = \sqrt{2M_3E_3}\cos\theta + \sqrt{2M_4E_4}\cos\phi \tag{14.28}$$

and

$$0 = \sqrt{2M_3E_3}\sin\theta - \sqrt{2M_4E_4}\sin\phi. \tag{14.29}$$

Now we can eliminate the angle ϕ from Equations 14.28 and 14.29 by squaring each equation and adding the results. The sum is

$$M_4E_4 = M_3E_3 + M_1E_1 - 2\sqrt{M_1M_3E_1E_3}\cos\theta.$$

Eliminating E_4 between this result and Equation 14.27, we obtain

$$Q = E_3\left(1 + \frac{M_3}{M_4}\right) - E_1\left(1 - \frac{M_1}{M_4}\right) - \frac{2}{M_4}\sqrt{M_1M_3E_1E_3}\cos\theta. \tag{14.30}$$

This is the conventional "Q equation" in laboratory coordinates. It is independent of the nature of the reaction, since it deals only with the initial and final products.

Example 14-10

Use Equation 14.30 to discuss a spontaneous reaction in which $Q > 0$.

Solution

When $Q > 0$, no bombarding energy is required, so $E_1 = 0$. Then Equation 14.30 reduces to

$$Q = E_3\left(1 + \frac{M_3}{M_4}\right), \tag{14.31}$$

and Equation 14.27 becomes

$$Q = E_3 + E_4.$$

It follows that

$$E_3 = \frac{QM_4}{M_3 + M_4} \text{ and } E_4 = \frac{QM_3}{M_3 + M_4}.$$

Example 14-11

Consider the reaction

$$^4_2\text{He} + {}^{10}_5\text{B} \rightarrow {}^1_1\text{H} + {}^{13}_6\text{C}.$$

If $Q = +4.0$ MeV and the incident alpha particle energy is 2 MeV, what is the energy of a proton ejected at 90°?

Solution

Since cos 90° = 0, Equation 14.30 becomes

$$4.0 \text{ MeV} = E_3(1 + 1/13) - 2.0 \text{ MeV} (1 - 4/13)$$

$$(14/13) E_3 = (4 + 18/13) \text{ MeV}$$

$$E_3 = 5 \text{ MeV}.$$

Example 14-12

The threshold energy for a reaction having a negative Q value can be expressed as follows in laboratory coordinates[1]:

$$(E_1)_{th} = -Q\left(\frac{M_1 + M_2}{M_2}\right). \tag{14.32}$$

Use this result to find the threshold proton energy in the laboratory for the reaction,

$$^1_1\text{H} + {}^3_1\text{H} \rightarrow {}^1_0 n + {}^3_2\text{He},$$

for which $Q = -0.764$ MeV.

Solution

$M_1 = 1$ u and $M_2 = 3$ u. Then, using Equation 14.32,

$$(E_1)_{th} = 0.764 \text{ MeV} \left(\frac{1+3}{3}\right) = 1.02 \text{ MeV}.$$

At the threshold energy, the product particles appear in the $\theta = 0$ direction in order to conserve the momentum $\sqrt{2M_1E_1}$ of the incident proton.

8. NUCLEAR FISSION

Heavy nuclei have less binding energy per nucleon than do intermediate nuclei, as a glance at Figure 13–3 will show. Thus it is sometimes energetically favorable for a heavy nucleus to split into two nuclei having nearly equal masses. This process is called *fission*. When fission occurs, the difference in total binding energy is released as kinetic energy of the fragments and as radiation. In order to get some estimate of the magnitude of this energy, one merely has to assume that roughly 200 nucleons give up about 1 MeV each, or a total energy difference of

---

[1] See R. D. Evans, *The Atomic Nucleus,* McGraw-Hill Book Co., New York, 1955, p. 414.

about 200 MeV per nucleus! This should be compared with chemical reactions, which release only a few eV per atom.

Uranium-235 undergoes fission after the capture of a thermal neutron. A typical fission event might be characterized by one of the following:

$$n + {}^{235}_{92}U \rightarrow {}^{141}_{56}Ba + {}^{92}_{36}Kr + 3n$$

$$n + {}^{235}_{92}U \rightarrow {}^{140}_{54}Xe + {}^{94}_{38}Sr + 2n.$$

A wide range of fission products is possible, but it has been found that the most probable values for the mass numbers of the two fission fragments are about 95 and 139 when two neutrons are emitted. The fission products are themselves radioactive and undergo a series of decays until stable nuclides are reached. An interesting feature of fission is that it produces additional neutrons along with the two new nuclides, so that these neutrons are than available to induce additional fission reactions. The control of the supply of neutrons of the optimum energy and the production of large quantities of fissionable nuclei were key problems that had to be solved in order to launch the nuclear era.

When a fission event occurs in a sample of fissionable material, under the right conditions fission can continue as long as the supply of neutrons and fissionable nuclei last. We call this process a *chain reaction*. If the rate of neutron production just equals the rate at which neutrons are lost through absorption or leakage through the walls of the container, the controlled chain reaction may be used as a source of power. When the reaction proceeds at a rate that produces excess neutrons, fission gets out of control and the device becomes an *atomic bomb*.

One way to describe the operating level of a reactor is to define the effective neutron multiplication factor, k_e, as

$$k_e = \frac{P}{A + L}.$$

Here P is the rate of neutron production, A is the neutron absorption rate, and L is the leakage rate. When $k_e < 1$ the reactor is subcritical and its power output level is dying down. When $k_e = 1$ the reactor is said to be *critical*, and it will be operating at a constant power output level. For $k_e > 1$ the reactor power level will increase rapidly, and it is said to be supercritical.

Among the factors that affect the number of neutrons that are available for fission are the neutron speeds, the geometry of the reactor, and the absorption cross sections of both fissionable and non-fissionable nuclei. Absorption cross sections are simply relative probabilities for absorption, and they in turn depend upon the neutron energies. In order to control the neutron multiplication factor, then, it is important (1) to optimize the geometry so as to minimize the neutron loss through leakage, (2) to regulate the neutron speeds, and (3) to devise a "variable" neutron absorber. The first of these is essentially fixed by the design of the reactor and its neutron reflector. The second is controlled by the use of a *moderator,* such as heavy water (D_2O) or graphite, which is used to slow down the neutrons through collisions. The moderator must not be a neutron absorber, and it should have a low mass number so that a neutron will give up a large part of its energy with each collision. The amount of moderator material and its geometrical arrangement should maximize the fraction of neutrons having the right energy for easy capture by the fissionable material. (In the case of ^{235}U the neutrons should have thermal energies, that is, about 1/40 eV.)

The variable neutron absorber is realized by using movable rods made of good neutron absorbers (such as boron or cadmium). These *control rods* are inserted in the fissionable material in order to reduce k_e, and their position is adjusted in order to achieve the desired operating level.

Nuclear reactors are not only vital sources of electrical power, but they also provide us with many useful radioactive isotopes. Furthermore, the reactors themselves have become extremely important research tools in science and medicine. A reactor can provide electrical power and at the same time produce an equivalent amount of fissionable material, which in turn can be used as a reactor fuel. Such a reactor is called a *breeder reactor*. An example of a breeder reactor is one that uses natural uranium as its fuel and produces plutonium-239, which is also fissionable.

The chief disadvantage of nuclear fission reactors is the difficulty of safely handling the long-lifetime radioactive wastes from spent reactor cores. This is a problem that will persist until fusion reactors become our principal source of power.

9. NUCLEAR FUSION

Turning again to Figure 13–3, note that nuclei with A less than 30 or 40 have smaller binding energies per nucleon than the heavier nuclei. This means that, in principle, the inverse process to fission is energetically feasible for the lighter nuclei. The process of forming a heavy nucleus from two or more lighter nuclei is called *fusion*. Stars, including our sun, obtain their energy in this manner. One set of reactions that probably occurs in the sun is as follows:

$$\left. \begin{aligned} {}^1_1\text{H} + {}^1_1\text{H} &\rightarrow {}^2_1\text{H} + \beta^+ + \nu \\ {}^1_1\text{H} + {}^2_1\text{H} &\rightarrow {}^3_2\text{He} + \gamma \\ {}^1_1\text{H} + {}^3_2\text{He} &\rightarrow {}^4_2\text{He} + \beta^+ + \nu. \end{aligned} \right\} \qquad (14.33)$$

The net effect of this cycle of three fusion reactions is the transformation of four protons into an alpha particle, two positrons, two neutrinos, and a gamma photon.

Example 14-13
Calculate the Q value for the following fusion reaction, which is the sum of Equation 14.33:

$$4\,{}^1_1\text{H} \rightarrow {}^4_2\text{He} + 2\beta^+ + 2\nu + \gamma.$$

Solution
The mass of the four hydrogen atoms includes that of two electrons, which are not needed for helium. These two β^- masses plus the two β^+ masses do not contribute to the Q value for the reaction, although they will presumably annihilate and produce gamma photons. Then,

$$Q = (\text{Mass of } 4{}^1_1\text{H} - \text{Mass of } {}^4_2\text{He} - \text{Mass of } 4\beta)c^2$$

$$= (4 \times 1.007\ 825 - 4.002\ 603 - 4 \times 0.000\ 549) \times 931.5 \text{ MeV}$$

$$= (0.0265 \times 931.5) \text{ MeV}$$

$$= 24.7 \text{ MeV}.$$

If the fusion of four protons releases nearly 25 MeV of energy, one might well ask why hydrogen is so abundant. Why is it that helium is not the most abundant element? The answer is that the Coulomb repulsion between protons prevents them from getting close enough together to fuse. In order to get within the range of the nuclear forces, protons must collide with kinetic energies greater than 10 keV. This would require temperatures of the order of 10^8 K in order for fusion to occur during collisions in hydrogen gas. Such temperatures are found in stellar interiors. Even higher temperatures are required for the fusion of nuclei with $Z > 1$.

Another set of fusion reactions that is believed to occur in our sun is known as the *carbon cycle*. Its end result is once again the conversion of four protons to an alpha particle, but in this case carbon serves as a catalyst for the process. This cycle can occur only in stars that are sufficiently hot to form the elements up through $Z = 8$ by fusion. The steps in the carbon cycle are:

$$\left.\begin{array}{l} {}^{12}_{6}\text{C} + {}^{1}_{1}\text{H} \rightarrow {}^{13}_{7}\text{N} + \gamma \\[4pt] \qquad {}^{13}_{7}\text{N} \rightarrow {}^{13}_{6}\text{C} + \beta^+ + \nu \\[4pt] {}^{13}_{6}\text{C} + {}^{1}_{1}\text{H} \rightarrow {}^{14}_{7}\text{N} + \gamma \\[4pt] {}^{14}_{7}\text{N} + {}^{1}_{1}\text{H} \rightarrow {}^{15}_{8}\text{O} + \gamma \\[4pt] \qquad {}^{15}_{8}\text{O} \rightarrow {}^{15}_{7}\text{N} + \beta^+ + \nu \\[4pt] {}^{15}_{7}\text{N} + {}^{1}_{1}\text{H} \rightarrow {}^{12}_{6}\text{C} + {}^{4}_{2}\text{He}. \end{array}\right\} \qquad (14.34)$$

The chain begins with ${}^{12}_{6}$C and ends with ${}^{12}_{6}$C, and during the process four protons are converted to one helium nucleus, two positrons, and two neutrinos as in Equation 14.33.

Controlled fusion reactors—or thermonuclear power plants—are probably the most promising sources of energy for the future. We have an abundant supply of hydrogen, and fusion is a relatively "clean" process compared to fission. The temperatures required to induce fusion can be achieved by bombarding the fuel with high-energy particles or with a high-power laser. A much more difficult problem is that of confining the heated plasma long enough to sustain the fusion reaction until all of the fuel is spent. Another serious problem is that of coupling the energy produced by fusion to an external load without large losses due to electromagnetic radiation.

At temperatures comparable to stellar temperatures, the atoms will be stripped of all electrons; the resulting gas of charged nuclei and electrons is called a plasma. In a star the plasma is confined by the enormous gravitational forces. For a terrestrial reactor, two confinement schemes are under development. One is *magnetic confinement,* often called the magnetic "bottle," in which external magnetic forces on the moving charges of the plasma prevent the plasma from dispersing. One geometry that is being seriously studied is the torus (doughnut); at the other extreme is a linear plasma, which could be as much as a kilometer in length but only a few centimeters in diameter. The other confinement scheme is called *inertial confinement*. In this case the fuel, in the form of a pellet of solid deuterium or tritium, is heated to the fusion temperature by an array of pulsed high-power lasers that bombard it from all directions in a few nanoseconds. The fusion process would be completed before the particles could disperse.

SUMMARY

The first evidence that nuclei undergo transformations was the discovery of natural radioactivity. The law of radioactive decay was found and the decay schemes of the naturally occurring radioelements were soon discovered. It was not long before nuclear reactions were produced in the laboratory, and man had finally achieved the ancient goal of transforming one element into another. Although we are not using this knowledge to produce gold, as the alchemists hoped to do, the applications of radioactivity to medicine, science, and engineering are virtually limitless today.

The radioactive elements found in the earth's crust emit alpha, beta, and gamma rays, which were later identified as helium nuclei, electrons or positrons, and photons, respectively. Alpha particles emitted from a given reaction are generally monoenergetic. This energy, E_α, is related to Q, the energy released in the reaction, by the expression

$$E_\alpha = Q \left(\frac{A - 4}{A} \right),$$

where A is the nucleon number of the parent nucleus and A-4 is the nucleon number of the daughter. The theoretical model used to explain alpha emission is quantum mechanical tunneling.

A beta particle, unlike both alpha particles and gamma rays, is emitted with an energy that is unpredictable except that the maximum energy it can have is limited to the Q value of the reaction. The emitted beta particle shares the reaction energy with a neutrino, which has no mass or charge but does have a spin of $\frac{1}{2}\hbar$. The neutrino is necessary in order to conserve spin, linear momentum, and energy in the reaction.

Gamma rays are photons having shorter wavelengths than x-rays. They are emitted when a nucleus decays from an excited state to a lower excited state or the ground state. The energy of the gamma photon is nearly equal to the energy difference between the states, the discrepancy being the recoil kinetic energy of the nucleus. Much of what we now know about nuclear energy levels has been learned from gamma ray spectroscopy.

Nuclear reactions may be initiated by the bombardment of nuclei with particles or photons. They may be grouped broadly into low-energy reactions where classical mechanics can be used in the calculations, and high-energy reactions where all calculations must use relativistic mechanics. Two extremely important reactions are fission and fusion, which have brought mankind into the nuclear era. They hold both the promise of solving earth's long-term energy needs and the potential for wholesale destruction. When a nucleus undergoes fission, it divides into two elements, each having nearly half the mass of the original nucleus. In doing so it releases nearly 1 MeV per nucleon or a total reaction energy of about 200 MeV per fission. Fusion is the inverse process, that is, the fusing of two nuclei to form a heavier nucleus. Fusion is energetically favorable only for light nuclei whose mass numbers are less than about 40. It is believed that fusion of hydrogen to form helium is the initial energy source of a young star. As its temperature rises, heavier elements are formed through fusion.

Fission is commonly induced by the capture of a neutron. In some cases the neutron must be energetic but in U-235, for example, the neutron should have an energy of only a fraction of an eV. Fusion, on the other hand, requires an energy of 10 to 100 keV in order to bring the two nuclei close enough together to fuse against the Coulomb repulsion.

Although fusion reactors will probably become the principal source of energy on earth, fission breeder reactors will continue to be an important source of radioisotopes for medicine and scientific research and of fissionable material to replace the dwindling supply of uranium.

Additional Reading

APS Study Group, "The Nuclear Fuel Cycle: An Appraisal," *Physics Today 30,* 32 (October 1977).

C. H. Blanchard, C. R. Burnett, R. R. Stoner, and R. L. Weber, *Introduction to Modern Physics,* Prentice-Hall, Inc., Englewood Cliffs, N.J., 1969, Chapters 14, 15, 16, and 17.

F. L. Culler, Jr., and W. O. Harms, "Energy from Breeder Reactors," *Physics Today 25,* 28 (May 1972).

W. C. Gough and B. J. Eastlund, "The Prospects of Fusion Power," *Scientific American 224,* 50 (February 1971).

W. F. Libby, *Radiocarbon Dating,* University of Chicago Press, Chicago, 1955.

M. J. Lubin and A. P. Fraas, "Fusion by Laser," *Scientific American 224,* 13 (June 1971).

"Magnetically Confined Fusion." Four articles in *Physics Today 32,* No. 5 (May 1979).

E. J. Moniz and T. L. Neff, "Nuclear Power and Nuclear Weapons Proliferation," *Physics Today 31,* 42 (April 1978).

J. Nuckolls, J. Emmett, and L. Wood, "Laser-Induced Thermonuclear Fusion," *Physics Today 26,* 46 (August 1973).

H. Semat and J. R. Albright, *Introduction to Atomic and Nuclear Physics,* 5th ed., Holt, Rinehart and Winston, Inc., New York, 1972, Chapters 14, 15, and 16.

G. T. Seaborg and J. L. Bloom, "Fast Breeder Reactors," *Scientific American 223,* 13 (May 1970).

C. M. Stickley, "Laser Fusion," *Physics Today 31,* 50 (May 1978).

P. A. Tipler, *Modern Physics,* Worth Publishers, Inc., New York, 1978, Chapter 11.

R. T. Weidner and R. L. Sells, *Elementary Modern Physics,* 3rd ed., Allyn and Bacon, Inc., Boston, 1980, Chapters 10 and 11.

PROBLEMS

14-1. Calculate the decay constant of $^{90}_{38}Sr$, whose half-life is 28.8 years.

14-2. Carbon-14 is a β^- emitter having a half-life of 5730 years. What percentage of the ^{14}C contained in a bone would disintegrate in 1000 years?

14-3. Radioactive ^{22}Na has a half-life of 2.58 years. What percentage of a sample would decay per year?

14-4. How many grams of $^{14}_6C$ would be required in order to have an activity of 5 curies?

14-5. What activity in disintegrations per minute per gram would be expected for carbon samples from bones that are said to be 2000 years old?

14-6. Calculate the mass of a millicurie of $^{234}_{92}U$, whose half-life is 2.48×10^5 years.

14-7. $^{238}_{92}U$ emits an alpha particle and decays to $^{234}_{90}Th$, which is itself unstable with a half-life of 24.1 days. If the ratio of ^{234}Th to ^{238}U found in a rock sample is 1.46×10^{-11}, what is the half-life of ^{238}U?

14-8. What is the line broadening due to recoil of a ^{60}Ni nucleus upon emission of a 1.33 MeV gamma photon?

14-9. Calculate the kinetic energy of the alpha particles emitted by $^{218}_{84}Po$.

14-10. What is the kinetic energy of alpha particles emitted by $^{232}_{92}U$?

14-11. Find the kinetic energy of the alpha particles emitted by the decay of $^{210}_{84}Po$.

14-12. Find the kinetic energy of the alpha particles emitted by the decay of $^{220}_{86}Rn$.

14-13. Find the threshold alpha particle energy in the laboratory for the reaction

$$^4_2He + {}^{14}_7N \rightarrow {}^1_1H + {}^{17}_8O,$$

for which $Q = -1.18$ MeV.

14.14. (a) Calculate the Q value for the reaction

$$^3_1H + {}^1_1H \rightarrow {}^3_2He + n + Q.$$

(b) What is the threshold energy in the laboratory for this reaction if tritium nuclei bombard stationary protons? (c) What is the threshold energy in the laboratory for this reaction if protons bombard stationary tritium nuclei?

14.15. (a) Calculate the Q value for the reaction

$$n + {}^3_2He \rightarrow {}^2_1H + {}^2_1H + Q.$$

(b) What is the threshold neutron kinetic energy for this reaction if ^3_2He is at rest in the laboratory?

14-16. Calculate the Q value for the reaction

$$n + {}^{235}_{92}U \rightarrow {}^{138}_{56}Ba + {}^{96}_{40}Zr + 2n + 4\beta^-.$$

14-17. Assume 200 MeV of available energy per fission. (a) How many nuclei would have to undergo fission per second in order to generate 1 megawatt of power? (b) How many grams of ^{235}U would be required per hour?

14-18. Calculate the Q value for each fusion reaction in Equation 14.33, and show that their sum equals the result obtained in Example 14-13.

14-19. Calculate the Q values for the following fusion reactions:
(a) $n + {}^1_1H \rightarrow {}^2_1H + \gamma$
(b) $^2_1H + {}^2_1H \rightarrow {}^3_1H + n + \beta^+$
(c) $^2_1H + {}^3_2He \rightarrow {}^4_2He + {}^1_1H$

14-20. Calculate the Q values for the following fusion reactions:
(a) $^2_1H + {}^3_1H \rightarrow {}^4_2He + n$
(b) $^1_1H + {}^{12}_6C \rightarrow {}^{13}_7N + \gamma$
(c) $^1_1H + {}^{13}_6C \rightarrow {}^{14}_7N + \gamma$

14-21. The sun radiates energy at the rate of 4×10^{23} kilowatts. If the fusion reaction of Example 14-13 were to account for all of the energy released by the sun, (a) how many protons would be fused per second, and (b) how much mass would be transformed per second?

14-22. Uranium-235 undergoes fission after absorbing a thermal neutron. The fission products are two nuclei of equal mass and two prompt neutrons. How much energy is released?

14-23. A 1 MeV neutron is emitted in a fission reactor. If it were to lose one half of its kinetic energy in each collision with an atom of the moderator, how many collisions must it undergo in order to achieve thermal energies?

14-24. How many grams of uranium-235 would be consumed in producing a megawatt-hour of electricity, if the conversion efficiency from heat to electricity is 25%?

Appendix A Velocity and Acceleration in Polar Coordinates

Consider the motion of a particle along a path such as that shown in Figure A–1. At the instant when the particle is at the point P, its location may be denoted by the position vector, $\vec{r} = r\,\hat{r}$, drawn to the point $P(r, \theta)$ from any point chosen as the origin of the coordi-

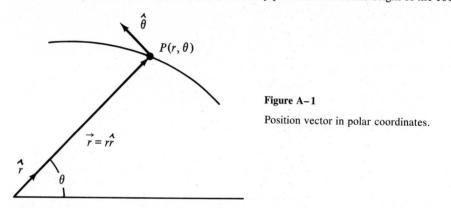

Figure A–1

Position vector in polar coordinates.

nates. Here $\hat{r}$ is a unit vector in the direction of $\vec{r}$, and $\hat{\theta}$ is a unit vector perpendicular to $\vec{r}$ in the sense of increasing polar angle θ. Since $\vec{r}$ may change in both magnitude and direction with time, both $\hat{r}$ and $\hat{\theta}$ may change direction with time. In order to find these time derivatives, let us look at Figure A–1. In (a) we see that the change in $\hat{r}$ may be written as

$$|d\hat{r}| = |\hat{r}|d\theta = d\theta. \qquad (A.1)$$

Since $d\hat{r}$ is evidently in the direction of $\hat{\theta}$, we then obtain

$$d\hat{r} = \hat{\theta}\,d\theta$$

and

$$\frac{d\hat{r}}{dt} = \hat{\theta}\,\frac{d\theta}{dt} = \dot{\theta}\,\hat{\theta}. \qquad (A.2)$$

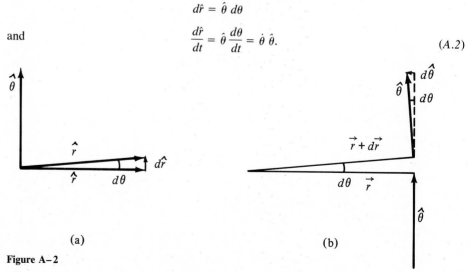

(a)

(b)

Figure A–2

Infinitesimal changes in direction of the unit vectors.

From (b) we write

$$|d\hat{\theta}| = |\hat{\theta}|d\theta = d\theta. \qquad (A.3)$$

But $d\hat{\theta}$ is directed toward the origin as shown. Therefore,

$$d\hat{\theta} = -\hat{r}\,d\theta$$

and

$$\frac{d\hat{\theta}}{dt} = -\dot{\theta}\hat{r}. \qquad (A.4)$$

Now we are ready to obtain the time derivatives of $\vec{r}$. Thus

$$\vec{r} = r\hat{r},$$

$$\dot{\vec{r}} = \dot{r}\hat{r} + r\frac{d\hat{r}}{dt} = \dot{r}\hat{r} + r\dot{\theta}\hat{\theta}, \qquad (A.5)$$

where Equation A.2 has been used. The first term of Equation A.5 is the radial velocity, and the second term is the angular velocity, of the position vector $\vec{r}$.

If we now take the second time derivative of the position vector, we obtain

$$\ddot{\vec{r}} = \ddot{r}\hat{r} + \dot{r}\frac{d\hat{r}}{dt} + \dot{r}\dot{\theta}\hat{\theta} + r\ddot{\theta}\hat{\theta} + r\dot{\theta}\frac{d\hat{\theta}}{dt}$$

$$= \ddot{r}\hat{r} + \dot{r}\dot{\theta}\hat{\theta} + \dot{r}\dot{\theta}\hat{\theta} + r\ddot{\theta}\hat{\theta} - r\dot{\theta}^2\hat{r}$$

$$= (\ddot{r} - r\dot{\theta}^2)\hat{r} + (2\dot{r}\dot{\theta} + r\ddot{\theta})\hat{\theta}. \qquad (A.6)$$

The first and second terms of Equation A.6 are the radial and angular accelerations, respectively, of the position vector $\vec{r}$.

Appendix B The Rutherford Scattering Equation

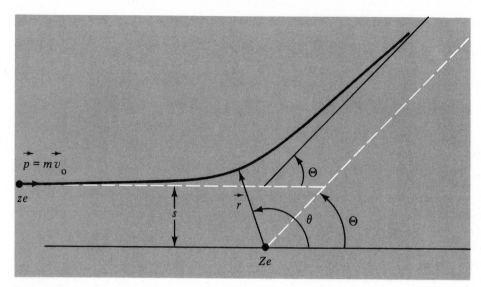

Figure B–1

Rutherford scattering.

Rutherford scattering is the kind of scattering that occurs when both the projectiles and the scattering centers are charged particles having the same sign. That is, the scattering interaction between the particles is simply the repulsive Coulomb force. It is for this reason that the terms "Coulomb scattering" and "Rutherford scattering" are used interchangeably.

Consider a projectile of mass m and charge ze incident upon a stationary scattering center such as a nucleus of charge Ze. For z and Z both positive, the Coulomb repulsion will scatter the incident particle along a hyperbolic trajectory, as shown in Figure B–1. Since the angular momentum about any point is constant for a central force, we may write

$$mv_0 s = mr^2\dot\theta = \text{constant}, \qquad (B.1)$$

where the nucleus has been chosen as the reference point. From Equation B.1 we obtain

$$r\dot\theta^2 = \frac{s^2 v_0^2}{r^3}. \qquad (B.2)$$

The radial part of the equation of motion is

$$F(r) = m(\ddot r - r\dot\theta^2), \qquad (B.3)$$

where we have used the radial term of Equation A.6. The central force is simply the Coulomb force, so Equation B.3 becomes

$$\frac{kzZe^2}{r^2} = m(\ddot r - r\dot\theta^2). \qquad (B.4)$$

It is convenient to change variables by making the substitution $r = 1/\mu$. Then, from Equation B.1,

$$\dot{\theta} = sv_0\mu^2. \qquad (B.5)$$

Using the chain rule of differentiation,

$$\dot{r} = \frac{dr}{d\mu}\frac{d\mu}{d\theta}\frac{d\theta}{dt} = \left(-\frac{1}{\mu^2}\right)(sv_0\mu^2)\frac{d\mu}{d\theta} = -sv_0\frac{d\mu}{d\theta}, \qquad (B.6)$$

and

$$\ddot{r} = -sv_0\frac{d^2\mu}{d\theta^2}\frac{d\theta}{dt} = -s^2v_0^2\mu^2\frac{d^2\mu}{d\theta^2}. \qquad (B.7)$$

Substituting these expressions into Equation B.4, we obtain

$$\frac{kzZe^2}{ms^2v_0^2} = -\mu - \frac{d^2\mu}{d\theta^2}. \qquad (B.8)$$

It is customary to define the collision diameter D as the closest distance of approach in the case of a head-on collision. That is, D is the distance from the target at which all of the kinetic energy of the projectile is converted to electrostatic potential energy. Then,

$$D = \frac{2kzZe^2}{mv_0^2}, $$

and the equation of the trajectory becomes

$$\frac{d^2\mu}{d\theta^2} + \mu = -\frac{D}{2s^2}. \qquad (B.9)$$

A solution of this equation is

$$\mu = A\cos\theta + B\sin\theta - \frac{D}{2s^2}, \qquad (B.10)$$

but we must use the boundary conditions of the problem in order to evaluate the unknown parameters. These boundary conditions are:

    (1) As $\theta \to \pi$, $r \to \infty$, $\mu \to 0$.
    (2) As $\theta \to \pi$, $\dot{r} \to v_0$.
    (3) As $\theta \to \Theta$, $r \to \infty$, $\mu \to 0$.

Substituting the first of these into Equation B.10, we obtain

$$0 = A\cos\pi + B\sin\pi - \frac{D}{2s^2}, $$

or

$$A = -\frac{D}{2s^2}. \qquad (B.11)$$

Substituting the second condition into Equation B.6 tells us that

$$\left(\frac{d\mu}{d\theta}\right)_{\theta=\pi} = -\frac{1}{s}. \qquad (B.12)$$

Putting this result into the derivative of Equation B.10, we get

$$\left(\frac{d\mu}{d\theta}\right)_{\theta=\pi} = -A\sin\pi + B\cos\pi = -B = -\frac{1}{s}, $$

or

$$B = \frac{1}{s}. \qquad (B.13)$$

If we incorporate Equations B.11 and B.13 in Equation B.10, our solution becomes

$$\mu = \frac{1}{s} \sin \theta - \frac{D}{2s^2} (1 + \cos \theta). \qquad (B.14)$$

Now we will use the third boundary condition, which involves the scattering angle Θ. Substituting it into B.14,

$$0 = \frac{1}{s} \sin \Theta - \frac{D}{2s^2} (1 + \cos \Theta),$$

from which we see that

$$s = \frac{D}{2} \frac{1 + \cos \Theta}{\sin \Theta} = \frac{D}{2} \cot \frac{\Theta}{2}. \qquad (B.15)$$

Equation B.15 is the result we seek, since it gives the relationship between the impact parameter s and the scattering angle Θ[1]. It is shown in Chapter 7 how the differential scattering cross-section is obtained from this.

---

[1] For the purpose of this derivation the scattering angle is defined as Θ since θ is used for the polar angle. Elsewhere in this book the scattering angle will be denoted by θ.

Appendix C The Fourier Integral and the Delta Function

Any periodic function, such as $f(t + T) = f(t)$, can be expanded in terms of sines and cosines provided that $f(t)$ is piecewise continuous and differentiable throughout the interval of the expansion. If the interval of expansion includes the end points of the period, then we add the assumption that the value of $f(t)$ at each end point is the arithmetic mean of its values at the right and left end points. Then, in the interval $-(T/2) \leq t \leq T/2$,

$$f(t) = \frac{a_0}{2} + \sum_{n=1}^{\infty} a_n \cos \omega_n t + \sum_{n=1}^{\infty} b_n \sin \omega_n t, \qquad (C.1)$$

where

$$a_n = \frac{2}{T} \int_{-T/2}^{T/2} f(t') \cos \omega_n t' \, dt',$$

$$b_n = \frac{2}{T} \int_{-T/2}^{T/2} f(t') \sin \omega_n t' \, dt',$$

$$\omega_n = \frac{2\pi n}{T}.$$

This may be written in a more convenient form[1] by means of the Euler identity:

$$f(t) = \sum_{n=-\infty}^{\infty} c_n e^{i\omega_n t} \text{ and } c_n = \frac{1}{T} \int_{-T/2}^{T/2} f(t') e^{-i\omega_n t'} \, dt', \qquad (C.2)$$

where $c_n = \frac{1}{2}(a_n - ib_n)$, $c_n^* = c_{-n}$ and $\omega_n = -\omega_{-n}$.

It is desirable to extend the interval of the expansion from T to ∞ so that non-periodic functions can also be represented by expansions in sines and cosines. To do this we must add the requirement that the integral $\int_{-\infty}^{\infty} |f(t)| \, dt$ exists. We rewrite Equation C.2 in the form

$$f(t) = \sum_{n=-\infty}^{\infty} \frac{1}{T} \int_{-T/2}^{T/2} f(t') e^{i\omega_n(t-t')} \, dt', \qquad (C.3)$$

and note that we must eliminate the factor $1/T$ before going to the limit. Since n is restricted to integral values,

$$\Delta\omega = \omega_{n+1} - \omega_n = \frac{2\pi(n+1) - 2\pi n}{T} = \frac{2\pi}{T},$$

or

$$\frac{1}{T} = \frac{\Delta\omega}{2\pi}. \qquad (C.4)$$

Substituting Equation C.4 into C.3,

$$f(t) = \frac{1}{2\pi} \sum_{n=-\infty}^{\infty} \Delta\omega \int_{-T/2}^{T/2} f(t')^{i\omega_n(t-t')} \, dt'.$$

In the limit as $T \to \infty$, the frequencies are distributed continuously instead of discretely and

$$f(t) = \frac{1}{2\pi} \int_{-\infty}^{\infty} d\omega \int_{-\infty}^{\infty} f(t') e^{i\omega(t-t')} \, dt',$$

if the integrals are absolutely convergent. The Fourier integrals are written in a more nearly symmetric form by defining a function $g(\omega)$ as follows:

$$f(t) = \frac{1}{\sqrt{2\pi}} \int_{-\infty}^{\infty} d\omega \, g(\omega) e^{i\omega t}$$

and

$$g(\omega) = \frac{1}{\sqrt{2\pi}} \int_{-\infty}^{\infty} dt' \, f(t') e^{-i\omega t'} \qquad (C.5)$$

The functions $f(t)$ and $g(\omega)$ are called Fourier transforms of one another. Any reasonably well-behaved function $f(t)$ can be represented by a superposition of harmonic functions with continuously varying frequency and weighting function $g(\omega)$. Conversely, $g(\omega)$ can be represented by a superposition of harmonic functions in time, each function multiplied by the weighting factor $f(t')$.

---

[1] Although the complex representation is introduced here as a mathematical device to represent real functions, its general utility will become evident when it is used to represent complex wave functions.

The above expansion is in the *time-frequency domain*. It is evident that an analogous expansion may be obtained in the position-wave-vector domain, or what is conveniently called the *coördinate-momentum domain*. Physically, this implies that our starting point was a *spatially* periodic function,

$$\psi(x + L) = \psi(x),$$

whose Fourier expansion is

$$\psi(x) = \sum_{n=-\infty}^{\infty} c_n e^{ik_n x},$$

where

$$c_n = \frac{1}{L} \int_{-L/2}^{L/2} \psi(x')e^{-ik_n x'} \, dx' \text{ and } k_n = \frac{2\pi n}{L}.$$

Following the same procedure as before, we eliminate the factor $1/L$ before taking the limit as $L \to \infty$ by the device that $\Delta k_n \sim 2\pi/L$ and $1/L \sim \Delta k_n/2\pi$. The results are:

$$\left. \begin{aligned}
\psi(x) &= \frac{1}{\sqrt{2\pi}} \int_{-\infty}^{\infty} dk\phi(k)e^{ikx} \\
\phi(k) &= \frac{1}{\sqrt{2\pi}} \int_{-\infty}^{\infty} dx'\psi(x')e^{-ikx'}
\end{aligned} \right\} \qquad (C.6)$$

Equations C.6 are readily generalized to any number of dimensions. For example,

$$\left. \begin{aligned}
\psi(\vec{r}) &= \left(\frac{1}{2\pi}\right)^{3/2} \iiint_{-\infty}^{\infty} d\vec{k}\phi(\vec{k})e^{i\vec{k}\cdot\vec{r}} \\
\phi(\vec{k}) &= \left(\frac{1}{2\pi}\right)^{3/2} \iiint_{-\infty}^{\infty} d\vec{r}'\psi(\vec{r}')e^{-i\vec{k}\cdot\vec{r}'},
\end{aligned} \right\} \qquad (C.7)$$

where $\vec{r}$ has components (x, y, z), $\vec{k}$ has components (k_x, k_y, k_z), $d\vec{r} = dx \, dy \, dz$, and $d\vec{k} = dk_x \, dk_y \, dk_z$. It is also possible to incorporate the time factor by writing

$$\Psi(\vec{r}, t) = \psi(\vec{r})e^{-i\omega t},$$

but the time will be omitted until it is specifically required.

The need for the Dirac delta function arises naturally from the Fourier integrals. Suppose we substitute $\phi(k)$ into $\psi(x)$ in Equations C.6 as follows:

$$\psi(x) = \frac{1}{2\pi} \int_{-\infty}^{\infty} dk \int_{-\infty}^{\infty} dx'\psi(x')e^{ik(x-x')}.$$

Interchanging the order of integration,

$$\psi(x) = \frac{1}{2\pi} \int_{-\infty}^{\infty} dx'\psi(x') \int_{-\infty}^{\infty} dk e^{ik(x-x')}.$$

Now define the delta function,

$$\delta(x - x') = \frac{1}{2\pi} \int_{-\infty}^{\infty} dk e^{ik(x-x')}. \qquad (C.8)$$

The delta function has the following properties:

$$\delta(x - x') = 0, \quad \text{if} \quad x' \neq x.$$

$$\int_a^b \delta(x - x') \, dx' = \begin{cases} 0, & \text{if} \quad x > b \text{ or if } x < a \\ 1, & \text{if} \quad a < x < b. \end{cases}$$

Since the delta function is zero everywhere but at the singular point, the only contribution to an integral occurs at that point. Then, using Equation C.8,

$$\psi(x) = \int_{-\infty}^{\infty} dx' \psi(x') \delta(x - x') \equiv \psi(x).$$

Thus the delta function may be thought of as a spike function having unit area but a non-zero amplitude at only one point, the point of singularity, where the amplitude becomes infinite. When integrated over all of space, its effect is to yield the remaining factors of the integrand evaluated at the singular point.

It is often convenient to place the origin at the singular point, in which case the delta function may be written as

$$\delta(x) = \frac{1}{2\pi} \int_{-\infty}^{\infty} dk e^{ikx}.$$

An alternative definition of the delta function may be obtained by integrating as follows:

$$\delta(x) = \lim_{a \to \infty} \frac{1}{2\pi} \int_{-a}^{a} e^{ikx} \, dk = \frac{1}{2\pi} \lim_{a \to \infty} \left(\frac{e^{iax} - e^{-iax}}{ix} \right) = \lim_{a \to \infty} \frac{\sin ax}{\pi x}, \qquad (C.9)$$

where a is positive and real. Let us examine the behavior of this function for both small and large x. First, consider the limit as x goes to zero:

$$\lim_{x \to 0} \frac{\sin ax}{\pi x} = \frac{a}{\pi} \lim_{x \to 0} \frac{\sin ax}{ax} = \frac{a}{\pi}.$$

Thus, $\delta(0) = \lim_{a \to \infty} a/\pi \to \infty$, or the amplitude becomes infinite at the singularity. For large $|x|$, $\sin ax/\pi x$ oscillates with period $2\pi/a$, and its amplitude falls off as $1/|x|$. But in the limit as $a \to \infty$, the period becomes infinitesimally narrow, so that the function approaches zero everywhere except for the infinite spike of infinitesimal width at the singularity. It now remains to show that the integral of Equation C.9 over all space is unity:

$$\int_{-\infty}^{\infty} \lim_{a \to \infty} \frac{\sin ax}{\pi x} \, dx = \lim_{a \to \infty} \frac{2}{\pi} \int_{0}^{\infty} \frac{\sin ax}{x} \, dx = \frac{2}{\pi} \cdot \frac{\pi}{2} = 1.$$

This establishes the validity of Equation C.9 as a representation of the delta function.

Appendix D

Ground State Electron Configurations for the Elements[1]

Z	Element	Shell: K 1 — s	L 2 — s	L 2 — p	M 3 — s	M 3 — p	M 3 — d	N 4 — s	N 4 — p	N 4 — d	N 4 — f	O 5 — s	O 5 — p	O 5 — d	O 5 — f	P 6 — s	P 6 — p	P 6 — d	Q 7 — s	Ground State Term
1	H hydrogen	1																		$^2S_{1/2}$
2	He helium	2																		$^1S_0$
3	Li lithium	2	1																	$^2S_{1/2}$
4	Be beryllium	2	2																	$^1S_0$
5	B boron	2	2	1																$^2P_{1/2}$
6	C carbon	2	2	2																$^3P_0$
7	N nitrogen	2	2	3																$^4S_{3/2}$
8	O oxygen	2	2	4																$^3P_2$
9	F fluorine	2	2	5																$^2P_{3/2}$
10	Ne neon	2	2	6																$^1S_0$
11	Na sodium	2	2	6	1															$^2S_{1/2}$
12	Mg magnesium	2	2	6	2															$^1S_0$
13	Al aluminum	2	2	6	2	1														$^2P_{1/2}$
14	Si silicon	2	2	6	2	2														$^3P_0$
15	P phosphorus	2	2	6	2	3														$^4S_{3/2}$
16	S sulfur	2	2	6	2	4														$^3P_2$
17	Cl chlorine	2	2	6	2	5														$^2P_{3/2}$
18	Ar argon	2	2	6	2	6														$^1S_0$
19	K potassium	2	2	6	2	6		1												$^2S_{1/2}$
20	Ca calcium	2	2	6	2	6		2												$^1S_0$
21	Sc scandium	2	2	6	2	6	1	2												$^2D_{3/2}$
22	Ti titanium	2	2	6	2	6	2	2												$^3F_2$
23	V vanadium	2	2	6	2	6	3	2												$^4F_{3/2}$
24	Cr chromium	2	2	6	2	6	5	1												$^7S_3$
25	Mn manganese	2	2	6	2	6	5	2												$^6S_{5/2}$
26	Fe iron	2	2	6	2	6	6	2												$^5D_4$
27	Co cobalt	2	2	6	2	6	7	2												$^4F_{9/2}$
28	Ni nickel	2	2	6	2	6	8	2												$^3F_4$
29	Cu copper	2	2	6	2	6	10	1												$^2S_{1/2}$
30	Zn zinc	2	2	6	2	6	10	2												$^1S_0$
31	Ga gallium	2	2	6	2	6	10	2	1											$^2P_{1/2}$
32	Ge germanium	2	2	6	2	6	10	2	2											$^3P_0$
33	As arsenic	2	2	6	2	6	10	2	3											$^4S_{3/2}$
34	Se selenium	2	2	6	2	6	10	2	4											$^3P_2$
35	Br bromine	2	2	6	2	6	10	2	5											$^2P_{3/2}$
36	Kr krypton	2	2	6	2	6	10	2	6											$^1S_0$
37	Rb rubidium	2	2	6	2	6	10	2	6			1								$^2S_{1/2}$
38	Sr strontium	2	2	6	2	6	10	2	6			2								$^1S_0$
39	Y yttrium	2	2	6	2	6	10	2	6	1		2								$^2D_{3/2}$
40	Zr zirconium	2	2	6	2	6	10	2	6	2		2								$^3F_2$

[1] McGraw-Hill Encyclopedia of Science and Technology, New York: McGraw-Hill Book Co., 1977, p. 571.

Appendix D (continued)

Z	Element		K 1 s	L 2 s	p	M 3 s	p	d	N 4 s	p	d	f	O 5 s	p	d	f	P 6 s	p	d	Q 7 s	Ground State Term
41	Nb	niobium	2	2	6	2	6	10	2	6	4	.	1								$^6D_{1/2}$
42	Mo	molybdenum	2	2	6	2	6	10	2	6	5	.	1								$^7S_3$
43	Tc	technetium	2	2	6	2	6	10	2	6	6	.	1								$^6S_{5/2}$
44	Ru	ruthenium	2	2	6	2	6	10	2	6	7	.	1								$^5F_5$
45	Rh	rhodium	2	2	6	2	6	10	2	6	8	.	1								$^4F_{9/2}$
46	Pd	palladium	2	2	6	2	6	10	2	6	10	.	.								$^1S_0$
47	Ag	silver	2	2	6	2	6	10	2	6	10	.	1								$^2S_{1/2}$
48	Cd	cadmium	2	2	6	2	6	10	2	6	10	.	2								$^1S_0$
49	In	indium	2	2	6	2	6	10	2	6	10	.	2	1							$^2P_{1/2}$
50	Sn	tin	2	2	6	2	6	10	2	6	10	.	2	2							$^3P_0$
51	Sb	antimony	2	2	6	2	6	10	2	6	10	.	2	3							$^4S_{3/2}$
52	Te	tellurium	2	2	6	2	6	10	2	6	10	.	2	4							$^3P_2$
53	I	iodine	2	2	6	2	6	10	2	6	10	.	2	5							$^2P_{3/2}$
54	Xe	xenon	2	2	6	2	6	10	2	6	10	.	2	6							$^1S_0$
55	Cs	cesium	2	2	6	2	6	10	2	6	10	.	2	6	.	.	1				$^2S_{1/2}$
56	Ba	barium	2	2	6	2	6	10	2	6	10	.	2	6	.	.	2				$^1S_0$
57	La	lanthanum	2	2	6	2	6	10	2	6	10	.	2	6	1	.	2				$^2D_{3/2}$
58	Ce	cerium	2	2	6	2	6	10	2	6	10	1	2	6	1	.	2				$^1G_4$
59	Pr	praseodymium	2	2	6	2	6	10	2	6	10	3	2	6	.	.	2				$^4I_{9/2}$
60	Nd	neodymium	2	2	6	2	6	10	2	6	10	4	2	6	.	.	2				$^5I_4$
61	Pm	promethium	2	2	6	2	6	10	2	6	10	5	2	6	.	.	2				$^6H_{5/2}$
62	Sm	samarium	2	2	6	2	6	10	2	6	10	6	2	6	.	.	2				$^7F_0$
63	Eu	europium	2	2	6	2	6	10	2	6	10	7	2	6	.	.	2				$^8S_{7/2}$
64	Gd	gadolinium	2	2	6	2	6	10	2	6	10	7	2	6	1	.	2				$^9D_2$
65	Tb	terbium	2	2	6	2	6	10	2	6	10	9	2	6	.	.	2				
66	Dy	dysprosium	2	2	6	2	6	10	2	6	10	10	2	6	.	.	2				$^5I_8$
67	Ho	holmium	2	2	6	2	6	10	2	6	10	11	2	6	.	.	2				$^4I_{15/2}$
68	Er	erbium	2	2	6	2	6	10	2	6	10	12	2	6	.	.	2				$^3H_6$
69	Tm	thulium	2	2	6	2	6	10	2	6	10	13	2	6	.	.	2				$^2F_{7/2}$
70	Yb	ytterbium	2	2	6	2	6	10	2	6	10	14	2	6	.	.	2				$^1S_0$
71	Lu	lutetium	2	2	6	2	6	10	2	6	10	14	2	6	1	.	2				$^2D_{3/2}$
72	Hf	hafnium	2	2	6	2	6	10	2	6	10	14	2	6	2	.	2				$^3F_2$
73	Ta	tantalum	2	2	6	2	6	10	2	6	10	14	2	6	3	.	2				$^4F_{3/2}$
74	W	wolfram (tungsten)	2	2	6	2	6	10	2	6	10	14	2	6	4	.	2				$^5D_0$
75	Re	rhenium	2	2	6	2	6	10	2	6	10	14	2	6	5	.	2				$^6S_{5/2}$
76	Os	osmium	2	2	6	2	6	10	2	6	10	14	2	6	6	.	2				$^5D_4$
77	Ir	iridium	2	2	6	2	6	10	2	6	10	14	2	6	7	.	2				$^4F_{9/2}$
78	Pt	platinum	2	2	6	2	6	10	2	6	10	14	2	6	9	.	1				$^3D_3$
79	Au	gold	2	2	6	2	6	10	2	6	10	14	2	6	10	.	1				$^2S_{1/2}$
80	Hg	mercury	2	2	6	2	6	10	2	6	10	14	2	6	10	.	2				$^1S_0$
81	Tl	thallium	2	2	6	2	6	10	2	6	10	14	2	6	10	.	2	1			$^2P_{1/2}$
82	Pb	lead	2	2	6	2	6	10	2	6	10	14	2	6	10	.	2	2			$^3P_0$
83	Bi	bismuth	2	2	6	2	6	10	2	6	10	14	2	6	10	.	2	3			$^4S_{3/2}$
84	Po	polonium	2	2	6	2	6	10	2	6	10	14	2	6	10	.	2	4			$^3P_2$
85	At	astatine	2	2	6	2	6	10	2	6	10	14	2	6	10	.	2	5			$^2P_{3/2}$
86	Rn	radon	2	2	6	2	6	10	2	6	10	14	2	6	10	.	2	6			$^1S_0$
87	Fr	francium	2	2	6	2	6	10	2	6	10	14	2	6	10	.	2	6	.	1	$^2S_{1/2}$
88	Ra	radium	2	2	6	2	6	10	2	6	10	14	2	6	10	.	2	6	.	2	$^1S_0$
89	Ac	actinium	2	2	6	2	6	10	2	6	10	14	2	6	10	.	2	6	1	2	$^2D_{3/2}$
90	Th	thorium	2	2	6	2	6	10	2	6	10	14	2	6	10	.	2	6	2	2	$^3F_2$

Appendix D (continued)

Z		Element	Shell: Subshell:	K 1 s	L 2 s	 p	M 3 s	 p	 d	N 4 s	 p	 d	 f	O 5 s	 p	 d	 f	P 6 s	 p	 d	Q 7 s	Ground State Term
91	Pa	protactinium		2	2	6	2	6	10	2	6	10	14	2	6	10	1	2	6	2	2	
92	U	uranium		2	2	6	2	6	10	2	6	10	14	2	6	10	3	2	6	1	2	$^5L_6$
93	Np	neptunium		2	2	6	2	6	10	2	6	10	14	2	6	10	4	2	6	1	2	
94	Pu	plutonium		2	2	6	2	6	10	2	6	10	14	2	6	10	6	2	6	.	2	
95	Am	americium		2	2	6	2	6	10	2	6	10	14	2	6	10	7	2	6	.	2	
96	Cm	curium		2	2	6	2	6	10	2	6	10	14	2	6	10	7	2	6	1	2	
97	Bk	berkelium		2	2	6	2	6	10	2	6	10	14	2	6	10	8	2	6	1	2	
98	Cf	californium		2	2	6	2	6	10	2	6	10	14	2	6	10	10	2	6	.	2	
99	Es	einsteinium		2	2	6	2	6	10	2	6	10	14	2	6	10	11	2	6	.	2	
100	Rm	fermium		2	2	6	2	6	10	2	6	10	14	2	6	10	12	2	6	.	2	
101	Md	mendelevium		2	2	6	2	6	10	2	6	10	14	2	6	10	13	2	6	.	2	
102	No	nobelium		2	2	6	2	6	10	2	6	10	14	2	6	10	14	2	6	.	2	
103	Lw	lawrencium		2	2	6	2	6	10	2	6	10	14	2	6	10	14	2	6	1	2	

Appendix E Gauss' Probability Integral and Others

$$I_0 = \int_0^\infty e^{-\alpha x^2}\, dx = \frac{1}{2}\sqrt{\frac{\pi}{\alpha}} \qquad \text{(Gauss' probability integral)}$$

$$I_1 = \int_0^\infty x e^{-\alpha x^2}\, dx = \frac{1}{2\alpha}$$

$$I_2 = \int_0^\infty x^2 e^{-\alpha x^2}\, dx = -\frac{dI_0}{d\alpha} = \frac{1}{4}\sqrt{\frac{\pi}{\alpha^3}}$$

$$I_3 = \int_0^\infty x^3 e^{-\alpha x^2}\, dx = -\frac{dI_1}{d\alpha} = \frac{1}{2\alpha^2}$$

$$I_4 = \int_0^\infty x^4 e^{-\alpha x^2}\, dx = \frac{d^2 I_0}{d\alpha^2} = \frac{3}{8}\sqrt{\frac{\pi}{\alpha^5}}$$

$$I_5 = \int_0^\infty x^5 e^{-\alpha x^2}\, dx = \frac{d^2 I_1}{d\alpha^2} = \frac{1}{\alpha^3}$$

$$\cdots\cdots\cdots\cdots\cdots\cdots$$

$$I_{2n} = (-1)^n \frac{d^n}{d\alpha^n} I_0$$

$$I_{2n+1} = (-1)^n \frac{d^n}{d\alpha^n} I_1.$$

Appendix F Table of Nuclear Masses

The masses given below are for neutral atoms of all stable and some unstable nuclides. They are taken from J. H. E. Mattauch, W. Thiele and A. H. Wapstra, *Nuclear Physics 67*, 1 (1965). Each mass is given in atomic mass units (u) and includes the mass of Z electrons. Natural abundances and half-lives are from the *Chart of the Nuclides*, 12th edition, General Electric Company, 1977.

ELEMENT	A	ATOMIC MASS (u)	NATURAL ABUNDANCE (%)	HALF-LIFE (IF UNSTABLE)
$_0n$	1	1.008 665		10.6 min
$_1H$	1	1.007 825	99.985	
	2	2.014 102	0.015	
	3	3.016 050		12.3 yr
$_2He$	3	3.016 029		
	4	4.002 603	99.999	
	6	6.018 891		805 ms
$_3Li$	6	6.015 123	7.5	
	7	7.016 004	92.5	
	8	8.022 487		844 ms
$_4Be$	7	7.016 930		53.3 d
	9	9.012 182	100	
	10	10.013 535		1.6×10^6 yr
$_5B$	10	10.012 938	20	
	11	11.009 305	80	
	12	12.014 353		20.4 ms
$_6C$	10	10.016 858		19.3 s
	11	11.011 433		20.3 min
	12	12.000 000	98.89	
	13	13.003 355	1.11	
	14	14.003 242		5730 yr
	15	15.010 599		2.45 s
$_7N$	12	12.018 613		11.0 ms
	13	13.005 739		9.97 min
	14	14.003 074	99.63	
	15	15.000 109	0.37	
	16	16.006 099		7.10 s
	17	17.008 449		4.17 s
$_8O$	14	14.008 597		70.5 s
	15	15.003 065		122 ms
	16	15.994 915	99.758	
	17	16.999 131	0.038	
	18	17.999 159	0.204	
	19	19.003 576		26.8 s
$_9F$	17	17.002 095		64.5 s
	18	18.000 937		109.8 min
	19	18.998 403	100	
	20	19.999 982		11.0 s

Appendix F (continued)

ELEMENT	A	ATOMIC MASS (u)	NATURAL ABUNDANCE (%)	HALF-LIFE (IF UNSTABLE)
	21	20.999 949		4.33 s
$_{10}$Ne	18	18.005 710		1.67 s
	19	19.001 880		17.2 s
	20	19.992 439	90.51	
	21	20.993 845	0.27	
	22	21.991 384	9.22	
	23	22.994 466		37.5 s
	24	23.993 613		3.38 min
$_{11}$Na	22	21.994 435		2.60 yr
	23	22.989 770	100	
	24	23.990 963		15.0 h
$_{12}$Mg	23	22.994 127		11.3 s
	24	23.985 045	78.99	
	25	24.985 839	10.00	
	26	25.982 595	11.01	
$_{13}$Al	27	26.981 541	100	
$_{14}$Si	28	27.976 928	92.23	
	29	28.976 496	4.67	
	30	29.973 772	3.10	
$_{15}$P	30	29.978 310		2.50 min
	31	30.973 763	100	
$_{16}$S	32	31.972 072	95.02	
	33	32.971 459	0.75	
	34	33.967 868	4.21	
	35	34.969 032		87.2 d
	36	35.967 079	0.017	
$_{17}$Cl	35	34.968 853	75.77	
	36	35.968 307		3.01×10^5 yr
	37	36.965 903	24.23	
$_{18}$Ar	36	35.967 546	0.337	
	37	36.966 776		34.8 d
	38	37.962 732	0.063	
	39	38.964 315		269 yr
	40	39.962 383	99.60	
$_{19}$K	39	38.963 708	93.26	
	40	38.963 999	0.01	1.28×10^9 yr
	41	40.961 825	6.73	
$_{20}$Ca	40	39.962 591	96.94	
	41	40.962 278		1.3×10^5 yr
	42	41.958 622	0.647	
	43	42.958 770	0.135	
	44	43.955 485	2.09	
	45	44.956 189		163 d
	46	45.953 689	0.0035	
	47	46.954 543		4.5 d
	48	47.952 532	0.187	
$_{21}$Sc	45	44.955 914	100	
$_{22}$Ti	46	45.952 633	8.25	
	47	46.951 765	7.45	

Appendix F (continued)

ELEMENT	A	ATOMIC MASS (u)	NATURAL ABUNDANCE (%)	HALF-LIFE (IF UNSTABLE)
	48	47.947 947	73.7	
	49	48.947 871	5.4	
	50	49.944 786	5.2	
$_{23}$V	48	47.952 257		16 d
	50	49.947 161	0.25	~10^{17} yr
	51	50.943 962	99.75	
$_{24}$Cr	48	47.954 033		21.6 h
	50	49.946 046	4.35	
	52	51.940 510	83.79	
	53	52.940 651	9.50	
	54	53.938 882	2.36	
$_{25}$Mn	54	53.940 360		312.5 d
	55	54.938 046	100	
$_{26}$Fe	54	53.939 612	5.8	
	56	55.934 939	91.8	
	57	56.935 396	2.1	
	58	57.933 278	0.3	
	59	58.934 878		44.6 d
$_{27}$Co	58	57.935 755		70.8 d
	59	58.933 198	100	
	60	59.933 820		5.3 yr
$_{28}$Ni	58	57.935 347	68.3	
	60	59.930 789	26.1	
	61	60.931 059	1.1	
	62	61.928 346	3.6	
	64	63.927 968	0.9	
$_{29}$Cu	63	62.929 599	69.2	
	64	63.929 766		12.7 h
	65	64.927 792	30.8	
$_{30}$Zn	64	63.929 145	48.6	
	66	65.926 035	27.9	
	67	66.927 129	4.1	
	68	67.924 846	18.8	
	70	69.925 325	0.6	
$_{31}$Ga	69	68.925 581	60.1	
	71	70.924 701	39.9	
$_{32}$Ge	70	69.924 250	20.5	
	72	71.922 080	27.4	
	73	72.923 464	7.8	
	74	73.921 179	36.5	
	76	75.921 403	7.8	
$_{33}$As	74	73.923 930		17.8 d
	75	74.921 596	100	
$_{34}$Se	74	73.922 477	0.9	
	76	75.919 207	9.0	
	77	76.919 908	7.6	
	78	77.917 304	23.5	
	80	79.916 520	49.8	
	82	81.916 709	9.2	

Appendix F (continued)

ELEMENT	A	ATOMIC MASS (u)	NATURAL ABUNDANCE (%)	HALF-LIFE (IF UNSTABLE)
$_{35}$Br	79	78,918 336	50.7	
	80	79.918 528		17.7 min
	81	80.916 290	49.3	
$_{36}$Kr	78	77.920 397	0.35	
	80	79.916 375	2.25	
	82	81.913 483	11.6	
	83	82.914 134	11.5	
	84	83.911 506	57.0	
	86	85.910 614	17.3	
$_{37}$Rb	85	84.911 800	72.2	
	87	86.909 184	27.8	4.9×10^{10} yr
$_{38}$Sr	84	83.913 428	0.6	
	86	85.909 273	9.8	
	87	86.908 890	7.0	
	88	87.905 625	82.6	
$_{39}$Y	89	88.905 856	100	
$_{40}$Zr	90	89.904 708	51.5	
	91	90.905 644	11.2	
	92	91.905 039	17.1	
	94	93.906 319	17.4	
	96	95.908 272	2.8	
$_{41}$Nb	93	92.906 378	100	
$_{42}$Mo	92	91.906 809	14.8	
	94	93.905 086	9.3	
	95	94.905 838	15.9	
	96	95.904 675	16.7	
	97	96.906 018	9.6	
	98	97.905 405	24.1	
	100	99.907 473	9.6	
$_{43}$Tc	99	98.906 252		2.1×10^{5} yr
$_{44}$Ru	96	95.907 596	5.5	
	98	97.905 287	1.9	
	99	98.905 937	12.7	
	100	99.904 217	12.6	
	101	100.905 581	17.0	
	102	101.904 347	31.6	
	104	103.905 422	18.7	
$_{45}$Rh	103	102.905 503	100	
$_{46}$Pd	102	101.905 609	1.0	
	104	103.904 026	11.0	
	105	104.905 075	22.2	
	106	105.903 475	27.3	
	108	107.903 894	26.7	
	110	109.905 169	11.8	
$_{47}$Ag	107	106.905 095	51.8	
	108	107.905 956		2.41 min
	109	108.904 754	48.2	
$_{48}$Cd	106	105.906 461	1.3	
	108	107.904 186	0.9	

Appendix F (continued)

ELEMENT	A	ATOMIC MASS (u)	NATURAL ABUNDANCE (%)	HALF-LIFE (IF UNSTABLE)
	110	109.903 007	12.5	
	111	110.904 182	12.8	
	112	111.902 761	24.1	
	113	112.904 401	12.2	9×10^{15} yr
	114	113.903 361	28.7	
	116	115.904 758	7.5	
$_{49}$In	113	112.904 056	4.3	
	115	114.903 875	95.7	5×10^{14} yr
$_{50}$Sn	112	111.904 823	1.0	
	114	113.902 781	0.7	
	115	114.903 344	0.4	
	116	115.901 743	14.7	
	117	116.902 954	7.7	
	118	117.901 607	24.3	
	119	118.903 310	8.6	
	120	119.902 199	32.4	
	122	121.903 440	4.6	
	124	123.905 271	5.6	
$_{51}$Sb	121	120.903 824	57.3	
	123	122.904 222	42.7	
$_{52}$Te	120	119.904 021	0.1	
	122	121.903 055	2.5	
	123	122.904 278	0.9	$\sim 1.2 \times 10^{13}$ yr
	124	123.902 825	4.6	
	125	124.904 435	7.0	
	126	125.903 310	18.7	
	128	127.904 464	31.7	
	130	129.906 229	34.5	
$_{53}$I	127	126.904 477	100	
	131	130.906 119		8.0 d
$_{54}$Xe	124	123.906 12	0.1	
	126	125.904 281	0.1	
	128	127.903 531	1.9	
	129	128.904 780	26.4	
	130	129.903 509	4.1	
	131	130.905 076	21.2	
	132	131.904 148	26.9	
	134	133.905 395	10.4	
	136	135.907 219	8.9	
$_{55}$Cs	133	132.905 433	100	
$_{56}$Ba	130	129.906 277	0.1	
	132	131.905 042	0.1	
	134	133.904 490	2.4	
	135	134.905 668	6.6	
	136	135.904 556	7.9	
	137	136.905 816	11.2	
	138	137.905 236	71.7	
$_{57}$La	138	137.907 114	0.1	1×10^{11} yr
	139	138.906 355	99.9	

Appendix F (continued)

ELEMENT	*A*	ATOMIC MASS (u)	NATURAL ABUNDANCE (%)	HALF-LIFE (IF UNSTABLE)
$_{58}$Ce	136	135.907 14	0.2	
	138	137.905 996	0.2	
	140	139.905 442	88.5	
	142	141.909 249	11.1	5×10^{16} yr
$_{59}$Pr	141	140.907 657	100	
$_{60}$Nd	142	141.907 731	27.2	
	143	142.909 823	12.2	
	144	143.910 096	23.8	2.1×10^{15} yr
	145	144.912 582	8.3	$>10^{17}$ yr
	146	145.913 126	17.2	
	148	147.916 901	5.7	
	150	149.920 900	5.6	
$_{61}$Pm	147	146.915 148		2.6 yr
$_{62}$Sm	144	143.912 009	3.1	
	147	146.914 907	15.1	1.1×10^{11} yr
	148	147.914 832	11.3	8×10^{15} yr
	149	148.917 193	13.9	$>10^{16}$ yr
	150	149.917 285	7.4	
	152	151.919 741	26.7	
	154	153.922 218	22.6	
$_{63}$Eu	151	150.919 860	47.9	
	153	152.921 243	52.1	
$_{64}$Gd	152	151.919 803	0.2	1.1×10^{14} yr
	154	153.920 876	2.1	
	155	154.922 629	14.8	
	156	155.922 130	20.6	
	157	156.923 967	15.7	
	158	157.924 111	24.8	
	160	159.927 061	21.8	
$_{65}$Tb	159	158.925 350	100	
$_{66}$Dy	156	155.924 287	0.1	$>1 \times 10^{18}$ yr
	158	157.924 412	0.1	
	160	159.925 203	2.3	
	161	160.926 939	19.0	
	162	161.926 805	25.5	
	163	162.928 737	24.9	
	164	163.929 183	28.1	
$_{67}$Ho	165	164.930 332	100	
$_{68}$Er	162	161.928 787	0.1	
	164	163.929 211	1.6	
	166	165.930 305	33.4	
	167	166.932 061	22.9	
	168	167.932 383	27.1	
	170	169.935 476	14.9	
$_{69}$Tm	169	168.934 225	100	
$_{70}$Yb	168	167.933 908	0.1	
	170	169.934 774	3.2	
	171	170.936 338	14.4	
	172	171.936 393	21.9	

Appendix F (continued)

ELEMENT	A	ATOMIC MASS (u)	NATURAL ABUNDANCE (%)	HALF-LIFE (IF UNSTABLE)
	173	172.938 222	16.2	
	174	173.938 873	31.6	
	176	175.942 576	12.6	
$_{71}$Lu	175	174.940 785	97.4	
	176	175.942 694	2.6	2.9×10^{10} yr
$_{72}$Hf	174	173.940 065	0.2	2.0×10^{15} yr
	176	175.941 420	5.2	
	177	176.943 233	18.6	
	178	177.943 710	27.1	
	179	178.945 827	13.7	
	180	179.946 561	35.2	
$_{73}$Ta	180	179.947 489	0.01	$>1.6 \times 10^{13}$ yr
	181	180.948 014	99.99	
$_{74}$W	180	179.946 727	0.1	
	182	181.948 225	26.3	
	183	182.950 245	14.3	
	184	183.950 953	30.7	
	186	185.954 377	28.6	
$_{75}$Re	185	184.952 977	37.4	
	187	186.955 765	62.6	5×10^{10} yr
$_{76}$Os	184	183.952 514	0.02	
	186	185.953 852	1.6	2×10^{15} yr
	187	186.955 762	1.6	
	188	187.955 850	13.3	
	189	188.958 156	16.1	
	190	189.958 455	26.4	
	192	191.961 487	41.0	
$_{77}$Ir	191	190.960 603	37.3	
	193	192.962 942	62.7	
$_{78}$Pt	190	189.959 937	0.01	6.1×10^{11} yr
	192	191.961 049	0.79	
	194	193.962 679	32.9	
	195	194.964 785	33.8	
	196	195.964 947	25.3	
	198	197.967 879	7.2	
$_{79}$Au	197	196.966 560	100	
$_{80}$Hg	196	195.965 812	0.2	
	198	197.966 760	10.0	
	199	198.968 269	16.8	
	200	199.968 316	23.1	
	201	200.970 293	13.2	
	202	201.970 632	29.8	
	204	203.973 481	6.9	
$_{81}$Tl	203	202.972 336	29.5	
	205	204.974 410	70.5	
$_{82}$Pb	204	203.973 037	1.4	1.4×10^{17} yr
	206	205.974 455	24.1	
	207	206.975 885	22.1	
	208	207.976 641	52.4	

Appendix F (Continued)

ELEMENT	A	ATOMIC MASS (u)	NATURAL ABUNDANCE (%)	HALF-LIFE (IF UNSTABLE)
	210	209.984 178		22.3 yr
$_{82}$Pb	214	213.999 764		26.8 min
$_{83}$Bi	209	208.980 388	100	$>2 \times 10^{18}$ yr
	212	211.991 267		60.6 min
$_{84}$Po	210	209.982 876		138 d
$_{84}$Po	214	213.995 191		0.16 ms
$_{84}$Po	216	216.001 790		0.15 s
	218	218.008 930		3.05 min
$_{85}$At	218	218.008 607		1.3 s
$_{86}$Rn	220	220.011 401		56 s
$_{86}$Rn	222	222.017 574		3.824 d
$_{88}$Ra	226	226.025 406		1.60×10^{3} yr
$_{90}$Th	228	228.028 750		1.9 yr
$_{90}$Th	230	230.033 131		7.7×10^{4} yr
	232	232.038 054	100	1.4×10^{10} yr
	233	233.041 580		22.2 min
$_{91}$Pa	233	233.040 244		27 d
$_{92}$U	232	232.037 168		72 yr
$_{92}$U	233	233.039 629		1.6×10^{5} yr
	234	234.040 947		2.4×10^{5} yr
	235	235.043 925	0.72	7.04×10^{8} yr
	238	238.050 786	99.28	4.47×10^{9} yr
$_{93}$Np	237	237.048 169		2.14×10^{6} yr
	239	239.052 932		2.4 d
$_{94}$Pu	239	239.052 158		2.4×10^{4} yr

Appendix G Nobel Prizes in Physics

DATE	NAME	COUNTRY	DISCOVERY
1901	Wilhelm Konrad Röntgen	Germany	Discovery of X-rays.
1902	Hendrik Antoon Lorentz	Netherlands	Influence of magnetism on electromagnetic radiation.
	Pieter Zeeman	Netherlands	
1903	Antoine Henri Becquerel	France	Discovery of radioactive elements.
	Pierre Curie	France	
	Marie Curie	France	
1904	John Strutt (Lord Rayleigh)	England	Discovery of argon.
1905	Philipp Lenard	Germany	Research on cathode rays.
1906	Sir Joseph John Thomson	England	Conduction of electricity through gases.
1907	Albert A. Michelson	USA	Spectroscopic and metrological investigations.
1908	Gabriel Lippmann	France	Color photography.
1909	Guglielmo Marconi	Italy	Development of wireless telegraphy.
	Karl Ferdinand Braun	Germany	
1910	Johannes Diderik van der Waals	Netherlands	Equations of state of gases and fluids.
1911	Wilhelm Wien	Germany	Laws of heat radiation.
1912	Nils Gustaf Dalén	Sweden	Coast lighting.
1913	Heike Kamerlingh-Onnes	Netherlands	Properties of matter at low temperatures; production of liquid helium.
1914	Max von Laue	Germany	Diffraction of X-rays in crystals.
1915	Sir William Henry Bragg	England	Study of crystal structure by means of X-rays.
	William Lawrence Bragg	England	
1917	Charles Glover Barkla	England	Discovery of the characteristic X-ray spectra of elements.
1918	Max Planck	Germany	Quantum theory of radiation.
1919	Johannes Stark	Germany	Discovery of the Doppler effect in positive ion beams and the splitting of spectral lines in an electric field.
1920	Charles Edouard Guillaume	France	Discovery of anomalies in nickel-steel alloys.
1921	Albert Einstein	USA	The theory of relativity and the photoelectric effect.
1922	Niels Bohr	Denmark	Quantum theory of the atom.
1923	Robert Andrews Millikan	USA	Measurement of the elementary electric charge and the photoelectric effect.
1924	Karl Manne Siegbahn	Sweden	Discoveries in the area of X-ray spectra.
1925	James Franck	Germany	Laws governing collisions between electrons and atoms.
	Gustav Hertz	Germany	
1926	Jean Perrin	France	Discovery of the equilibrium of sedimentation.
1927	Arthur H. Compton	USA	Explanation of the scattering of X-rays by electrons and atoms.
	Charles T. R. Wilson	England	Cloud chamber for particle detection.
1928	Sir Owen Williams Richardson	England	Discovery of the law of thermionic emission of electrons.
1929	Prince Louis-Victor de Broglie	France	Wave nature of electrons.
1930	Sir Chandrasekhara Raman	India	Diffusion and scattering of light by molecules.
1932	Werner Heisenberg	Germany	Creation of quantum mechanics.
1933	Paul Adrien Maurice Dirac	England	Extensive development of quantum mechanics.
	Erwin Schrödinger	Austria	
1935	James Chadwick	England	Discovery of the neutron.
1936	Victor F. Hess	Austria	Discovery of cosmic radiation.
	Carl David Anderson	USA	Discovery of the positron.

Appendix G (continued)

DATE	NAME	COUNTRY	DISCOVERY
1937	Clinton Joseph Davisson George P. Thomson	USA England	Discovery of diffraction of electrons by crystals.
1938	Enrico Fermi	Italy	Artificial radioactivity induced by slow neutrons.
1939	Ernest O. Lawrence	USA	Invention of the cyclotron.
1943	Otto Stern	Germany	Detection of magnetic moment of protons.
1944	Isidor Isaac Rabi	USA	Magnetic moments by beam methods.
1945	Wolfgang Pauli	Austria	Work on nuclear fission.
1946	Percy Williams Bridgman	USA	High-pressure physics.
1947	Sir Edward Appleton	England	Discovery of ionospheric reflection of radio waves.
1948	Patrick Maynard Stuart Blackett	England	Discoveries in cosmic radiation.
1949	Hideki Yukawa	Japan	Theoretical prediction of the meson.
1950	Cecil Frank Powell	England	Photographic method of studying atomic nuclei; research on mesons.
1951	Sir John Douglas Cockcroft Ernest Thomas Sinton Walton	England Ireland	Transmutation of atomic nuclei by accelerated atomic particles.
1952	Felix Bloch Edward Mills Purcell	USA USA	Measurement of magnetic fields in atomic nuclei.
1953	Fritz Zernike	Netherlands	Development of phase contrast microscopy.
1954	Max Born	Germany	Contributions to quantum mechanics.
	Walther Bothe	Germany	Use of the coincidence method in studies of cosmic radiation.
1955	Willis E. Lamb, Jr. Polykarp Kusch	USA USA	Atomic measurements.
1956	John Bardeen Walter H. Brattain William B. Schockley	USA USA USA	Invention and development of the transistor.
1957	Chen Ning Yang Tsung Dao Lee	China, USA China, USA	Overthrow of principle of conservation of parity.
1958	Pavel A. Cerenkov Ilya M. Frank Igor Y. Tamm	USSR USSR USSR	Interpretation of radiation effects and development of a cosmic-ray counter.
1959	Owen Chamberlain Emilio Gino Segre	USA USA	Demonstrating the existence of the antiproton.
1960	Donald A. Glaser	USA	Invention of bubble chamber.
1961	Robert L. Hofstadter	USA	Electromagnetic structure of nucleons from high-energy electron scattering.
	Rudolph L. Mössbauer	Germany	Discovery of recoilless resonance absorption of gamma rays in nuclei.
1962	Lev D. Landau	USSR	Theories of condensed matter.
1963	Eugene P. Wigner Maria Goeppert-Mayer J. Hans D. Jensen	USA USA Germany	Nuclear shell structure.
1964	Charles H. Townes Nikolai G. Basov Alexander M. Prokhorov	USA USSR USSR	Amplification by laser-maser devices.
1965	Richard P. Feynman Julian S. Schwinger Sin-Itiro Tomonaga	USA USA Japan	Quantum electrodynamics.
1966	Alfred Kastler	France	Atomic energy levels.
1967	Hans A. Bethe	Germany	Nuclear theory and stellar energy production.
1968	Luis W. Alvarez	USA	Elementary particles.
1969	Murray Gell-Mann	USA	Theory of elementary particles.

Appendix G (continued)

DATE	NAME	COUNTRY	DISCOVERY
1970	Hannes Alfvén	Sweden	Astrophysics
1971	Dennis Gabor	USA	Holography
1972	John Bardeen	USA	Theory of superconductivity
	Leon Cooper	USA	
	John Schrieffer	USA	
1973	Leo Isaki	Japan	Discoveries in semiconductor physics.
	Ivar Giaever	USA	
	Brian Josephson	England	Prediction of Josephson effects in superconductivity.
1974	Anthony Hewish	England	Discovery of pulsars.
	Martin Ryle	England	Radioastronomy.
1975	Aage Bohr	Denmark	The collective model of the nucleus.
	Ben Mottelson	Denmark	
	James Rainwater	USA	Measurements of nuclear electric quadrupole moments.
1976	Burton Richter	USA	Discovery of the psi/J particle.
	Samuel C. C. Ting	China, USA	
1977	Phillip W. Anderson	USA	Theoretical investigations of the electronic structure of magnetic and disordered systems.
	Sir Neville Mott	England	
	John H. Van Vleck	USA	
1978	Robert W. Wilson	USA	Discovery of cosmic microwave background radiation.
	Arno Penzias	USA	
	Peter L. Kapitza	USSR	Discoveries in low temperature physics.
1979	Sheldon L. Glashow	USA	Theory of the unified weak and electromagnetic interaction between elementary particles.
	Abdus Salam	Pakistan	
	Steven Weinberg	USA	
1980	James W. Cronin	USA	Discovery of violations of fundamental symmetry principles in the decay of neutral K mesons.
	Val L. Fitch	USA	
1981	Nicolaas Bloembergen	USA	Contributions to the development of laser spectroscopy.
	Arthur Schawlow	USA	
	Kai Siegbahn	Sweden	Contribution to the development of a high-resolution electron spectroscopy.

Appendix H Answers to Odd-Numbered Problems

Chapter 1

1. 3.27×10^{-25} kg
3. 7.31×10^{-26} kg
5. half of its original value
7. 2.69×10^{22}/liter
9. 2.98 kcal/kmole-K for He; 4.97 kcal/kmole-K for N_2
17. 412 m/s
19. 6.21×10^{-21} J; 2.72×10^3 m/s
21. 913 Å

Chapter 2

1. 9.648×10^7 coul
3. 1.6×10^{-18} J; 1.88×10^6 m/s
5. 8942 coul; 8.3 hr.
9. $r_p = 14.4$ cm; $r_d = 20.4$ cm; $r_\alpha = 20.4$ cm
11. 0.48 m
13. (a) 5.89×10^{-4} m/s; (b) 2.97×10^5 volts/m
15. 47,620 volts/m

Chapter 3

1. 20 m/s; 30 m/s
5. $L' = (1 - \frac{1}{2}\beta^2)^{1/2}$; $\theta = $ arc tan γ
7. (a) $0.44c$; (b) 16.2 m
9. $0.98c$
11. $\dfrac{\beta\gamma}{\gamma + 1}$, where $\beta = v/c$ and $\gamma = (1 - \beta^2)^{-1/2}$
13. (a) $0.89c$; (b) $0.66c$; (c) $0.82c$ at 68.5°
15. $\beta = \dfrac{\beta_1 + \beta_2}{1 + \beta_1\beta_2}$
17. (a) 50 s; (b) 1.47×10^7 km
19. (a) $\beta = 0.6$; (b) $\tau_0 = 2.67 \times 10^{-5}$ sec
21. $\beta = 0.75$

Chapter 4

1. $v = 0.943c$
5. $\omega = q\mathcal{B}/\gamma\, m_0$; $\omega_{class-} = q\mathcal{B}/m_0$
7. (a) $v = 0.98c$; (b) $p = 4.9E_0/c$
9. (a) 1.33×10^8 m/s; (b) $0.943c$; (c) $0.9987c$
11. $p' = 11.3\, m_0c$; $T' = 10.3\, m_0c^2$
13. $0.4\, m_0c^2$
17. 139.1 MeV
19. (a) 6120 Å, 5880 Å, 5999 Å; (b) 12,240 Å, 3060 Å, 5094 Å

Chapter 5

1. 5000 Å
3. (a) 2.9 A; (b) 7.13×10^{21} W/m²
5. 5800 K
7. (a) 2500 Å, 1.2×10^{15} Hz; (b) 1.25×10^{-2} Å, 2.4×10^{20} Hz
9. 6.6×10^{-25} J; 30 cm
11. 1.6×10^{30} photons/s
13. 250 W
15. 10^{-2260}

Chapter 6

1. 1840 Å
3. (a) 0.97 eV; (b) 0.01 Å
5. 2 eV
7. 1.37 eV
9. hc/λ_0
11. 3.15 mW
13. 1.56 volts
17. 1.77 MeV
19. 41.5°; 0.68 MeV
21. 6.6×10^{-25} kg-m/s
23. 6.6×10^{-22} J

Chapter 7

1. 2.30×10^{-3}
3. 69 particles/s
5. 122.4 volts
7. $\lambda_{min} = 912$ Å; $\lambda_{max} = 1215$ Å
9. $n = 3$
11. 4.58×10^{14} Hz; $\lambda = 6550$ Å
17. 1.79 Å; 2.38 Å
19. 3×10^{-15} m; -18.9 MeV
21. 5.9 keV
23. 1.56×10^{-11} s

Chapter 8

1. 146 eV; $10.7\, w_0$
3. 1.44×10^{-4} Å
5. $2\phi = 37.1°$; $\theta = 71.5°$
7. (a) 1.44 Å; (b) 0.54 Å
9. 124 MeV
15. 0.26 Å
17. 1 MHz
19. $\Delta p \sim \sqrt{3}\ h/2a$; $E_{min} \sim 3h^2/8ma^2$

21. $v_g = 0.925c$; $u_{ph} = 1.08c$; 0.836 MeV
23. $A = B = 1/a^{1/2}$
25. $N = (K\sqrt{\pi})^{-1/2}$

Chapter 9

5. $A = \left(\dfrac{\omega m}{\pi\hbar}\right)^{1/4}$

7. 0, $\frac{1}{2}\omega m\hbar$
11. $C = 2\alpha$, $q = \alpha^2$
13. 0, $3/2$, 0, $3\hbar^2/2$

15. $\left(\dfrac{1}{\sigma\sqrt{\pi}}\right) \exp\left(-\dfrac{x^2}{\sigma^2}\right)$

17. $\hbar^2 k^2$; $\dfrac{\hbar^2 k^2}{2m}$

Chapter 10

1. $\psi_1 = \left(\dfrac{2}{L}\right)^{1/2} \sin\dfrac{\pi x}{L}$; $\psi_2 = \left(\dfrac{2}{L}\right)^{1/2} \sin\dfrac{2\pi x}{L}$; $\psi_3 = \left(\dfrac{2}{L}\right)^{1/2} \sin\dfrac{3\pi x}{L}$

3. $\langle x \rangle_1 = \langle x \rangle_2 = \langle x \rangle_3 = L/2$
5. $(h/2L)^2$
9. $\omega\hbar/4$
11. $n = 1$; $E = 3\omega\hbar/2$
13. $(5)^{-1/2}$
15. $29h^2/72mL^2$
17. $A = (\alpha/4\pi)^{1/4}$; $\Psi(x, t) = (\alpha/4\pi)^{1/4}$ $(e^{-i\omega t/2} + \sqrt{2\alpha}\, xe^{-i3\omega t/2})e^{-\alpha x^2/2}$
19. $17E_1/5$

Chapter 11

3. $-6\hbar^2$; $-w_0/9$
5. $9a_0$
7. $5a_0$
9. $-w_0/2$
11. $w_0/4$
13. 7.0×10^9 Hz
15. 5.66×10^{-2} Å
17. 5.1×10^{-2} tesla
19. 10 GHz
21. 5×10^{-3} rad
23. $^8S_{7/2}$; $g = 2.00$; $7.94\,\mu_B$
25. $g = 1.20$; $9.6\,\mu_B$

Chapter 12

1. (a) $\times 10^{-26}$: 3.82, 6.49, 22.1, 10.5, 17.9, 32.7
   (b) $\times 10^{28}$: 2.54, 1.33, 0.85, 8.53, 5.87, 5.90
3. 7.1 eV, 5.5 eV, 5.5 eV
5. 4
7. 1.39×10^6 m/s
9. 0.017 eV; 0.31%
11. (a) p, (b) n, (c) n, (d) p, (e) p, (f) n
13. 0.029 eV; 20.5 Å
15. $0.42\,m_0$
17. 2.9×10^{-6}; 3800 K (above the melting point)
19. 11,360 Å

Chapter 13

1. 1.24 GeV
3. 92.16 MeV; 7.68 MeV
5. $^3_1$H
7. 11.1 MeV
9. $\Delta E_C = 4.1$ MeV; $\Delta E_B = 3.5$ MeV
11. 36.7 MeV; 24.5 MeV
13. 243 MeV
15. 74.2 MeV; 78.3 MeV
17. 13.0 MeV
19. 8.5 MeV
21. (a) 15.6 MeV; (b) 8.33 MeV; (c) 71.4 MeV; (d) 87 MeV

Chapter 14

1. 2.41×10^{-2} yr^{-1}
3. 24%
5. 11.8 disint/min/g
7. 4.51×10^9 yr
9. 6.0 MeV
11. 5.3 MeV
13. 1.52 MeV
15. -3.27 MeV; 4.36 MeV
17. 3.12×10^{16} s^{-1}; 43.8 mg/hr
19. 2.22 MeV; 3.27 MeV; 18.35 MeV
21. 4×10^{38} s^{-1}; 4.4×10^9 kg/s
23. 13

Index

(Page numbers in italics refer to illustrations.)

302